Tides: A scientific history

Throughout history, the prediction of earth's ti
tant. This book provides a history of the study of the tides over two millennia, from the primitive ideas of the Ancient Greeks to the present sophisticated geophysical techniques which require advanced computer and space technology.

Tidal physics has puzzled some of the world's greatest philosophers, scientists and mathematicians: amongst many others, Galileo, Descartes, Bacon, Kepler, Newton, Bernoulli, Euler, Laplace, Young, Whewell, Airy, Kelvin, G. Darwin, H. Lamb, have all contributed to our understanding of tides. The problem of predicting the astronomical tides of the oceans has now been, in essence, completely solved, and so it is a perfect time to reflect on how it was all done from the first vague ideas to the final results. The volume traces the development of the theory, observation and prediction of the tides, and is amply illustrated with diagrams from historical scientific papers, photographs of artefacts, and portraits of some of the subject's leading protagonists.

The history of the tides is in part the history of a broad area of science, and the subject provides insight into the progress of science as a whole: this book will therefore appeal to all those interested in how scientific ideas develop. It will particularly interest those specialists in oceanography, hydrography, geophysics, geodesy, astronomy and navigation whose subjects involve tides.

After graduating in Natural Sciences at Cambridge University and in Mathematics at London University, David Cartwright worked from 1954 to 1973 as a researcher in Marine Physics at the then UK National Institute of Oceanography. His subjects were initially sea waves and the motion of ships, but later developed into tides, surges and variations in sea level. During a year at the Scripps Institute of Oceanography at La Jolla, California, he and Walter Munk revolutionised the whole concept of tidal spectroscopy and prediction, which later extended to the analysis of storm surges in sea level. He was appointed Assistant Director of the UK Institute of Oceanographic Sciences soon after its formation in 1973. Here, his team developed a pioneering programme of tide measurements in the Atlantic Ocean, from pressure variations on the ocean floor to topography of the ocean surface measured from satellites. On retirement from IOS in 1987, he accepted a Senior Research Associateship at the NASA–Goddard Space Flight Center in USA, where he pursued his interest in satellite altimetry and where, with Richard Ray, he made the first near-complete mapping of the tides in the global ocean from the US Navy's Geosat spacecraft. He was also a leading figure over several years in the international scientific planning team for the highly successful Topex/Poseidon satellite, launched by the USA and France in 1992.

While working at the leading edge of modern oceanic tide research, David Cartwright became increasingly aware of and interested in the long and neglected history of the science. On his retirement from NASA, he was in a unique position to write such a history: this book is the result.

The pathfinders: Sir Isaac Newton (1642–1727) and Marquis Pierre Simon de Laplace (1749–1827). (Both by permission of the President and Council of the Royal Society, London.)

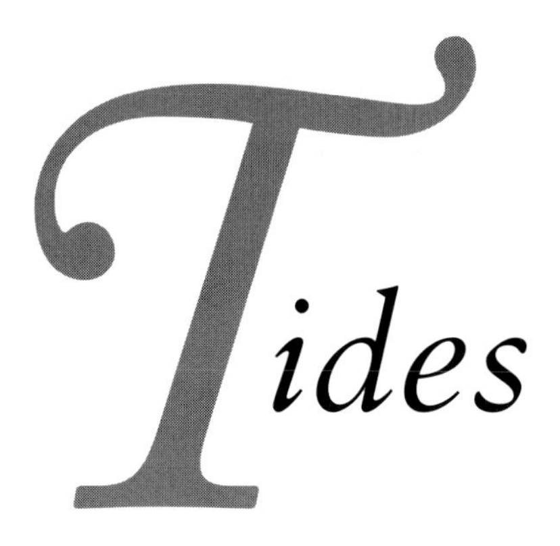

A SCIENTIFIC HISTORY

David Edgar Cartwright FRS

PUBLISHED BY THE PRESS SYNDICATE OF THE UNIVERSITY OF CAMBRIDGE
The Pitt Building, Trumpington Street, Cambridge, United Kingdom

CAMBRIDGE UNIVERSITY PRESS
The Edinburgh Building, Cambridge CB2 2RU, UK http://www.cup.cam.ac.uk
40 West 20th Street, New York, NY 10011-4211, USA http://www.cup.org
10 Stamford Road, Oakleigh, Melbourne 3166, Australia
Ruiz de Alarcón 13, 28014 Madrid, Spain

First published 1999
Reprinted 2000
First paperback edition 2000

Printed in the United Kingdom at the University Press, Cambridge

Typeface Monotype Sabon 10½/13pt *System* QuarkXpress™ [SE]

A catalogue record for this book is available from the British Library

Library of Congress Cataloguing in Publication data

Cartwright, David Edgar, 1926–
Tides : a scientific history / David Edgar Cartwright.
p. cm.
Includes bibliographical references and index.
ISBN 0 521 62145 3 (hb)
1. Tides. 2. Tides – History. I. Title.
GC301.2.C37 1998
551.47′08–dc21 98-4660 CIP

ISBN 0 521 62145 3 hardback
ISBN 0 521 79746 2 paperback

Contents

Preface

This book describes the growth of scientific understanding of a phenomenon which is superficially familiar but subtly complex, starting with primitive ideas in the remote past and leading up to sophisticated geophysical relationships which require modern computers and space technology for their evaluation. It is not an encyclopedic review of modern tidal knowledge, but it is the story of how such knowledge came about. It is not a catalogue of every book and paper that has been written about tides, but it includes at least a brief account of every major discovery and most branches of related research, whether they involve physical observations, theory, analysis of data or prediction technique.

Those readers who are themselves historians of science should realise that, unlike the history of physics or astronomy, the history of tidal research during the last two centuries has been very largely neglected. The present work not only summarises the early history which has been treated, at least piecemeal, by others, but it attempts to provide for the first time a basic framework and perspective for the vast area of all post-Newtonian tidal science, whose history is practically unwritten. This framework is mostly based on the published record of scientific papers and reviews, interpreted with specialist understanding of the subject from many years' personal experience of all aspects of tide research. I must, however, leave the *minutiae* of biographies, private correspondence and notebooks to the specialist historian who wishes to focus on a particular period, subject or personality.

Inevitably the subject requires the use of at least some of the terminology of elementary mathematical physics, but I have kept such terminology to the minimum necessary to make work in a wide range of disciplines intelligible to the general scientific reader who understands mathematical language, without trivialising the thought behind it. The essence of what I wish to convey is the historical growth of ideas over the centuries. In general, I have used equations only to express relationships between physical quantities which would be hard to express succinctly in words. I trust that readers who find a particular section 'difficult' will practise the art of skipping to a more accessible page or chapter in the interest of following the march of scientific history, perhaps to return later when the context has become clearer. I have added appendices and diagrams to some chapters in order to explain unavoidable jargon or to enlarge on some point of theory for those who would appreciate it, without interrupting the main text.

I feel that a few direct quotations and photographic excerpts from the original writings of classic sages convey valuable insight of the 'period flavour' of ideas and the way they were originally expressed. I have used standard English translations where the original writings were in Latin. The one exception is Figure 5.2, photographed directly from Newton's *Principia*, where the illustrated passages formed the first major turning point in our understanding of the tides. Sufficient explanation in English is given in the Figure caption and in the main text. The second major turning point was Laplace's enunciation of his equations of tidal dynamics, and I have displayed in Figure 7.1 an even longer excerpt from Laplace's seminal paper of 1776, in the hope that more readers are able to appreciate the classic French style than understand Latin. French scientists supplied important initiatives at several points in the history of tidal science, especially in the 18th and 19th centuries, and I have seen fit to include photographs of other classic pages of French (Figures 5.4, 6.1–6.3, 10.1, 13.1), which are rarely if ever referred to in modern scientific literature, English or French. At the suggestion of a referee, I have added my own English translation to all French quotations in the text. In return, I hope that English readers will accept the legitimate spelling 'tide-gage' to rhyme with 'greengage', instead of the more conventional spelling which suggests the sound of 'gauze' or 'gaunt'.

Chapters 1–5 follow progress more or less chronologically, but in Chapters 6–10 I have found it more appropriate to allocate a specific subject to each Chapter, with looser chronological sequence. Chapters 11–15 are firmly in the 20th century (with occasional retrospect) and the increasing pace of research forces narrower time limits with consideration of several subjects making contemporary progress. However, the time intervals stated in some Chapter headings are interpreted with a good deal of elasticity; they are intended merely as a rough guide to the central epoch of the Chapter. The subject matter reaches to about the end of 1995, but I have endeavored to avoid imitating the style of the modern scientific *Review*, with its comprehensive survey of every item of published literature.

Following the usual historical convention, dates of birth and death of scientists are quoted once, where known, at first or nearly their first mention. I am disinclined to quote the names of scientists still living and active at the time of writing (including of course my own name) in an essentially historical context. Chapters 12–15 therefore progressively ascribe the origins of research to institutions rather than to individuals, with a few practically unavoidable exceptions.

The names of the individual authors involved are retrievable from the numbered references.

I have for long felt the historical background to tidal studies to be unjustifiably neglected. I hope that the present work will fill a gap in the literature available to the modern science historian, and that it will also be interesting to those specialists in oceanography, hydrography, geophysics, geodesy, astronomy and navigation whose subjects involve tides, and of course to more general amateurs of earth science.

Acknowledgements

Research for this book has been aided by a one-year Research Grant from the Royal Society of London, with the Proudman Oceanographic Laboratory at Bidston Observatory, Birkenhead (a distant successor to the Liverpool Tidal Institute of 1919–1968) named as 'Host Institution'. The Library and Archives at POL provided valuable material for Chapters 11 and 12 in particular, and various staff members gave advice and technical assistance for line-drawings. I have also made extensive use of the National Oceanographic Library at Southampton and of its predecessor at Wormley, Surrey (originally part of the National Institute of Oceanography of 1949–1973 under Sir George Deacon).

Outside oceanography, I have found much useful material in the British Library, Bloomsbury, and in the libraries of the Royal Astronomical Society, London and of the National Maritime Museum, Greenwich. However, my most important source of historic scientific literature before about 1850 has been the Library and Archives of the Royal Society of London, whose staff have always been so helpful. More papers on tides have appeared in the *Philosophical Transactions* and the *Proceedings* of the Royal Society from Volume 1 (1665) onwards than in any other scientific journal in the world.

On the personal level, Dr. Philip Woodworth of POL gave helpful opinions on Chapters 7, 10 and 14; likewise, Professor George Platzman of Chicago on Chapter 15. Professor Willard J. Pierson jr. of CUNY made useful suggestions regarding the early days of satellite altimetry as an oceanographic tool, with which he was closely associated, and general encouragement for the book as a whole. But I am especially grateful to my ex-colleague Richard D. Ray at NASA/Goddard Space Flight Center, who volunteered to read through and criticise the entire text from the point of view of science history. The comments from all these people and from Professor Carl Wunsch of MIT as referee resulted in improvements and additions to the text. Finally, I must thank my wife, Anne-Marie, for cheerfully supporting some 40 Moons of asocial behavior from her spouse.

David E. Cartwright,
Petersfield, Hampshire

1

Introduction – the overall pattern of enquiry

Excepting Mediterranean and Baltic communities, the daily pulse of the tide has always been familiar to those who live by a sea exposed to the oceans. Both the rise and fall at the shoreline and the swing of the currents offshore are obvious. The close alliteration of the Anglo-Saxon words *Zeit*, *Gezeiten, Time* and *Tide*, (which 'wait for no man'), testifies to early awareness of the tide's regularity by North Sea coastal dwellers, fishermen, navigators and pirates. 'Rules-of-thumb' expressing a relationship between times of High Water and lunar phases, based on careful observation, were embodied in medieval tide-tables and clocks (Figures 3.1, 3.2) and in archaic terminology such as 'What moon maketh a Full Sea?' (Chapter 3). For many centuries, then, or at least throughout the Christian era, the tide has been a commonplace phenomenon, apparently as predictable as sunrise and Full Moon. What, some may ask, has made it a subject for serious research up to the present day?

Leaving aside the fact that, to a professional astronomer the accurate prediction of the times of sunrise and Full Moon demands sophisticated computations, research on the oceanic tides (and later, tides of air and earth) has been driven by practical needs and by the variety of fundamental questions which have been posed. Practical needs originally stemmed from harbor management, protection from coastal and estuarine flooding, coastal navigation and surveying, but more recently have been additionally concerned with amphibious military operations, harnessing tidal power, and precise corrections to measurements to and from artificial satellites. In all such cases, easy solutions which suffice for rough purposes have proved inadequate, while every improvement in accuracy of measurement and prediction has led to further fundamental research into previously hidden details.

From the dawn of scientific enquiry, basic questions about the *mechanism* of how the moon and sun drive the tides and how the ocean responds to the driving forces have inspired distinguished philosophers and earth-scientists. Descartes and Galileo (Chapter 4) proposed theories which were later proved untenable by Isaac Newton. Pierre Simon, Marquis de Laplace, the pioneer of tidal dynamics, declared the subject to be '*... ce problème, le plus épineux de toute la mécanique céleste ...*' (*the spiniest problem in all celestial mechanics.*) Solution of Laplace's tidal equations, even in seas of idealised shape, taxed mathematicians for well over a century until the advent of modern computers. Even then, some decades were to elapse before computers were large enough to represent the global ocean in sufficient detail, and techniques had improved sufficiently to give realistic results.

From a more empirical viewpoint, ever since primitive measurements from the western shores of the Atlantic became known, natural philosophers from Sir Francis Bacon onwards became curious to know how the tides behave in mid-ocean. Observing, for example, that High Water on the Atlantic coast of Florida occurs at the same time as at the Canary Islands, what happens in between? Does the tide progress northwards everywhere as observed along the coast of western Europe, or does it behave differently in the east and west Atlantic with a region of low amplitude in between? This important question was not seriously tackled until the 19th century, and then only after much speculation, – see Chapter 9 and later Chapters.

The 19th century polymath, William Whewell of Trinity College, Cambridge, became obsessed with this very question, and while unable to solve it, continually stressed the need for worldwide coastal measurements and berated the astronomers for having abandoned this fundamental science for more celestial pursuits.[(1)] Similar accusations have been leveled at oceanographers in the 20th century, and for analogous reasons. The fact is, that when a scientific problem does not yield to currently available tools, scientists tend to turn to other subjects which, if no easier, at least have the attraction of novelty. The tides have been an 'old subject' for a long time.

From time to time a new idea has arisen to cast fresh light on the subject. While such events have spurred some to follow up the new ideas and their implications, they have also had a negative effect by appearing superficially to solve all the outstanding problems. Newton's gravitational theory of tides (Chapter 5) explained so many previously misunderstood phenomena that British scientists in the 18th century saw little point in pursuing the subject further. The initiative passed to the French Académie Royale, culminating in the work of Laplace (Chapter 7) who took over where Newton stopped, at the dynamic response of the ocean to Newton's correctly defined force field. Similarly, William Thomson's idea of harmonic analysis, which stemmed from Laplace's theory, was so successful (after development by George Darwin) in providing for accurate predictions at any site where the tide had been measured for a long enough period of time, that one of the mainstays of research, namely from the commercial and naval producers of tide-tables, was transferred to routine computing activity. It was left to altruistic bodies like the British Association for the Advancement of Science (who had in fact promoted the development of the 'harmonic method' of prediction) to encourage further research into the *spatial* and *global* properties of tides and their currents.

When the new subject of *geophysics* began to develop towards the end of the 19th century, its investigators soon found that many of its problems involved the large-scale properties of the tides of ocean, atmosphere and the elastic earth.[(2)] Increasingly, these problems were seen to be global in character and solutions to them few or lacking. Of central and lasting interest was the total rate of dissipation of energy by the oceanic tides, and its implications for the apparent acceleration of the moon's longitude and (later) the rate of increase in the length of the day. Progress was now made, not by the tide-table experts, the naval hydrographers and academic mathematicians, who had hitherto kept tidal lore to themselves, but by geophysicists and by certain oceanographers inclined to mathematical physics.

The central problem of the 20th century, essential to the understanding of global energy dissipation as well as to a host of other geophysical problems, has remained that of determining the behavior of tides in the deep ocean. This is essentially the same problem as had bothered Whewell in the previous century, but at an altogether more refined level of precision than Whewell ever imagined. Persistently this problem defied formal mathematical analysis and measurement technique, and final or nearly final solutions have had to await modern technology.

Research on oceanic tides in the modern sense, then, has spanned at least four centuries. It has involved scientists from disciplines ranging from astronomy and satellite geodesy to ocean instrument technology, and activity from mathematical analysis and computing to sea-going expeditions. Relatively few people have been involved at any one time, but the subject seems to have had a peculiar fascination for 'lone workers'. As one worker has 'shot his bolt' or retired, another has taken up the challenge from a different viewpoint or discipline. Schools of expertise in different countries have led certain aspects of the field at different times, chiefly in Britain, France, Germany, Russia and USA. Only in the last decades of the 20th century, with the enormously increased power of computation and space geodesy, have the major goals been achieved. This book is therefore unusual in being concerned with the history of a science which has both a recognisable beginning and an 'end', or at least a temporary plateau, taking the reader from the earliest writings to the most recent research. The present is particularly timely for such a history.

Previous historians of tidal science have concentrated on the minutiae of limited, rather distant epochs of research, chiefly in the 16th to 18th centuries. The papers of the late E.J. Aiton in *Annals of Science*[(4)] (see Chapters 4 and 5) are particularly well studied in depth and have made useful guides to their epochs. Other historians from whose writings I have learnt much are Margaret Deacon on the debates about tides led by the early Royal Society of London (Chapters 4 and 6) and David Kushner on Sir George Darwin and the controversies over lunar acceleration among 19th century astronomers (Chapter 10).

Some textbooks have included a Chapter or part-Chapter on early history. Most comprehensive of these is in Book I, Part I of Rollin Harris's 5-volume monograph,[3] written very nearly a century ago and now long out-of-print. Harris's history has been widely acknowledged as the most thorough guide to early ideas on tides from antiquity to near the end of the 19th century. However, it was written as an introduction to Harris's own painstaking construction of a world map of tidal times. Ironically, he too probably thought the science was reaching the end of an era, but the dynamical theory on which his constructions were based was unsound, and was soon criticised on rigorous standards by George Darwin, though later accepted by Henri Poincaré as a reasonable compromise (Chapter 9). As one who has played a modest part in the international activity in tidal science during most of the last third of the 20th century, I feel confident that the year 1996 has a stronger claim to have reached the end of a long epoch of research, with the achievement of centuries-old objectives.

The above statements should not, however, be taken to imply that research on the tides is likely to come to a halt. Some details of the physical mechanism of energy dissipation are still unclear, and at the time of writing, *acoustic tomography* and *satellite altimetry* are revealing unexpected features of *internal waves* of tidal periodicity.[7] Tidal motion at diurnal and semidiurnal periods has also been observed for the first time in the rotation of the earth, (see Chapter 15). The small-scale dynamics of tidal motion on continental shelves, not to mention coastal engineering problems, will no doubt continue to demand attention.[5,6] This history is more concerned with the *global* aspects of tidal science, which do indeed seem to have reached a state of near-culmination.

Notes and references

1. Deacon, Margaret. *Scientists and the Sea, 1650–1900*, Academic Press, London, 445pp., 1971 – reprinted 1997. (Especially Chapters 5 and 12.)
2. Darwin, Sir George. *The Tides and Kindred Phenomena in the Solar System,* 3rd edn., John Murray, London, 437pp., 1911. (The 1st edition, lacking some later Supplements, was published in 1898.)
3. Harris, Rollin A. *Manual of Tides* – Appendices to Reports of the U.S. Coast & Geodetic Survey, (Parts 1–5, 1894–1907), Book 1, Part 1, Chapters 5–8. Govt. Printing Office, Washington, D.C., 1897.
4. Aiton, E.J. *Ann. Sci.*, **10**, 44–57, 1957; **11**, 206–233 and 237–248, 1955; *ISIS*, **56**, 56–61, 1965.
5. Pugh, D.T. *Tides, Surges and Mean Sea-Level; a handbook for engineers and scientists,* John Wiley, Chichester (UK), 472pp., 1987.
6. Parker, B.B., (Ed.). *Tidal Hydrodynamics*, John Wiley, New York, 883pp., 1991.
7. Dushaw, B.D., B.D. Cornuelle, P.F. Worcester, B.M. Howe, D.S. Luther. Barotropic and baroclinic tides in the central north Pacific Ocean, determined from long-range reciprocal acoustic transmissions. *J. Phys. Oceanogr.*, **25**, 4, 631–647, 1995. *Also*: Ray, R.D & G.T. Mitchum. Surface manifestation of internal tides generated near Hawaii. *Geophys. Res. Letters*, **23**, (16), 2101–2104, 1996.

2

Early ideas and observations

Seafarers of all kinds have always made their own observations of the tides along the coasts they frequent, and have devised their own practical 'rules-of-thumb' for rough predictions. Their observations would have included the relationship to lunar phase but they would not have been concerned with the *cause* of the relationship. The information would have been passed from father to son and from master to apprentice as part of the lore of the sea, but no written record would be made. The subject of this Chapter is therefore somewhat fragmentary and partly based on secondary sources and legends.

The Megalithic Age

There is evidence, though of a controversial nature, that the people who erected huge stones ('megaliths') near the western shores of Europe around 2500–1000 BC chose sites for them from which the rising and/or setting of the sun and moon could be observed against distant landmarks on the horizon.[1,2] Such observations could, it is supposed, have enabled the *cognoscenti* of the Megalithic Age to keep track of the seasons and to note the variations of the moon's orbit, including conditions leading to lunar eclipses. It has even been suggested[1] that forecasting the tides could have been one of several motives for daily observing, since many of the supposed 'lunar observatories' are sited close to seas whose tides are a major hazard to navigation.

Sceptics[2] have doubted this suggestion, on the grounds that horizon positions (equivalent to declinations) give insufficient information *per se* for tide prediction, and few scholars now accept the whole theory. Against the sceptical view, one may argue that forecasting the equinoxes and solstices would be relevant to tides, and the trained observers would make more accurate assessment of lunar phase and time of transit than an average seaman. If, in addition, they could sense the *rate of change* of phase or the change in apparent diameter, indicating the approach of lunar perigee, this would be a distinct advantage, provided they knew how to use it. But, as with all 'megalithic science', applying one's own ingenuity to deduce the possible ingenuity of prehistoric man tends to weaken the logic. In any case, the complete absence of any intelligible written record makes all such ideas purely speculative.

Early Indian and Arabic civilisations

The earliest concrete evidence for mankind's interaction with the rise and fall of the tide, implying some understanding of their nature, is the discovery by Indian archeologists of a *tidal dock* off the Gulf of Cambay near Ahmedabad, dating from roughly 2000 BC.[3] A tidal dock is a large wall-sided basin with a narrow entrance to the sea, closable by a sluice-gate which is opened only at High Water to let ships in and out, then closed to prevent ships inside from grounding at Low Water. Photographs[3] show the dock to be a remarkable engineering feat for its period.

The authors of (3) quote from various early texts of religious origin which show recognition of the moon's influence on the tides. A *Puràna* document (300–400 BC) likens the ocean to 'water in a cauldron which in consequence of the heat expands, so the waters of the oceans swell with increase of the moon'. This curious notion of the moon imparting heat to the sea also occurs in an Arabic document of the 13th century AD, quoted in full by Darwin.[4] An old Icelandic document, also quoted by Darwin, refers to heating by the sun as well as the moon, in a confused attempt to account for spring tides occurring at both Full and New Moons. Explanations for the tides involving the 'pulse' or the 'breathing' of a monstrous sea god may also be found in certain Indian and oriental texts.[3,4]

In view of the long tradition of Arabian astronomy, it is very probable that port astronomers of the early Arabic civilisations concerned with navigation took note of the subtleties of local tide behavior in relation to lunar phases and the seasons. However, I have not seen any account of written Arabic documents on tides dating from before the 9th century AD. That knowledge existed before then is evident from the account by Posidonius, transcribed by Strabo,[6] of singular properties of the tides related to him or written by 'Seleucus' the astronomer, a native of the Persian Gulf. The phenomena described by Seleucus are discussed in the following section of this Chapter.

The influential 9th century treatise on astrology, including tides, by Albumasar – Chapter 3 – bespeaks a tradition of tidal knowledge at least as advanced and probably older than that generally ascribed to the pre-Christian Greeks and Romans, although by the time of Albumasar it is hard to distinguish original from derived knowledge. Indeed, the knowledge described by the early Mediterranean cultures was itself derivative from observers at oceanic ports outside the Mediterranean Sea.

The ancient Greek, Babylonian and Roman civilisations

We pass to the decidedly enlightened enquiries of members of the early Greek civilisation. The Greeks had little direct experience of tides in their home waters, but they knew of their existence outside the 'Temple of Hercules' (Gibraltar) and from the roughly contemporaneous voyages of Pytheas of Marseille to northern Europe and of Alexander the Great to the mouth of the Indus River on the Arabian sea coast, (330–324 BC).

Aristotle (384–322 BC) wrote influential books on *Physics* and *Meteorology* which cover most aspects of the then known physical world, but he had little to say about tides, probably on account of lack of personal experience of them. He commented on tidal phenomena reported from 'Gades' (Cadiz), but attributed them vaguely to the rocky nature of the coast. Curiously, during his last years, when Aristotle lived at Khalkis, he was said to be greatly perplexed by his inability to understand the so-called *Tide of the Euripus* there – oscillatory currents in the narrows between the long island of Euboea and the mainland of Greece. He was also said to be drowned in the Euripus.[5] The oscillatory currents are now known to be caused by differences in the very small tidal elevations to the north and south of the strait, a fact first suggested by Eratosthenes the geodesist (c.276–194 BC) according to writings on tides attributed to him by Strabo.[6,7] These 'tides' are highly *irregular* in the sense used by Seleucus – see below – and they are disturbed by local resonances at nontidal periods induced by weather. Aristotle may be forgiven for failing to recognise any simple rythmic pattern or causality in what is regarded even today as a very complex dynamic regime.

The Stoic, Posidonius (135–51 BC), gave the first reasonably correct and detailed account of the tides at Gades from personal observations and accounts by unnamed local people. All his original writings were lost in the fire at the library of Alexandria (47 BC), but their essence is preserved in the 17 books of the *Geography* of Strabo (64 BC to 21 AD), who quotes extensively from Posidonius.[6] Thus Strabo writes:

> 'Now he (Posidonius) asserts that the motion of the sea corresponds with the revolution of the heavenly bodies and experiences a diurnal, monthly and annual change, in strict accordance with the motion of the moon.'
>
> For the diurnal motion: '... when the moon is elevated one sign of the zodiac [30 degrees] above the horizon, the sea begins sensibly to swell and cover the shores, until she has attained her meridian; but when that satellite begins to decline, the sea again retires by degrees until the moon wants merely one sign of the zodiac from setting; it then remains stationary until the moon has set and also descended one sign of the zodiac below the horizon, when it again rises until she has attained her meridian below the earth; it then retires again until the moon is within one sign of the zodiac of her rising above the horizon, when it remains stationary until the moon has risen one sign of the zodiac above the earth, and then begins to rise as before.'
>
> For the monthly revolution: '[he says] that the spring tides occur at the time

of the new moon, when they decrease until the first quarter; they then increase until full moon, when they again decrease until the last quarter, after which they increase until new moon; [he adds] that these increases ought to be understood both of their duration and speed.'

For the annual motion: 'he says that he learned from the statements of the Gaditanians, that both the ebb and flow tides were at their extremes at the summer solstice; and that hence he conjectured that they decreased until the [autumnal] equinox; then increased to the winter solstice; then decreased again until the vernal equinox; and finally increased until the summer solstice.'

The above accounts of the diurnal (twice daily) and monthly (twice monthly) variations are quite correct. One may assume that by 'duration and speed' Posidonius or Strabo means the duration of exceeding a high level and the speed of succession of High Waters, respectively; the 'speed' is slightly greater at spring tides. But the account of the annual variation is curiously wrong, for from modern knowledge, in any semidiurnal tide regime spring-tides reach their *highest* levels around the equinoxes and their *lowest* levels at the solstices, as may be verified by study of a modern tide table for Cadiz or nearby Gibraltar. (In fact, one has to take rather careful averages to detect any systematic difference at all.)

It is conceivable that the original observations on which the account told to Posidonius was based were made in only one year which happened to be a year when 'perigean spring tides' (that is, spring tides coincident with perigee – the closest approach of the moon) occurred near the summer solstice, as happens about every $4\frac{1}{2}$ years. An unusually high perigean spring tide would be followed 15 days later by an unusually low 'apogean' spring tide, but perhaps this subtlety was overlooked. Darwin's[(4)] remarks on this issue are ambiguous, but he writes: 'I doubt whether there is any foundation for that part [of Posidonius' account] which was derived from hearsay.'

Nevertheless, the Roman sage Pliny the Elder (AD 23–79), writing less than a century after Strabo, and like him quoting from existing texts, writes in his *Natural History*:[(8)]

> ... and all these (tidal) effects are likewise increased by the annual changes of the sun, the tides rising up *higher* at the equinoxes and more so at the autumnal than the vernal, while they are *lower* about the winter solstice and still more so at the summer solstice ... *not exactly at the Full nor at the New Moon but after them.* [my italics]

Pliny's source thus gives the correct account of the equinoctial/solstitial inequality, contradicting the account ascribed to Posidonius, and in addition has taken note of the *age* of the tide – the delay of a day or two of spring tides after Full or New Moon. The reported inequality between the vernal and autumnal equinoxes is not true of tides today, but it was indeed correct 2000 years ago, when *perihelion* – closeness to the sun – occurred some 35 days earlier than its present date of about 2 January, and so closer to the autumnal equinox.

Returning to Posidonius, the Stoic was also the first to take note of the *diurnal inequality*, that is, the difference in heights between successive High Waters or successive Low Waters. The diurnal inequality is quite low at Cadiz, as it is in most of the North Atlantic, but it is very noticeable in the Arabian and Red Seas, and in the Persian Gulf. In fact, Posidonius obtained his information from Seleucus of Babylon, an astronomer, 'native of the country next the Erythraean (Arabian) Sea'.[6] Seleucus, Strabo writes,

> states [according to Posidonius] that the regularity and irregularity of the ebb and flow follow the different positions of the moon in the zodiac; that when she is in the equinoctial signs the tides are regular, but that when she is in the signs next the tropics, the tides are irregular both in their height and force; and that for the remaining signs the irregularity is greater or less, according as they are more or less removed from the signs just mentioned. Posidonius adds that during the summer solstice and whilst the moon was full, he himself passed many days in the Temple of Hercules at Gades, but could not observe anything of these annual irregularities.

Here, the term *irregularity* evidently means diurnal inequality, and the description of how it would appear at equinoxes and solstices accords exactly with that observed in all seas adjacent to modern Saudi Arabia, while being barely perceptible at Cadiz.

In general, the most remarkable feature of all these accounts is, as observed by Lord Kelvin, the interest, experimental ability and persistence of people 2000 years ago to record the heights of the tide throughout the year, to note quite small changes, and to relate them to the ephemerides of the moon and sun.

No writings of Seleucus survive, but from secondary sources[7] we learn that he postulated that the moon causes tides by pressing or resisting the atmosphere, assumed to extend to the moon's orbit. He supposed the pressure or resistance to set up winds which would transfer the disturbance to the ocean. Like later hypotheses involving a heating or lighting effect (Chapter 3) Seleucus's idea failed to account for the tide which is manifest when the moon is below the horizon. However, replacing the atmosphere by the 'aether', supposed to fill all space, it has some affinity with Descartes' *vortex theory* of more than 17 centuries later, which by subtle argument did purport to account for the twice daily influence of the moon (Chapter 4).

The stark facts related by Posidonius also stirred primitive superstitions and vaguely religious feelings. Pliny[8] attributes to Aristotle the enigmatic remark: 'no animal (on the ocean coast of Gaul) dies except when the tide is ebbing' – an aphorism echoed after two millennia in Charles Dickens' description of the death of Barkis, *David Copperfield*, Chapter XXX, – and is then inspired to poetic thoughts:

> Hence we may certainly conjecture that the moon is not unjustly regarded as the star of our life. This it is that replenishes the earth; when she approaches it, she fills all bodies, while when she recedes she empties them. From this cause it is that shellfish grow with her increase, and those animals that are without

blood more particularly experience her influence; also, that the blood of man is increased or diminished to the quantity of her light, and the leaves and vegetables generally, as I shall show in the proper place, feel her influence, her power penetrating all things.

From the time of Posidonius onwards, there seems to be general agreement among those writers of the civilised world who made passing reference to the tides, about their apparent relation to the motions of the moon and sun. But more than a thousand years were to pass before one finds any evidence of a written table of tide predictions, or of any constructive thought applied to understanding the *cause* of the tides. These long intervals may be cited as the first example of the peculiarity noted in the previous Chapter, that substantial advances in understanding have tended to be followed by a lull during which nobody seemed inclined to learn any more about the subject.

The 'Dark Ages'

However, the present example has special circumstances related to human history. Athens and Rome fell successively to uncivilised invaders, and the climate of leisured philosophical enquiry, for which early Greece especially is remembered, collapsed; for many, mere survival became the essence of life. By AD 500 all the countries which now comprise modern Europe were overrun by barbarous warriors, civil strife abounded, and there was no chronicler even to write down a comprehensive history, still less engage in natural philosophy. The 'Dark Ages' had begun. Only isolated groups of Christian monks managed to preserve and teach what learning there was. We shall see in Chapter 3 that the Monastic institutions of Europe did indeed produce writers who occasionally commented on the tide in their accounts of the natural world, and who sometimes added their own observations of the phenomenon.

Otherwise, occasional references to the effects of the tide on men's affairs appear in the sparse historical fragments from the Dark Ages. The Battle of Maldon (Essex) in AD 991 between defending Anglo-Saxons and Viking invaders was held up by the tide, as related with dramatic effect in an anonymous Anglo-Saxon poem.[9] The popular legend of King Canute or Knut of England, Denmark and Norway (995–1035) is probably an exaggerated embellishment of some real altercation between the king and his courtiers. Whatever the true facts, there is historical evidence that such an altercation took place close to the tidal shore of present-day Southampton where it is commemorated by a wall-plaque (Figure 2.1). The legend is also associated by some with the tidal port of Bosham, West Sussex.

Understanding of tides in ancient China

In volume 3 of his monumental treatise, *Science and Civilisation in China*, Joseph Needham (1900–1995) devotes eleven pages to accounts related to sea tides in ancient Chinese writings.[10] Needham exaggerates when he says:

Figure 2.1. Building in Southampton bearing a plaque to commemorate the site where, according to popular legend, King Canute (Knut) demonstrated his inability to control the tides. The plaque's wording suggests a more general interpretation. (Photograph by author's son, Timothy.)

> Until modern times there was, on the whole, more knowledge of, and interest in, the phenomena of tides in China than in Europe.

From his own accounts one could fairly say that Chinese thought on the subject was roughly parallel with contemporary European thought, give or take a century or two, until the upsurge of the European Scientific Revolution in the 17th century. At any rate, it is clear from the early writings cited by Needham (many collected by the Revd. A.C. Moule,[11] a China scholar) that a relationship with the moon was recognised in China in late pre-Christian times, as it was in Greece.

Much of the interest (or written accounts of it) was centered on the spectacular tidal bore in the Chhien-Thang River (Needham's spelling) near Hangchow, which, according to legend, was caused by the spirit of a certain unjustly murdered minister or general named Wu Tsu-Hsü, whose body had been thrown into the river. This legend is also related by Darwin[4] from other sources. But Needham, following Moule, quotes in full a logical demolition of it by Wang Chhung, a well known sceptic of the 1st century AD, which concludes with the revealing statements: 'Finally the rise of the wave follows the waxing and waning moon, smaller and larger, fuller or lesser, never the same. If Wu Tsu-Hsü makes the waves, his wrath must be governed by the phases of the moon!'

The construction of a Table for predicting the Chhien-Thang bore in the 10th or 11th century will be described in Chapter 3. Here, it suffices to note that from an early age the bore focussed attention on the regularity of *spring tides*, their association with Full and Change of the moon, and their seasonal variation in strength. The experience was different from that of the daily tides in Europe and of course on the coasts of China. Moule and Needham cite the construction in the 1st century AD of high sea walls to protect the river valley from floods caused by the bore.

However, China had no equivalent of the later European natural philosophers. By the 18th century, Chinese writers were still discussing supernatural legends to account for the tides, while at the same time, Chinese mariners were evidently still relating tides to lunar phase as a simple everyday rule-of-thumb. On this last point, one may add that in all times and countries there has been a dichotomy of understanding of tides between philosophers and practical seamen. Even in 20th century England, the writer Hillaire Belloc, no theorist himself but a keen amateur yachtsman, wrote:[12] 'When they [theoreticians] pontificate on the tides it does no great harm, for the sailorman cares nothing for their theories but goes by real knowledge.' Belloc's irony was probably *tongue-in-cheek*, but it expresses a real and deep mistrust between seamen and theorists which has always been hard to bridge.

Notes and references

1. Thom, A. *Megalithic Lunar Observatories,* Oxford Univ. Press, 127pp., 1971.
2. Heggie, D.C. *Megalithic Science; Ancient Mathematics and Astronomy in northwest Europe,* Thames & Hudson, London, 256pp., 1981.
3. Pannikar, N.K. and T.M. Srinivasan. The concept of tides in ancient India. *Indian J. Hist. Sci.*, **6**, 36–50, 1971.
4. Darwin, Sir George. *The Tides and Kindred Phenomena in the Solar System,* 3rd edn., John Murray, London, 437pp., 1911.
5. Gill, Adrian, 'Walter, Aristotle and the tides of the Euripus'. *A Celebration in Geophysics and Oceanography – 1982,* (Walter Munk's 65th birthday volume). Scripps Inst. of Oceanography, Ref. Ser. 84–5, 96–99, 1984. An intensive study by M.Tsimplis, based on measured currents and sea levels, has recently appeared in *Estuarine, Coastal and Shelf Science*, **44**, 91–101, 1997.
6. Strabo, *Geography*, (17 vols.) transl. H.L. Jones; extracts quoted from Bk. 3, Ch. 5, Loeb, 1917–1933. For Eratosthenes on tides, see Bk. 1, Ch. 1 & 3.
7. The astronomer Seleucus of Babylon, (not to be confused with the kings of the Seleucid dynasty, 305–64 BC), was active around 150 BC. He is best known as the only supporter in ancient times of the pioneering heliocentric hypothesis of Aristarchus, 310–230 BC. A summary of Seleucus's explanation of the cause of tides, together with secondary Greek references and mention of contemporary solar theories, may be found in pp. 305–307 of: Sir Thomas Heath, *Aristarchus of Samos, the Ancient Copernicus.* Oxford, 1913; reprinted Dover pubs., New York, 1981, 425pp. (Heath describes Seleucus as 'mathematician'.)
8. Pliny the Elder (Plinius Secundus). *Natural History,* (37 books), Engl. transl. H. Rackham; extracts from Bk. 2, Ch. 99–102, Loeb, 1942–1963.
9. Cooper, Janet, (Ed.) *The Battle of Maldon – fiction and fact.* Hambledon Press, London, 265pp., 1993. (See also: Cartwright, D.E. & C.P. Conway. Maldon and the tides, *The Cambridge Review*, **112**, 180–183, 1991.)
10. Needham, Joseph, (with Wang Ling). *Science and Civilisation in China*, **3**, 21, 483–494, Cambridge University Press, 877pp., 1959.
11. Moule, A.C. The Bore on the Chhien-Thang River in China, *T'oung Pao*, (Archives concernant l'Histoire ... de l'Asie Orientale), **22**, 135–188, Brill, Leyden, 1923.
12. Belloc, H. *The Cruise of the Nona,* Constable, London, 1925.

3

What moon maketh a full sea?

The Venerable Bede and Gerald of Wales

By far the most influential of the early Christian monks was the Venerable Bede (672–735) of Jarrow Abbey, Northumbria. His *History of the English Church and People* (731) is one of the few reliable sources for Dark Age history. Earlier, Bede had written two works on 'Time and its Reckoning'. They are mainly concerned with the lunar determination of Easter, but they also established our system of reckoning dates from the birth of Christ. Section XXIX of *Opera de Temporibus* (703), entitled 'On the harmony of the sea and the moon', starts:[1]

> But the most admirable thing of all is this union of the ocean with the orbit of the moon. At every rising and every setting of the moon the sea violently covers the coast far and wide, sending forth its surge, which the Greeks call *reuma*; and once this same surge has been drawn back it lays the beaches bare, and simultaneously mixes the pure outpourings of the rivers with an abundance of brine, and swells them with its waves. As the moon passes by without delay, the sea recedes and leaves these outpourings in their original state of purity and their original quantity. It is as though it is unwittingly drawn up by some breathings of the moon, and then returns to its normal level when this same influence ceases.

The observations on rivers and brine are unusual, and 'breathings of the moon' is a conceptual advance on the breathing of a sea-god. At least it shows some wonder at the true mechanism, and is not so far removed from Descartes' theory of 'vortices' (Chapter 4). Otherwise, Bede's account of the behavior in time follows the pattern of the Greek and Roman statements, but without factual errors and in a more numerate form. For example, he points out that in 12 lunar months of 354 days the tide rises and falls 684 times.

But Bede's most original contribution was to notice, from local accounts of tides round Britain, that High Water does not occur simultaneously on all coasts:

> For we who inhabit the various coasts of the British Sea know that wherever one tide begins to swell, another one simultaneously recedes.

Further on:

> ... but on one and the same shoreline those who live to the North of me will see every sea tide both begin and end much earlier than I do, whilst indeed those to the South will see it much later.

This recognition of the progressive wave-like character of tides, more evident in the shelf seas round Britain than around the oceanic coast of the Iberian peninsula, invalidates the assumption of some Greek and Latin authors, that the waters swell up simultaneously in all seas. It also points to the question: when *does* High Water occur at a given place, relative to the moon's passage?

In his *History of Cosmology from Plato to Copernicus*, Pierre Duhem has a Chapter on the comments on tides by various 12th century clerics in which he devotes four pages to the archdeacon Giraud de Barri, otherwise known as Giraldus Cambrensis, or in English, Gerald of Wales.[(2)] Gerald (1147–1220+) was a prolific, and by all accounts, entertaining writer, mainly on Welsh and Irish history. His comments on tides occupy just five paragraphs of a treatise on Irish geography and history, in a sub-section entitled *Of the Wonders and Miracles of Ireland*.[(3)] At the start of the sub-section he assures his readers that all his informants were 'most trustworthy men', but some of the facts he relates have been dismissed as fanciful, or even false. I judge his facts about tides to be basically correct within allowable limits of accuracy. Misunderstandings may have been caused by inaccurate translation of the Latin text.

Gerald relates *inter alia* that, while the tides at Wexford (southeast Ireland) have the same timing as those of Milford (southwest Wales), the tides at Wicklow and Dublin, only 50–70 miles further north, have quite different timing, being at lowest ebb when Milford and Wexford are nearly in full flood. When the moon crosses the meridian, he says (contrary to some astrologers) that the tide is *at ebb* in northern Britain while at flood near Dublin. If one interprets 'northern Britain' as the coasts of Ireland and Scotland between Malin Head and Cape Wrath, all these facts accord with modern observations. From the last quotation above, similar facts were also in general known to Bede. Gerald's statement (in a standard translation): 'What is still more remarkable, there is a rock in the sea not far from Arklow where the tide comes in on one side while it ebbs on the other.' may sound literally impossible, but if one interprets the two 'sides of the rock' as loose parlance for the sea within about 30 miles north and south of the region of Arklow (about half way between Wexford and Dublin) it accords with the now recognised region of low amplitude and rapid phase change along that part of the coast.

He goes on to describe the increase and decrease of tidal range with the waxing and waning of the moon and their effects on humans and animals, in terms reminiscent of Pliny and Bede. With greater originality, he expresses wonder at the ability of the 'western ocean' (i.e. the Atlantic) to sustain such large tidal activity in contrast to the Mediterranean. He suggests the wider extent for free course of the tides in the ocean as a major cause, as is indeed correct. Altogether, Gerald's brief contribution to 12th century knowledge of the tides contains a surprising amount of interesting detail from one whose main interest seems to have been in human affairs.

Albumasar, Grosseteste, and Chaucer

As the Dark Ages lightened and more literary works emerged, it was apparent that this sort of basic tidal lore had become part of common learning among those who used the sea in maritime Europe. A 9th century work by the Muslim Astrologer Abu Ma'shar (or 'Albumasar') appeared in two 12th century translations into Latin. Although mainly astrological, it included an article on the tides, which, while accepting the common observational knowledge, showed a significant distinction in trying to explain their causes, not by invoking supernatural agencies but in terms of the perceived physical nature of the universe. The philosophy may be said to stem from the enigmatic but seminal idea expressed by Seleucus of Babylon (Chapter 2).

Robert Grosseteste, Oxford scholar and later Bishop of Lincoln,[(4)] was impressed by Albumasar's work, and his own account of the tides, *Questio de fluxu et refluxu maris*,[(5)] is almost a paraphrase of the earlier work, but with differences in interpretation clearly attributed to Grosseteste, (or as some authorities aver, to one of his pupils). I shall briefly describe their main features as presented by their English author. Pierre Duhem gives a full account of the discourse presented by Albumasar, together with their underlying astrological philosophy which assigned occult powers to all celestial bodies, in Volume 2 of *Système du Monde*.[(2)]

Water and air are supposed to respond to movements of the heavenly bodies because they may be rarified (i.e. made less dense) by heating, unlike earth. The moon is the principal agent of rarefaction, partly because of the suggestive similarity in timing, but also because its light imparts heat to the sea. (The importance of light is due to Grosseteste, but he shared with Albumasar a firm belief that in some sense the tide starts to increase at moonrise and returns to its lowest point again at moonset.) The hypothesis immediately calls into question the role of heating for the tide which occurs when the moon is below the horizon. 'This requires further investigation', says Grosseteste.

The sun is seen to strengthen the moon's tides when in conjunction (i.e. at New Moon), but the effect weakens as the moon distances itself from the sun towards the First Quarter. Then on approaching Full Moon, the greater luminosity of the moon causes the tide to become stronger again, hence accounting for the twice-monthly spring-neap cycle. To account for other variations in the strength of the tide, Grosseteste follows Albumasar in listing eight causes, one of which is that the moon is less effective when furthest from the earth (Apogee), and *vice versa*. A second cause of variation is the moon's 'declination'; northern declination produces higher tides in northern seas, and *vice versa*. Again, the sun 'helps' the moon to produce higher tides at the solstices than at the equinoxes – an echo of the false statement handed down by Strabo (Chapter 2). Finally, the tide may be increased by the action of the wind.

Another section of Grosseteste's essay explains why some seas have tides while others do not, or do not seem to – on account of lack of observers in the right places. But I have described enough to give the tenor of these theories, without deigning to criticise them in the light of modern knowledge. The hypotheses include some ideas which have since been proved false, together with others which in essence have stood the test of time. Their real interest lies in the evident desire of medieval philosophers to *explain* phenomena hitherto taken as granted, without calling on the supernatural, and their appreciation of what celestial configurations are important in controlling the tide. Such factors as lunar and solar longitude, lunar distance and declination, and the wind, are still essentially relevant in modern theory; only the medieval *physics* is inadequate.

The Albumasar/Grosseteste treatises were accepted and quoted as the standard texts for many years, serving as the basis for other 13th century writers, notably St. Thomas Aquinas and the Franciscan philosopher Roger Bacon, in their references to tides. By this time, every serious writer had to show that the behavior of tides was part of his understanding of the world system.

While such philosophers were groping for convincing theories, the common sailor developed his own practical understanding of the tides by dint of necessity, without searching for explanations. The seamen who now regularly plied all the coasts of Europe felt safe in relating any observed tides and currents to the moon, and acquired knowledge of the moon's bearings at floods and ebbs wherever they went. In 1386, the poet Geoffrey Chaucer writes of his Skipper in the Prologue to *The Canterbury Tales* (otherwise a hard-drinking rough diamond):[(6)]

> As for his skill in reckoning his tides,
> Currents and many other risks besides,
> Moons, harbors, pilots, he had such dispatch,
> That none from Hull to [Cartagena] was his match.

It was from this sort of knowledge that permanent *tide-tables* arose, (that is, printed almanacs which could be used by a mariner to estimate the state of the tide at any time at a particular port or ports) partly as a guide to seamen entering unfamiliar waters, partly for landsmen wanting to know when ships might arrive at or sail from their ports.

Early tide prediction – London Bridge and the Chinese bore

The earliest tide-tables, of roughly the 11th–15th centuries, were not ephemerides as of today, pertaining to a particular period and no other, but they expressed a simple set of rules for the local time of 'High Water', sometimes called 'Full Sea' or 'Flood', according to the age or the phase of the moon. The moon's age (0–29 days, sometimes written 1–30 days) was a more important factor in everyday life than it is to most people today, and it could be looked up in Almanacs. In basic terms, the assumed rule was that Flood at Full or Change of the moon occurs at a certain constant sun time at each place; on other days four-fifths of an hour (48 minutes) was added for each day of the moon's age. The time at Full or Change was termed the *Establishment* of the port in 19th century English literature, (after the earlier French *Etablissement*), but that term was disliked by seafarers and is in any case now obsolete.

The said rule, expressed in various forms to suit the understanding of the user, is very rough, because the moon's times of southing ('transit') do not advance by exactly 48 minutes from day to day but fluctuate, and the time of Flood does not follow the moon by a constant interval. But the rule served the rough needs of most seamen and had the great virtue of simplicity. It must be remembered that in the days before clocks few people could assess the time with much accuracy, and the measurements on which the 'constant' was based were themselves rough.

The first known written tide-table for European waters is for 'flod at london brigge' (High Water at London Bridge), from a 13th century manuscript now held in the British Museum (Codex Cotton, Julius DXVII, page 45b). It originally belonged to the Abbey of St. Albans, and is thought to be largely the work of Abbott John of Wallingford who died in 1213. (John is not to be confused with Richard of Wallingford (1292–1336), also Abbott of St. Albans, a pioneer in the design of astronomical clocks.[(7)]) The existence of this table was first brought to the attention of tidalists by John Lubbock.[(8)] The table, hardly worth reproducing here, simply consists of a column of 30 times in hours and minutes, each time corresponding to a day of the moon's age – 'aetas lunae'. The 'constant' for London Bridge was reckoned as 3 hours, so the time listed for day 1 was 3h 48m, and so on in steps of 48 minutes, subtracting 12 hours wherever necessary. To the assumed accuracy a.m. and p.m. are interchangeable, and the 'age' could be reckoned from either New or Full Moon.

The form of table just described was intended mainly for landsmen. Mariners preferred to think in terms of the compass bearing of the moon at High Water. For example, since a Full Moon bears south at local midnight, so for London, at the Flood Tide of 3 a.m. the moon bears roughly SW (or at 3 p.m., NE). Hence the statement in a 15th century manuscript:[(7)]

> For schulen undirstonde that a southwest Moone or a northeest Moone maken an high flood Lundoun brigge ...

Similarly, a port with High Water at Full or Change at about 9 a.m. or p.m. (e.g. Honfleur) would be described as requiring a southeast or a northwest moon; and similarly for other points of the compass, counting 3 hours for every 2 points. The fact that the moon does not uniformly change its bearing by 4 points (90 degrees) in 6 hours added another source of inaccuracy or misunderstanding. Four-fifths of an hour was to be added for each day of the moon's age, as before.

Abbott John's table for London Bridge was for long considered in the English speaking world to be the earliest known tide-table. However, It is now known to be predated by at least a century by the Chinese Table for the bore on the Chhien-Thang river, to which I alluded in Chapter 2. The table was described in a Chinese document dated 1056 AD, but the document was not fully translated into English until Moule's paper of 1923.(9) Needham(10) makes use of several parts of Moule's article, but does not reproduce the table itself. It consists of 15 rows of information, each row pertaining to the age of the moon in days from Full or New Moon, treated equally. Each row gives the predicted time of the bore to about 40 minutes and a qualitative estimation of its predicted strength, for each of the four seasons of the year. There is also a distinction between 'day-hour' and 'night-hour'. Altogether, then, the table embodies assessments from observation of the daily, monthly and annual inequalities of the tide as exemplified by the bore. The information was carved in the stone walls of a pavilion overlooking the river at a place where the bore was most spectacular.

Moule (and later, Needham) was strongly motivated to correct the impression given by Sir George Darwin(11) and other English writers, that Chinese understanding of the bore was limited to superstitious legends. In his preface, Moule writes: 'If he [Darwin] were alive, he would perhaps modify his opinions after reading this chapter, which was written 3 or 4 centuries before Kepler or Newton, and is based on yet earlier authorities.'

Recent experiments in comparing the 1000-year old tide-table to modern data in the Chhien Thang River(12) show good agreement at the highest tides of perigee, less so at apogee. However, the very existence of this table is a remarkable testimony to an advanced tidal lore in 11th century China and probably before, previously unsuspected from the superstitious legends more usually associated with the subject.

Brouscon's Almanacs

Howse(13) describes and illustrates several historic diagrams for tide prediction based upon the compass notation, dating from c.1375 to 1598. Here, I illustrate only two pairs of the colorful diagrams from Brouscon's Almanac of 1546. These were prepared by Guillaume Brouscon, a Breton man from the port of Le Conquet, near Brest. Brouscon's Almanacs were at one time widely used by French and English, as well as by Breton seamen. They were pocket-sized in vellum for handy use at sea, and they contained other information besides tides, including the dates of religious festivals which depend on Full Moon. There were two types of tide-diagram, one comprising a series of small charts of different coastal areas, with the 'Flood-Moons' of each port indicated by lines joined to the appropriate part of a compass rose. The upper panels of Figure 3.1

Figure 3.1. Four diagrams from Brouscon's Almanac (c.1546). *Upper panel*: Left: The 'compass bearings' of High Waters at several ports round the Gascon coast (Bay of Biscay), all of which have 'Southwest Moons'. Right: *ditto* along the north coast of France from Brittany to Flanders, including Dover, with Southwest to West Moons. *Lower panel*: Tidal diagrams for calculating High Water times according to the moon's Age, at ports with SSW Moons (left) and with SW Moons (right) – see text for explanation of code. The Almanac from which these pages were photographed bears the signature 'F.Drak', thought to be that of Sir Francis Drake. (Courtesy, Derek Howse and the National Maritime Museum, Greenwich.)

show two such charts, for the Bay of Biscay (left) and for the north coast of France (right). The names are in Breton, but recognisable; in the right hand chart, for example, Lannion, Dinan and St. Malo appear near the bottom, Rouen half way up in larger letters, Calais and Ostende near the top. Note that the North of the compass rose points downwards and to the left, and is unrelated to the rough orientation of the map itself. All ports in the left hand chart were reckoned to have 'Southwest Flood-Moons'.

The second type of Brouscon's tide diagrams was a simple calculator for the time of flood for any age of the moon, one diagram for each compass bearing. The lower panels of Figure 3.1 show the diagrams for 'southwest' ports (left) and 'southeast' ports (right). (The central letters 'OU' and 'S' are irrelevant, being three of the eight letters of the author's name.) The outside ring indicates lunar age in a Breton version of Roman numerals (+ for ×) in 29 segments, the lowest segment being split between 1 and 30 days. The next ring inwards illustrates the principal phases of the moon, and also the days of spring tides (segment full of water) and neap tides (dark semicircle), which occur about 1–2 days after the relevant phase. These signs would be understood by illiterate fishermen. The 3rd and 4th rings show the times in hours and quarters (short red strokes for quarters) of High and Low Waters for each day of the moon's age. For example, in the right hand chart we see that at 30d (=0d) and at 15d, (+v), HW is at 3h, LW is at 9h, fittingly for a 'SW' port. At age 11d, (+i), HW is at 0h, LW at 6h, and so on.

Tide clocks

More sophisticated methods for predicting tide times from the 17th century onwards will be described in later Chapters, but the old simple rules, as embodied in clockwork mechanisms, continued to be used by the common people for several centuries. As clock making developed during the 17th and 18th centuries, some of the more elaborate clocks were fitted with a lunar dial, driven by a gearing ratio of 59:2, and indicated the age or phase of the moon. With a simple adjustment such a mechanism could easily be adapted to show the time of High Water at a chosen port, usually but not always London. A remarkable example of a *public* tidal clock on a church tower is shown in Figure 3.2. The dial depicted is a 20th century restoration of the original 1681 clock, constructed by one Thomas Tue, which was severely damaged when the church spire collapsed in 1741 and was then allowed to disintegrate over many years. The restored clock is seen mounted on the Southwest Tower of St. Margaret's Church in King's Lynn (originally 'Lynn'), a seaport in Norfolk, England. The twelve letters LYNN HIGH TIDE mark the even hours in a 24-hour system with 'L' marking noon (top) and 'G' marking midnight (bottom). The pointer, a green dragon with a red cross issuing from its mouth, the emblem of St. Margaret, shows the time of Flood at Lynn on that time system, and so revolves

Figure 3.2. Thomas Tue's 1681 tide clock, reconstructed to its original design under the supervision of Colin Shewring, city architect, on the southwest tower of St. Margaret's Church at the port of King's Lynn, Norfolk, and photographed by him on 24 July 1994, a day or two after Full Moon. The pointer performs a complete rotation in 29.5 days. At the date depicted, it indicated High Water at about 7 a.m. and p.m., GMT. At lunar phases other than Full Moon, the 'face' gives way to a tapering white segment. Note: A postcard on sale locally shows the Full Moon 'face' with the 'dragon' pointing incorrectly to 3 a.m./p.m. (Courtesy, C. Shewring, King's Lynn.)

round the dial in $29\frac{1}{2}$ days. Thus, for example, both the 'T' and the first 'N' would indicate Flood at 4a.m. and 4p.m. (Curiously, the use of a dragon's head or tail is also found in much older astronomical clocks,[7] to indicate the 19-year 'Draconic' cycle of the moon's nodes.)

The blue inner disc is in fact rigidly attached to the tide pointer; there is no need for a separate indicator of solar time, since a normal clock has always been displayed on the Northwest Tower. The large white disc with a fanciful 'face' represents the Full Moon; a tapered white shape is progressively covered and uncovered by the circular hole in the blue disc to give the appearance of other lunar phases. High Water at King's Lynn actually occurs at about 6.15 on days of Full Moon, so the dragon pointer is in the expected position for 1–2 days after Full Moon, when the 'face' would be slightly more central to the orifice.

A more elaborate tide clock, mounted in a wooden case for interior use, is on display in the Museum at Rye, East Sussex. This is of quite recent (1971) manufacture by G.H. Bell of Winchester, from a 1773 design by James Ferguson, FRS.[14] As befits an interior clock, it contains more moving parts and scales than

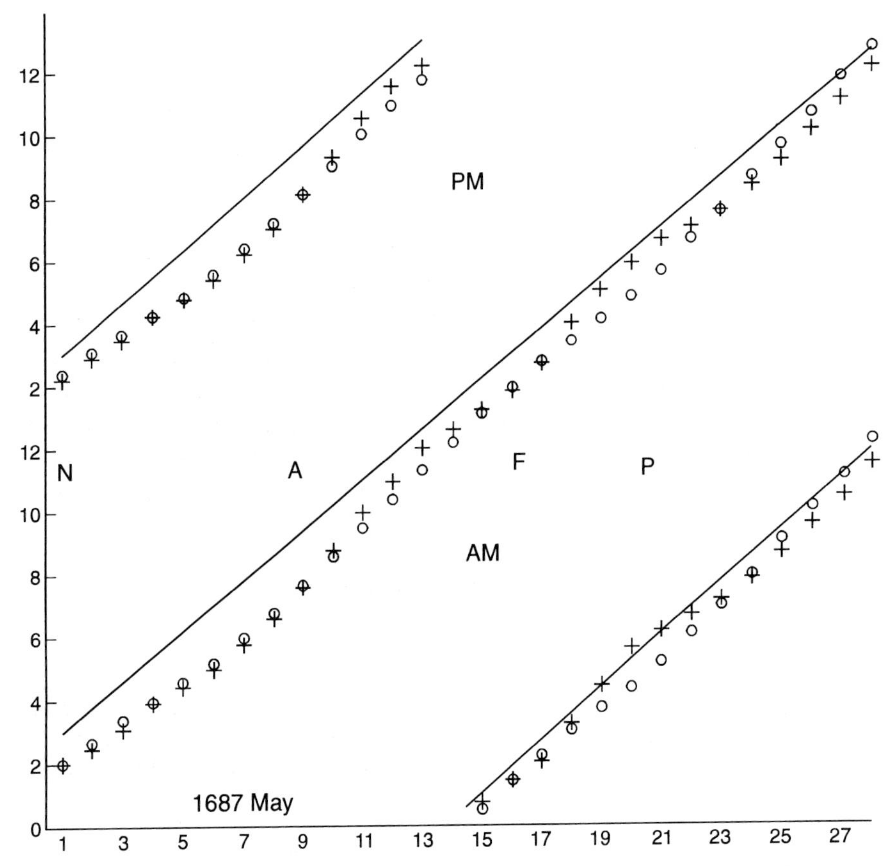

Figure 3.3. Comparison of early predictions of High Water times (hours a.m. and p.m.) at London Bridge for 1–28 May 1687. The straight lines represent 'Abbot John's Rule'; circles are Flamsteed's predictions based on a lunar ephemeris computed in 1686 with direct observations made in 1682;[(15)] crosses are improved predictions based on Bernoulli's tables of 1740, (Figure 5.4). The discontinuity occurs where Flamsteed's prediction for 1144 p.m./13 is succeeded by 1211 p.m./14 and 0039 a.m./15.

the church clock at King's Lynn, including independent sun and moon pointers, and a sea horizon which rises and falls between engraved rocks.

Medieval and 17th century predictions compared

Finally, in case the reader may wonder why anyone should want to improve on this admirably simple method of tide prediction, Figure 3.3 was constructed to compare such predictions as nearly as possible with real measurements. Here, circles indicate predictions which are close to the measured times of High Water at London Bridge for the first 28 days of May 1687, as recorded under the direction of John Flamsteed, the first Astronomer Royal[(15)] – see Chapter 6. New Moon occurred near noon on May 1st (Julian Calendar), so the appropriate 'prediction' by Abbott John's rule may be written

$$3 + 0.8 \times (\text{May date} - 1) \text{ hours},$$

which is represented by the straight lines shown. The predictions are not bad towards the end of the month, but they differ seriously from the measurements at other times, up to 1.8h around the 1st Quarter. This characteristic error was first pointed out by Henry Philips in 1668.[16]

Figure 3.3 suggests that the constant term may be too great, but Philips wrote that mariners still used 3 hours in 1668, and that it seemed appropriate to his measurements at Full and Change, but not at other times. The 'constant' offset probably varies with changing sedimentation in the estuary. If such a rule were to be applied to London Bridge today it would require a constant of only 1.7 hours, largely on account of dredging and other works in the Thames Estuary. It was to take some two centuries before sensibly improved tide predictions could be computed. We shall return to this subject in Chapters 5–8, but it is now appropriate to review more abstract researches which started seriously in the 16th century.

Notes and references

1. Bede (Bedae), the Venerable. *Opera de Temporibus,* Art 29, ed. C.W. Jones, Cambridge, Mass., 1943, (originally AD 703). Extracts translated by Mark Haywood, Univ. Liverpool – courtesy D.T. Pugh.
2. Duhem, Pierre. *Le Système du Monde – (Histoire des doctrines cosmologiques de Platon à Copernic).* Paris, 1913–1917. Vol. 2, Ch. 13, Secn. 14: La théorie des marées selon les Arabes – Abou Masar; Vol. 3, Ch. 3, Secn. 10: La théorie des marées au XIIe siecle ... – Giraud de Barri.
3. Opera Giraldi Cambrensis, (ed. J.F. Dimock). Vol V – *Topographia Hibernica,* Distinctio 2, Cap 2, 3, pp. 77–80, London, 1867 (originally 1187). English transl. in: *The Historical Works of Giraldus Cambrensis – The Topography of Ireland,* ed. T. Wright, pp. 57–61, London, 1863.
4. Laird, E.S. Robert Grosseteste, Albumasar and mediaeval tidal theory. *Isis,* **81**, 684–694, 1990. A recent paper (*Br. J. Hist. Sci.*, **29**, 385–401, 1996) discusses a rare early 16th c. manuscript by the Paduan, Frederico Chrisogono, who accepted (without explanation) roughly equal twice-daily influences from sun and moon, reinforcing each other at syzygies, largely cancelling at quadratures. Disregarded by Galileo, Wallis and Descartes (Chapter 4), this notion was eventually rationalised as an element of Newton's tidal theory (Chapter 5).
5. Dales, R.C. The text of Robert Grosseteste's *Questio de fluxu et refluxu maris* with an English translation, *Isis*, **57**, (4), 455–474, 1966.
6. Chaucer, G. *The Canterbury Tales,* (Modern English transl. N. Coghill), Century Hutchinson, London, 288pp., 1986.
7. North, J.D. *Richard of Wallingford,* 3, Appendix 31, Oxford, 1976.
8. Lubbock, J.W. On the Tides. *Phil. Trans. R. Soc. London,* **127**, 97–104, 1837. (In his remarks on the historic tide-table for London, p. 103, Lubbock misinterprets 'Age = 1 day' as 'Age = 0', here equivalent to 'Age = 30 days', and thereby wrongly deduces a secular change of 48 minutes in the Establishment.)
9. Moule, A.C. The Bore on the Chhien-Thang River in China, *T'oung Pao,* (Archives concernant l'Histoire ... de l'Asie Orientale), **22**, 135–188, Brill, Leyden, 1923. (*T'oung Pao* is the name of a quarterly Review of oriental history with articles in French and English.)
10. Needham, Joseph, (with Wang Ling). *Science and Civilisation in China,* **3**, 21, 483–494, Cambridge Univ. Press, 877pp., 1959.
11. Darwin, Sir George. *The Tides and Kindred Phenomena in the Solar System,* 3rd edn., John Murray, London, 437pp., 1911.

12. Yang, Z., K.O. Emery, Y. Xui. Historical development and use of 1000-year old tide-prediction tables, *Limnol. & Oceanogr.*, **34**, 5, 953–957, 1989.
13. Howse, Derek. Some early tidal diagrams, *The Mariner's Mirror*, **79**, 1, 27–43, 1993. (As late as 1667, the 2nd edition of the authoritative treatise *Hydrographie* by Georges Fournier, Ist edition, Paris, 1643, still described tides in terms of compass directions.)
14. Ferguson, James. A clock showing the apparent daily motion of the Sun and Moon, the Age and Phases of the Moon, with the time of her coming to the Meridian, and the Times of High and Low Water, by having only two Wheels and a Pinion added to the common Movement, pp.11–19 in *Select Mechanical Exercises …*, 272pp., London 1773.
15. Flamsteed, J., A correct TIDE TABLE showing the true times of High-Waters at London Bridge to every day in the year 1687, *Phil. Trans. R. Soc. London*, **16**, 232–235, 1686. (One of a series of similar annual tables starting **13**, 10–15, 1683, where their origin is partially explained.)
16. Philips, H. Time of the tides observed at London, *Phil. Trans. R. Soc. London*, **2**, 656–659, 1668.

4

Towards Newton

The Copernican Revolution

The European Renaissance movement brought a deeper level of enquiry into the nature of the physical world. Thinkers were concerned less with practical problems than with finding fundamental mechanisms which might put the observed phenomena into a more rational framework. For progress, it was necessary to take a sceptical view of time-honored physical ideas, mostly originating from Aristotle and sanctioned by the Church as part of religious dogma. The first and most far-reaching result of this new outlook was the hypothesis of Nikolaus Copernicus (1473–1543), that the earth was not the center of the universe but revolved round the sun together with the other planets. The 'Copernican Revolution' gained ground only slowly. It presented a simpler and more rational interpretation of the observed celestial motions than the ancient earth-centered hypothesis due to the Greek astronomer Ptolemy (2nd Century AD), but was at odds with Aristotelian mechanics. Galileo's new science of mechanics, which introduced inertia, opened the way for the Copernican Revolution in the 17th century. The Vatican, however, saw the new theory as contrary to Biblical teaching, and for many years banned the writings of Copernicus and his followers. Giordano Bruno (1548–1600) dared to suggest that not only did the earth revolve round the sun, but that the sun itself was merely a close star. He refused to retract from his beliefs and was burned at the stake as a heretic. We shall see that the attitude of the Vatican and the Inquisition to all thinkers who came within their reach was to have a profound effect on Galileo's thinking about the tides.

Let me say at the outset, that all the tidal theories to be described in this chapter, with the exception of the theory of Kepler who did not develop his ideas on tides to any extent, were all completely replaced in the light of Sir Isaac Newton's gravitational theory presented in 1687. It would be tempting to dismiss them with brief comments, if it were not that their authors were all quite distinguished in other fields and had for a time popular following.

On the other hand, it seems unnecessary to enlarge on some quite unscientific theories of the tides, involving for example animal breathing or heat radiated from the moon, which still persisted in some 16th century writings. I alluded to such ideas in Chapter 2 in the context of primitive thought, but some of them were refurbished by later writers and teachers. Shea[1] cites several such writers known to, and scorned by Galileo.

William Gilbert and Francis Bacon

The first plausible-seeming physical theory of the tides was postulated by William Gilbert (1544–1603), the pioneer of experimental research into magnetism and static electricity. Gilbert's greatest discovery, that the earth acts like a large magnet, was published during his lifetime. In a posthumous work published in 1651, entitled (in translation) *A New Philosophy of our Sub-Lunar World,* he went further, to suggest that the planets are held in their orbits round the sun by their mutual magnetic attraction, and that the tides were a manifestation of the magnetic attraction between earth and moon. (Being a member of the established Church of England, divorced from the Roman Church, Gilbert had no inhibitions about believing in a Copernican universe.) Unlike the later tidal theories of Galileo and Descartes, the suggestion of magnetic attraction comes remarkably close to the Newtonian law of gravity. But Gilbert had no experience of kinematics and so was unable to suggest what prevents the planets from falling into the sun, or the moon from falling into the earth, not to mention the twice-daily periodicity of the tides.

Gilbert's contemporary, Sir Francis Bacon (1561–1626), knew and respected his work on magnetism but never read his *New Philosophy*, so his speculations about tides are quite independent. Bacon is best known as a statesman, 'Lord Chancellor of England' under both Elizabeth I and James I. As well as his prowess as a statesman, Bacon was also a polymath and amateur of scientific enquiry. His *Essays*, written in his later years, cover a wide range of philosophical subjects, including natural knowledge. The Essay 'On the ebb and flow of the sea'[2] is somewhat verbose, but interesting for its disinclination to accept old theories which seemed implausible in the light of the known measurements. Where the measurements were insufficient, Bacon clearly advised where they should be supplemented.

Bacon first noted the progressive character of the tide from Gibraltar to the North Sea (as Bede had noted on a smaller scale), and the impossibility of accounting for this behavior in terms of a simultaneous expansion or contraction of the world's seas:

> That [the rise and fall] should be done so quickly, namely twice a day; as if the earth, according to that foolish conceit of Appolonius, were taking respiration, and breathing water every six hours and then taking it in again; is a very great difficulty.

After citing a supposed fact, that High Water in Florida occurred at about the same time as on the coast of Europe, Bacon proposed that the tide may progress northward over the entire Atlantic Ocean.

It is not clear where he obtained this fact. Along the Atlantic coast of Florida, High Water occurs about 1 hour before Gibraltar, 4 hours before, say, Galway. A later time could have been obtained from the region of Cape Sable at the extreme southwest tip of Florida. Otherwise simultaneity would have been valid between Florida and the Canary Islands whose tide is about 1 hour earlier than Gibraltar. But there were then no data south of Gibraltar. Bacon in fact exhorted mariners to record some times along the coast of west Africa. For similar reasons he advocated comparison of tidal times between China and Peru, that is, across the Pacific Ocean.

Bacon's speculation of a northward progressing tide wave happens to be correct over the entire eastern half of the Atlantic from South Africa to Norway, but its behavior in the western Atlantic is much more complex, for reasons far beyond contemporary knowledge of that period. The idea of oceanic tides as *progressive waves*, first suggested by the limited observations of Bede, was to dominate the researches of William Whewell in the early 19th century, but it was followed by a swing of opinion towards *stationary waves*, as discussed in Chapter 9. Stationary waves are also implicit in Bacon's ideas on resonance, below. Bacon postulated that the monthly and annual inequalities would be nearly simultaneous everywhere, as is largely correct. He suggested that measurements should be made to verify this fact. It is unusual to find someone of that epoch without professional experience of the sea taking such a global view of the ocean tides.

Bacon's suggestion for the *cause* of the tides was, however, informed by his adherence to a Ptolemaic, earth-centered universe. He observed that every celestial object travels from east to west relative to the earth, and supposed this to be a universal trend extending right down to the atmosphere and the sea. As evidence from the atmosphere, Bacon cited the westward tendency of the Trade Winds; he believed that the ocean would partake of the same westward flow if it were not obstructed by the great continents of 'the old and the new world' which cause the daily flow to be reflected. Reverberation between Africa and South America, he claimed, would somehow result in a twice daily oscillation, as observed:

> we think it necessarily follows that these two ramparts impart and communicate the character of double reaction to the entire mass of waters. Whence arises that motion in the quarter of a day, so that the waters being cooped in both sides, the ebb and flow of the sea would become visible twice a day, since there is a double advance and also a double recoil.

Although he did not share Gilbert's or Galileo's Copernican sympathies in this notion, Bacon evidently shared with Galileo a belief that the relation of the tide to the passage of the moon was a mere coincidence, the observed periodicity being due to some vague resonance of the ocean basins to an essentially daily (solar or sidereal) impulse.

Galileo's theory of tides

The much discussed theory of the tides by Galileo Galilei (1564–1642) was developed for the express purpose of providing a convincing proof of the Copernican hypothesis, that the earth not only rotates on its axis but also revolves in an orbit about the sun. He was convinced of the truth of this by his own telescopic observations of celestial rotations, but the Vatican was not so persuaded. His ideas were first aired in 1616 as *Discorso sopra il Flusso e Reflusso del Mare* (Discourse on the flow and ebb of the sea), but this discourse was not published in his lifetime. He incorporated the same ideas 18 years later (with some subtle changes of wording) in his long *Dialogo sopra i due massimi sistemi del mondo, Tolemaico e Copernicano* (Dialogue on the two main world systems, Ptolemaic and Copernican) of which tides were the subject of the 'Fourth Day'.[3] With this, Galileo hoped to end the Vatican's suppression of Copernican astronomy. Unsuccessful after long arguments, he was condemned by the Inquisition to complete retirement, and his works were banned from publication for the remainder of his life, under threat of direr retribution.

Essentially, Galileo held that the tides are the direct result of the accelerations due to the combined rotations of the earth on its axis and about the sun. In Figure 4.1, B is a basin near the earth's equator, center E, which rotates about the sun at S. To paraphrase Galileo's argument with modern notation, when B is at B_1, furthest from the sun, the basin has eastward velocity

$$R\,\mathrm{d}\phi/\mathrm{d}t + r\,\mathrm{d}(\theta+\phi)/\mathrm{d}t = R\,\Omega + r\,\omega,$$

where Ω, ω are rotation rates of one cycle per year and one cycle per sidereal day, respectively. (The obliquity of the ecliptic is ignored, for simplicity.) However, at B_2, nearest the sun, the basin's eastward velocity is

$$R\,\Omega - r\,\omega,$$

less than at B_1. Therefore B is periodically accelerated and decelerated, a state of affairs likened by Galileo to a tank of water being carried in a barge whose speed is changed. The water in the barge executes a stationary wavelike motion; hence the tide in the sea basin B.

The above argument may seem superficially persuasive, but as Aiton[4] points out, it is confused in mixing the frames of reference relative to the sun and to the earth. In more exact terms, the acceleration of B in an inertial framework is the vector sum of the centripetal vectors $r\omega^2$ along BE and $R\Omega^2$ along ES. Both vectors have components along BE which slightly alter the effective mass of the water without disturbing its equilibrium, but the net eastward *horizontal*

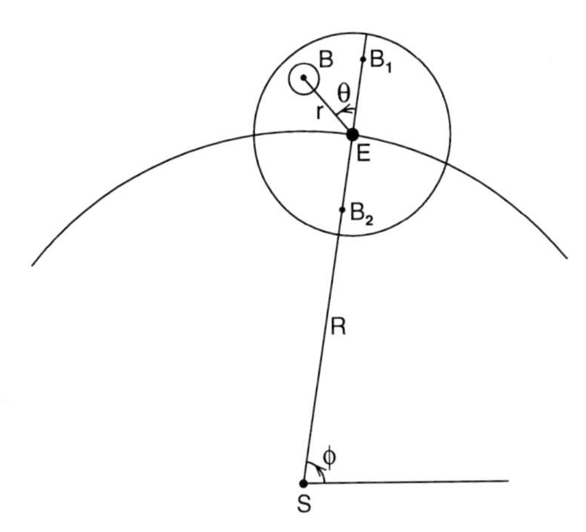

Figure 4.1. Diagram to illustrate Galileo's theory of the tides. B is a small sea basin rotating with the earth, center E, which itself rotates about the sun at S (Copernican system). The velocity of B relative to S is greater when at position B_1 than at position B_2. Therefore (in Galileo's reasoning), the sea experiences a daily cycle of acceleration which causes waves (tides) like the waves in an accelerated water-barge.

acceleration, $R\Omega^2 \sin\theta$, always has a component directed towards the sun. Ignoring inertial forces within the basin, the latter acceleration will cause the water in B to tilt upwards to the west or east, according to the sign of $\sin\theta$, so that it is balanced by horizontal pressure against the sides of the basin. An observer standing on the western shore of the basin would indeed see the water rise and fall with $\sin\theta$, that is, with a daily period. Galileo's main point was that, within the framework of his kinetics, that is, without invoking any attractive force between earth and sun, there would be no motion at all at B if E did not rotate about S.

Two obvious questions occur to the modern reader: why daily instead of half daily? and: what about the moon? But Galileo had a partial answer to both criticisms. In the first place, he was not convinced of a worldwide periodicity of half a lunar day. At Venice, the tide does indeed have a strong diurnal component (actually a sidereal day, but this could be taken for a solar day over a limited period of observation); at the Euripus, the so-called 'Tide' sometimes oscillates with a period of only a few hours. Still thinking of his water-barge analogy, Galileo maintained that each sea basin responds to the external disturbance with its own natural period, which happened to be about $12\frac{1}{2}$ hours in the Atlantic Ocean, he supposed, but which was different in other sea basins. (In reality, of course, resonances do play an important part in modern tide theory, but they do not alter the period of the external disturbance; they only magnify certain spectral components which are already present in the external stresses.)

Most critics castigate Galileo for excluding the moon from his tidal theory, but as Burstyn,[5] his kindest critic, insists in his discussion with Aiton, Galileo did at least point out the correct fact, that E in Figure 4.1 is really intermediate between the centers of earth and moon, so that the earth does indeed perform a small epicycle about the moon in a lunar month. However, this motion is far too small for any significant effect, and Galileo invokes it only to account (incorrectly) for the monthly inequality of the tides. His attempt to account for the annual inequality in terms of Declination had similar defects.

Galileo's tidal theory has attracted a surprising number of modern commentators,[1,4,5] most of whom are impressed by its originality of thought and occasional correct elements. If only he had stumbled on the idea that the dynamics of orbiting bodies require a mutual *attraction* to balance the accelerations, he might have been able to go further, to see that his diurnal effect disappears in a free orbit, leaving only a differential twice-daily attraction due to the different distances SB. But this insight had to await Newton's law of gravitation.

John Wallis

The development most akin to Galileo's concept was the Essay of John Wallis, Professor of Geometry at Oxford.[4,6] This essay was one of the earliest of many discussions on the tides to be published by the newly formed *Royal Society of London for Improving Natural Knowledge* (1663). The formation of such a Society, together with that of the *Académie Royale des Sciences* (1666) in Paris, marked an important step forward in the scientific enlightenment of the mid 17th century.

Wallis was prepared to go along with Galileo in recognising the significance of changes in absolute velocity along an orbit with axial rotation, but he was seriously concerned, where Galileo was not, in relating the tides to the moon:

> The sea's ebbing and flowing has so great a connexion with the moon's motion, that in a manner all philosophers have attributed much of its cause to the moon, which either by some occult quality, or particular influence which it has on moist bodies, or by some magnetic virtue, drawing the water towards it, which should therefore make the water highest where the moon is vertical, ... that it would seem very unreasonable to separate the consideration of the moon's motion from that of the sea.

Wallis attempted to involve the moon by observing (correctly) that it was the *Center of Gravity* of earth and moon which orbited the sun, and that this entailed a monthly perturbation in the velocity of all points in the earth. Galileo had suggested a very similar idea in trying to account for the monthly inequality, but Wallis was concerned here with the *daily* motion of the tides. However, by his arguments Wallis could only justify a period of one lunar day, where Galileo had a solar day. Wallis's hypothesis raised a good deal of controversy among the Society's Fellows, as recorded in the *Philosophical Transactions*. He replied with counter-objections, but finally advocated more direct observations of the tides as a means of settling the controversy.[6] We shall return to the observational activity of this period in Chapter 6.

Johannes Kepler

Later, in justifying against criticism his concept of an orbiting Center of Gravity, Wallis came close to postulating a gravitational attraction between all orbiting bodies.[6] He was not the first to do so, since as early as 1609 Johannes Kepler (1571–1630) had made similar suggestions, without developing them into a coherent theory.[7] Kepler's (eight) axioms for the 'true theory of gravity' include the following:

> If the moon and the earth were not retained by animal force, or some other equivalent force, each in its orbit, the earth would ascend toward the moon a fifty-fourth part of their distance, and the moon would fall toward the earth by 53 parts of this distance, where they would be united ...

(The moon/earth mass ratio was then reckoned by geometric reasoning to be about 1/54, on an assumption of equal density.) Further:

> If the earth ceased to attract (to itself) the waters of the sea, they would rise and pour themselves over the body of the moon.

And again:

> The sphere of the moon's drawing power extends to the earth and incites the waters under the Torrid Zone ... insensibly in enclosed seas, but sensibly where there are very broad beds of the ocean and abundant liberty for reciprocation of the waters.

Kepler's interests, of course, lay not in the tides, but in accounting for the laws of behavior of planetary orbits about the sun, for which he is justly famous, and which ultimately led directly to Newton's law of gravitation.

René Descartes

The last theory which we shall consider in this chapter, that of René Descartes, (1596–1650), is quite independent of any other theory, and the author makes no reference to any of them. Descartes accepted a heliocentric universe, but recognised that a theory for the tides must be strongly related to the moon. He provided for interaction between close celestial bodies through an ethereal substance which he supposed occupied the whole of space. The action of this 'ether' formed the basis for his *Theory of Vortices* exposed in Part 3 of his *Principes de la Philosophie*.[8] The theory of vortices supposes that the sun's rotation imparts a vortex motion to its surrounding ether, spreading outwards with decreasing angular velocity by some unspecified form of viscosity to the whole solar system, and carrying with it the planets in their orbits.

The theory makes no attempt to account for Kepler's laws of planetary motion, which would require considerable qualification of the solar vortex. Indeed, Newton[9] later proved that vortex motion requires the periods of

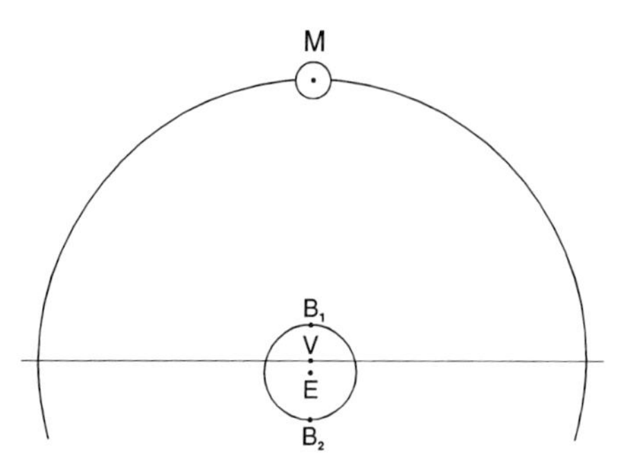

Figure 4.2. Diagram to illustrate Descartes' theory of the tides. The moon M is carried round the earth, center E, by a Cartesian vortex of 'ether' whose center is at V, where the distance VM is slightly less than EM. The 'ether' partaking in the vortex motion is slightly compressed at B_1 on account of the moon's proximity, and at B_2 on account of being further from the vortex center V. According to the hypothesis, the sea therefore has a tendency to Low Water at both B_1 and B_2; hence two tides per lunar day.

revolution at different radial distances to be proportional to the *square* of the radial distance, not to its power of $\frac{3}{2}$ according to Kepler's planetary laws. In the second of the two theorems cited,[9] Newton also showed that any planet carried round periodically by such a vortex must have the same density and motion as the fluid of the vortex. It is practically impossible to reconcile Descartes' theory with these requirements.

To continue, however, along Descartes' line of reasoning, the earth manifestly spins on its axis much faster than it revolves about the sun, so the earth too activates its own Cartesian vortex in the local ether, and this in turn is supposed to carry the moon round in its monthly orbit. The resulting stresses between the ether and the earth's surface cause interactions between the moon and the earth, producing tides, as explained in Part 4 of *Principes*,[8,10] and illustrated in Figure 4.2.

In Figure 4.2, E represents the earth's center and V, slightly displaced from E, is the center of the vortex in which the moon M rotates. It is supposed that when a point B on the earth's surface is at B_1 opposite M, the ether between B_1 and M is restricted and hence compressed by the presence of the moon; the increased downward pressure manifests itself in a Low Tide at B_1 and a small displacement of E from the center of the vortex in the direction EV. (The argument would be more convincing in a two-dimensional context of parallel cylinders.) The idea of moving E away from V, somewhat akin to Wallis's theory of the lunar-terrestrial center of gravity, is supposed to increase the pressure of the ether also at B_2, on the side of the earth remote from M. This, it was held, accounted for there being two tides each day.

For the monthly inequality (i.e. the spring–neap cycle), Descartes cited the small solar perturbation in the moon's orbit known as the *variation*, which had been observed astronomically by Tycho Brahe some 50 years previously. The *variation* has a period of half a synodic month, which is suggestive, but it amounts to only about 1/120 of the mean lunar distance, which is much smaller than the elliptic monthly inequality (about 1/18), ignored in this theory. (The *variation* was shown by G.H. Darwin more than two centuries later to account for only a very minor harmonic term in the tide, known as 'μ_2'.)

Descartes' tidal theory had wide popularity among his disciples in France, and it even survived the first few decades following Newton's gravitational theory. This was partly due to Descartes' deserved fame as a philosopher, and also to the popularity of the *Geography* of Varenius (or Bernhard Varen), which appeared in Latin in 1650 and ran into many editions. Varen adopted Descartes' as the least implausible of the then existing tidal theories, but he was not altogether uncritical. When the *Geography* first appeared in an English translation by Dugdale (1733),[11] it included critical notes made before his death by Sir Isaac Newton, and also by a disciple of Newton's, Dr. Jacob Jurin of Trinity College, Cambridge. While faithfully reproducing Varen's account of Descartes' theory, Jurin appended to it a summary of the physical defects in Descartes' reasoning and a brief account of Newton's theory which superseded it. On the vortex theory, Jurin writes:

> The Flux and Reflux of the Sea, which des Cartes [sic] has endeavoured to explain by an imaginary Plenum and Vortex may be more easily and fully explained upon other Principles ... for these [Descartes' assumptions] are mere Fictions and no way agreeable to Nature and Motion, as appears from the following Arguments ...

Jurin's following arguments are, in brief; 1. that with a 'Plenum' (ether) filling the universe, no body could move without moving every other body in the universe; 2. that comets are observed to move freely in all directions about the sun; 3. the formal dynamic objections to the vortex theory raised by Newton,[9] as mentioned above.

Notes and references

1. Shea, W.R.J. Galileo's claim to fame: the proof that the earth moves from the evidence of the tides. *Brit. J. Hist. Sci.*, 5, 18, 111–127, 1970.
2. Bacon, Francis. *De fluxu et refluxu maris,* (1623), transl. in: *Works,* (ed. J. Spedding, R. Ellis, D. Heath), 5, 443–458, Longman, London, 1858.
3. Galilei, Galileo. *Dialogo sopre i due massimi sistemi del mondo*, (1632), Opere (Edit. Nazionale, Firenze), 7, 21–250, 1890–1909. Engl. transl. Stillman Drake, 'Fourth Day', pp. 416–465 of *Dialogue ...*, Univ. California, Berkeley, xxvii + 496pp., 1962.
4. Aiton, E.J. Galileo's theory of the tides. *Ann.Sci.*, 10, 44–57, 1954.
5. Burstyn, H.L. Galileo's attempt to prove that the earth moves, *Isis,* 53, 2, 161–185, 1962. This paper was followed by a lengthy controversy between Aiton and Burstyn in *Isis,* 54, 265 and 400, 1963; 56, 56–63, 1965.
6. Wallis, John. Hypothesis on the flux and reflux of the sea, *Phil. Trans. R. Soc. London*, 1, 263–281, 281–289, 297–298, 1666. The article on pp. 281–289 is by way of 'Answers' to objections raised by various people; pp. 297–298 is a short note calling for more direct measurements.
7. Kepler, Johannes. *De fundamentis astrologiae certioribus,* (Pragae Boemorum, 1602), Translated in: A. Koyre, *The Astronomical Revolution* (Engl. transl. R.E.W. Maddison), esp. pp. 190–196, Methuen, London, 531pp., 1973.

8. Descartes, René. *Principes de la Philosophie,* (1647) in *Oeuvres,* ed. V. Cousin, **3**, 371–377, 1824.
9. Newton, Isaac. *Principia,* (1687), 3rd edn. (1726), Engl. transl. A. Motte, revised F. Cajori; **2**, Props. 52, 53, Univ. California Press, Berkeley, 1962.
10. Aiton, E.J. Descartes' theory of the tides. *Ann. Sci.*, **11**, 337–348, 1955.
11. Varenius, B., (Engl. transl. Dugdale). *A compleat system of general Geography,* 'improved and illustrated by Sir Isaac Newton and Dr. Jacob Jurin', London, 1733. Especially: Ch. XIV, 226–274, 'Of the motion of the Sea in general, and of its flux and reflux in particular'.

5

Newton and the Prize Essayists – the Equilibrium Theory

The *Principia* – Book I

It is hardly necessary to state that the *Philosophae Naturalis Principia Mathematica* of Isaac Newton (1642–1727) is one of the great milestones in the history of science. It is also well known that Newton himself was reluctant to publish his researches, but the genesis of the *Principia* was due to the encouragement and efforts of Edmond Halley, who published it at his own expense for the Royal Society of London in 1687.(1) In its three Books, Newton expounded the fundamental laws of dynamic motion and his concept of universal gravitational attraction between all massive bodies. He justified these by calculations and geometrical constructions whose results accorded with all then observed phenomena in the solar system, including Kepler's Laws of planetary motion (see Chapter 4). The ocean tides were of course included among the phenomena considered by Newton, but his account of them occupied only a few pages of Books 1 and 3, and in the Supplement entitled *The System of the World*.

Once grasped, Newton's theory of the tides is, unlike the theories described in the previous chapter, obviously correct in its basic principles. However, some of his deductions from these principles were incorrect, owing to an apparent misconception of the effects of vertical forces on the surface level of the ocean. For corrected accounts, we have to supplement Newton's basic ideas with those of Colin Maclaurin (1698–1746), Leonhard Euler (1707–1783), and Daniel Bernoulli (1700–1782), all of whom wrote seminal essays on the theory of tides in response to a Prize Competition offered in 1738 by the *Académie Royale des Sciences* of Paris. A summary of these Prize Essays, especially that of Bernoulli, is given at the end of the present Chapter.

It will be recalled that in his theory, René Descartes (Chapter 4) invoked the observed *variation* in the motion of the moon, that is, its perturbation by the sun with a period of half a synodic month, in an attempt to account for the monthly inequality or 'spring–neap' cycle. The *variation* of the moon also

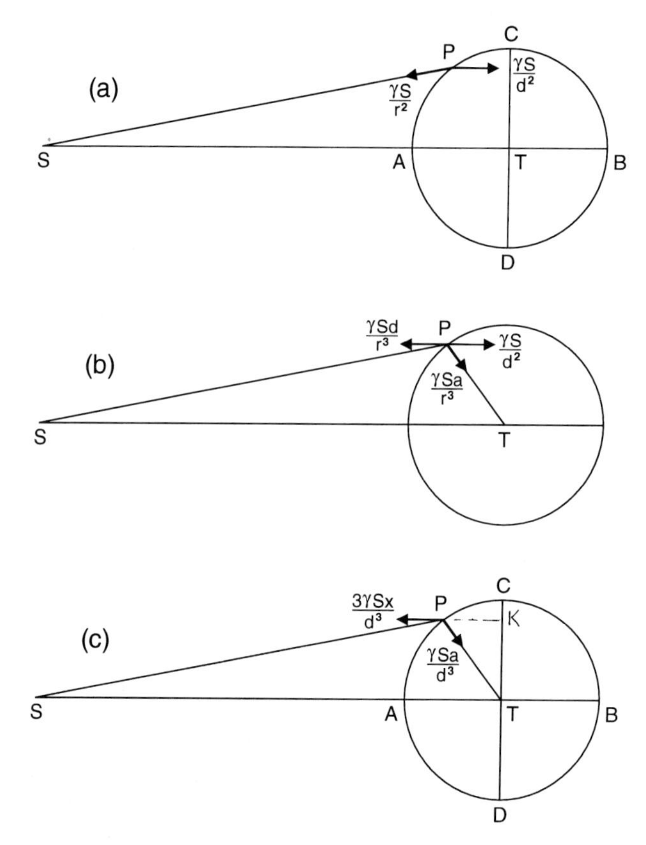

Figure 5.1. Three Figures adapted from Proudman's commentary,[2] based on Newton's own diagrams in Book I, Proposition, 66,[1] to illustrate the accelerative attractions between the sun (S), the earth (T), and a satellite P (e.g. the moon) orbiting around T in the direction ADBC. All points are supposed coplanar. The forces on P are illustrated by vectors in three equivalent ways in panels (a), (b), (c), with the notation $r=SP$, $d=ST$, $a=PT$, $x=KP$. The force vector at P parallel to TS is the variable tide-like component, whose sign reverses in the quadrants BC, DB. When P is considered to be the moon, this force produces the solar *variation* in the orbit. When considered as an element of fluid in a canal round the earth's circumference, it causes the tide – see text.

appears as the starting point for Newton's tidal investigations, but he adapted the concept in an entirely original way to account for the very existence of the tides. He started (Book I, Proposition 66) by explaining in terms of his gravitational theory how the sun perturbs the moon's orbit in this particular way.

In order to aid the reader to follow Newton's verbal arguments, I have reproduced in Figures 5.1(a, b, c) the three diagrams used by J. Proudman in his commentary of 1927 on the same subject.[2] Figures 5.1 embody the essential elements and lettering of Newton's own diagram, reproduced on five pages of the *Principia*, (Book I, Prop.66), but they are easier for a modern reader to follow. In these Figures, S and T are the centers of gravity of the sun and earth respectively; P is (initially) the position of a satellite body such as the moon in its unperturbed easterly orbit ADBC. A and B are the positions of the *syzygies* or of *conjunction*, and C and D the positions of *quadrature*, at right angles to the line of syzygies. The sectors AD, DB, BC, CA are referred to as the 1st, 2nd, 3rd and 4th *quadrants*, respectively, and the first 45° sector from A towards D is called the first *octant*, and so on. (See Appendix A for definitions of specialised astronomical jargon.)

The annotated arrows, added by Proudman,[2] summarise the vector components of the perturbing forces per unit mass at P, relative to T. Here, γ is Newton's universal constant of gravitation, and S is used to denote the sun's mass, with $r = \mathrm{SP}$, $d = \mathrm{ST}$, $a = \mathrm{PT}$. Figure 5.1(a) shows the inverse-square attraction of S on P partially balanced by the orbital acceleration of T. The attraction between P and T is supposed constant, and balanced by the orbital acceleration of P towards T. The three force components shown in Figure 5.1(b) are exactly equivalent to those in 5.1(a), and follow from the rule for force triangles. The third version in Figure 5.1(c), in which $x = \mathrm{PK}$, reduces the same forces to just two components, using the expansion of r^{-3} to first order of the small quantity (a/d). The dependence of the disturbing forces on the inverse *cube* of the distance $d = \mathrm{ST}$ is deduced verbally by Newton in Proposition 66, Corollary14.

From the configuration of the forces depicted in Figure 5.1(c), one sees (Prop. 66, Cor. 2, 3) that the eastward angular motion of the satellite is accelerated in the quadrants CA, DB, and retarded in the quadrants AD, BC. It therefore moves fastest at the syzygies A, B and slowest at the quadratures D, C. These are exactly the characteristics of the *variation*, first observed by Tycho Brahe around 1585. Newton also simply deduced (Cor. 4, 5) that the satellite is deflected closer to the earth at the syzygies and farther from the earth at the quadratures, a now wellknown characteristic. The salient fact for our interest, however, is that the perturbations have fundamentally half the synodic period of P in its rotation about T, namely 14.8 days.

Newton's translation of these concepts into the theory of tidal motion (original Latin text reproduced in Figure 5.2) was a stroke of genius. He replaced the body P by a fluid particle accompanied by a string of similar particles, in the limit coalescing into a rotating ring of fluid. The radius of the solid body with center T was then expanded to reach that of the fluid ring, which thus became a canal of fluid supported (in a circular groove) by a spherical solid body. With the spherical body further supposed to rotate about an axis normal to ADBC, we have in effect an equatorial canal, or canal following any parallel, round the earth, whose particles are disturbed by the sun's gravity to exhibit (solar) tidal motion. The principal difference from the previous role for P as a freely orbiting satellite is that the earth's rotatory period, which is logically shared by the fluid in the canal, is much slower than that required for a free orbit at the equatorial radius. The fluid is therefore constrained to the circular form of the canal, except for a slight deformation of the free surface.

The same perturbing forces apply as in Figure 5.1(c) with $a =$ earth radius, and, the synodic period being now a solar day, we have tidal stresses applied

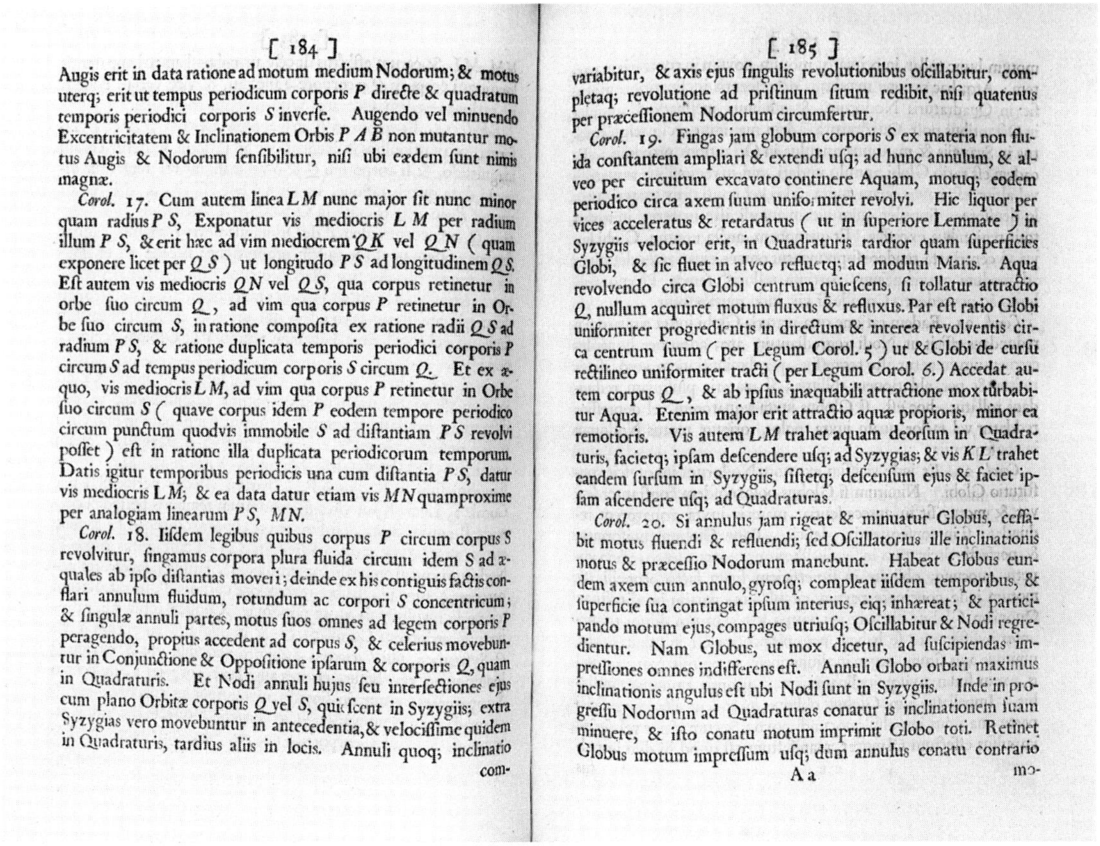

[184]

Augis erit in data ratione ad motum medium Nodorum; & motus uterq; erit ut tempus periodicum corporis *P* directe & quadratum temporis periodici corporis *S* inverſe. Augendo vel minuendo Excentricitatem & Inclinationem Orbis *P A B* non mutantur motus Augis & Nodorum ſenſibilitur, niſi ubi eædem ſunt nimis magnæ.

Corol. 17. Cum autem linea *L M* nunc major ſit nunc minor quam radius *P S*, Exponatur vis mediocris *L M* per radium illum *P S*, & erit hæc ad vim mediocrem *Q K* vel *Q N* (quam exponere licet per *Q S*) ut longitudo *P S* ad longitudinem *Q S*. Eſt autem vis mediocris *Q N* vel *Q S*, qua corpus retinetur in orbe ſuo circum *Q*, ad vim qua corpus *P* retinetur in Orbe ſuo circum *S*, in ratione compoſita ex ratione radii *Q S* ad radium *P S*, & ratione duplicata temporis periodici corporis *P* circum *S* ad tempus periodicum corporis *S* circum *Q*. Et ex æquo, vis mediocris *L M*, ad vim qua corpus *P* retinetur in Orbe ſuo circum *S* (quave corpus idem *P* eodem tempore periodico circum punctum quodvis immobile *S* ad diſtantiam *P S* revolvi poſſet) eſt in ratione illa duplicata periodicorum temporum. Datis igitur temporibus periodicis una cum diſtantia *P S*, datur vis mediocris L *M*; & ea data datur etiam vis *M N* quamproxime per analogiam linearum *P S*, *M N*.

Corol. 18. Iiſdem legibus quibus corpus *P* circum corpus S revolvitur, fingamus corpora plura fluida circum idem S ad æquales ab ipſo diſtantias moveri; deinde ex his contiguis factis conflari annulum fluidum, rotundum ac corpori *S* concentricum; & ſingulæ annuli partes, motus ſuos omnes ad legem corporis *P* peragendo, propius accedent ad corpus *S*, & celerius movebuntur in Conjunctione & Oppoſitione ipſarum & corporis *Q*, quam in Quadraturis. Et Nodi annuli hujus ſeu interſectiones ejus cum plano Orbitæ corporis *Q* vel *S*, quieſcent in Syzygiis; extra Syzygias vero movebuntur in antecedentia, & velociſſime quidem in Quadraturis, tardius aliis in locis. Annuli quoq; inclinatio com-

[185]

variabitur, & axis ejus ſingulis revolutionibus oſcillabitur, completaq; revolutione ad priſtinum ſitum redibit, niſi quatenus per præceſſionem Nodorum circumfertur.

Corol. 19. Fingas jam globum corporis *S* ex materia non fluida conſtantem ampliari & extendi uſq; ad hunc annulum, & alveo per circuitum excavato continere Aquam, motuq; eodem periodico circa axem ſuum uniformiter revolvi. Hic liquor per vices acceleratus & retardatus (ut in ſuperiore Lemmate) in Syzygiis velocior erit, in Quadraturis tardior quam ſuperficies Globi, & ſic fluet in alveo refluetq; ad modum Maris. Aqua revolvendo circa Globi centrum quieſcens, ſi tollatur attractio *Q*, nullum acquiret motum fluxus & refluxus. Par eſt ratio Globi uniformiter progredientis in directum & interea revolventis circa centrum ſuum (per Legum Corol. 5) ut & Globi de curſu rectilineo uniformiter tracti (per Legum Corol. 6.) Accedat autem corpus *Q*, & ab ipſius inæquabili attractione mox turbabitur Aqua. Etenim major erit attractio aquæ propioris, minor ea remotioris. Vis autem *L M* trahet aquam deorſum in Quadraturis, facietq; ipſam deſcendere uſq; ad Syzygias; & vis *K L* trahet eandem ſurſum in Syzygiis, ſiſtetq; deſcenſum ejus & faciet ipſam aſcendere uſq; ad Quadraturas.

Corol. 20. Si annulus jam rigeat & minuatur Globus, ceſſabit motus fluendi & refluendi; ſed Oſcillatorius ille inclinationis motus & præceſſio Nodorum manebunt. Habeat Globus eundem axem cum annulo, gyroſq; compleat iiſdem temporibus, & ſuperficie ſua contingat ipſum interius, eiq; inhæreat; & participando motum ejus, compages utriuſq; Oſcillabitur & Nodi regredientur. Nam Globus, ut mox dicetur, ad ſuſcipiendas impreſſiones omnes indifferens eſt. Annuli Globo orbati maximus inclinationis angulus eſt ubi Nodi ſunt in Syzygiis. Inde in progreſſu Nodorum ad Quadraturas conatur is inclinationem ſuam minuere, & iſto conatu motum imprimit Globo toti. Retinet Globus motum impreſſum uſq; dum annulus conatu contrario mo-

A a

Figure 5.2. The original Latin text of Book I, Proposition 66, Corollaries 17–20 of Newton's *Principia*, in which the essential steps were made to define the tidal forces. The main theorem of Proposition 66 concerns the interactions between two coplanar bodies orbiting about a third body, all subject to an inverse-square law of attraction. In Cor. 17, Newton refers to a diagram related to Figure 5.1(c), in which the variational force is represented by a vector LM. The fact that the magnitude of LM involves $1/d^3$ is established by him in Cor. 14, (not shown here). He goes on in Cor. 17 to show that the force is proportional to the square of the ratio of the orbital periods of P about T and of T about S. The key transition from an orbiting body P to a circular canal of fluid on the surface of the earth is made in Cor. 18–20, where he considers the effects of replacing P by a large number of contiguous fluid particles forming a ring (Cor. 18) to which the radius of the solid earth is expanded, (Cor. 19). The same variations to which the orbit was subjected then apply in exact analogy to cause tides in the water of the circular channel. In Cor. 20 Newton observes that the pressure at the bottom of the canal takes the place of the centrifugal force of the orbit. (The passage from Cor. 20 whose English translation is quoted in the text occurs on the page following the above excerpt.) (Photograph courtesy of the Royal Society.)

with a period of 12 hours. If the sun at S is replaced by the moon we similarly get a tidal period of half a lunar day, 12.42 hours. As before, the fluid is accelerated eastwards in the 2nd and 4th quadrants, (DB and CA), westwards in the 1st and 3rd quadrants, (AD and BC).

Now, as observed by Aiton,[3] if we take these accelerations at their face value, we get maximum eastward *velocity* at A and B and minimum *velocity* at C and D. By simple application of mass continuity this implies Low Water at the upper and lower transits of the sun and High Water when S is on the horizon. Neglecting inertial forces, this result is now known to be correct for an equatorial canal of ocean-like depth. But Newton did not follow that line of reasoning; he chose instead to consider only the *vertical* force along PT as the sole activator of tidal motion. This leads to quite different results for the positions of High and Low Water. His arguments being somewhat obscure, they are presented here in his own wording (Prop. 66, Cor. 20, Motte–Cajori translation of the original Latin):

> ... for the water is now no longer sustained and kept in its orbit, but by the channel in which it flows. And besides, the force (PT) attracts the water downwards most in the quadratures, and the force (along KP) upwards most in the syzygies. And these forces conjoined cease to attract the water downwards, and begin to attract it upwards in the octants before the syzygies; and cease to attract the water upwards, and begin to attract the water downwards in the octants after the syzygies. And thence the greatest height of the water may happen about the octants after the syzygies; and the least height about the octants after the quadratures; excepting only so far as the motion of ascent or descent impressed by these forces may by the inertia of the water continue a little longer, or be stopped a little sooner by impediments in its channel.

In the light of later knowledge this passage seems to contain two misleading notions; (1) the assumption that the water level is controlled by the *vertical* component of the tidal force; (2) vagueness as to how the water responds to varying vertical forces. Regarding (1), it was later shown by Euler[5] and by Laplace[6] that tidal motion is entirely controlled by the *horizontal* stress relative to gravity. Newton (Cor. 20 – see Figure 5.2) mentions that the pressure on the sea bed replaces the centrifugal force of the orbiting configuration, but, as Proudman[2] observed, he did not fully realise its implication in cancelling the effect of the vertical component of the *tidal* force.

Regarding (2), let us first note that, on removing the common factor $\gamma S/d^3$ from the two components shown in Figure 5.1(c), the upwards vertical force is proportional to $(3x^2/a - a)$, which is maximum at A and B where $x = \pm a$, minimum at C and D where $x = 0$, and zero where $x^2 = a^2/3$, about 55° before and after A and B. However, Newton was almost certainly thinking of the *variable* part of the upward vertical force, proportional (in later terminology) to the cosine of twice the angle from A or B which is zero at the ends of octants 1, 3, 5, 7. Evidently, inertia and acceleration were not considered to exercise a major control over the canal's response to these forces.

If, on the other hand, we were to suppose that the water level were to rise in exact proportion to the upward vertical force, as in the sea's response to barometric pressure, then statically speaking, High Water would occur at the syzygies A and B, Low Water at the quadratures. This would accord with Bernoulli's *Equilibrium Tide*;[(7)] but Newton postulates in the above passage that the water continues to rise as long as the upward vertical force remains positive, corresponding perhaps to motion controlled mainly by friction. The latter supposition does produce maxima and minima at the octants, as Newton suggests. In summary, Newton seems to be considering, not what he later called an 'Equilibrium' theory, but a motion controlled by friction which increases with velocity, and therefore maximum when the applied force is zero, but slightly modified by inertial forces proportional to acceleration.

Principia – Book III

Whatever mechanism Newton had in mind, the apparent result, that oceanic High Tide lags the syzygies by roughly 3 hours, seemed at the time to be supported by much of the then available data from oceanic coasts (as well as London), while later delays could, he supposed, be accounted for by propagation through shallow water. Thus, at the start of Book III, Proposition 24:

> ... it appears [from Book I] that the waters of the sea ought twice to rise and twice to fall every day, as well lunar as solar; and that the greatest height of the waters in the open and deep seas ought to follow the approach of the luminaries to the meridian of the place by a less interval than six hours; as happens in all that eastern tract of the Atlantic and Ethiopic seas between France and the Cape of Good Hope, and on the coasts of Chile and Peru in the South Sea; in all which shores the flood falls out about the 2nd, 3rd or 4th hour, unless where the motion propagated from the deep ocean is by the shallowness of the channels through which it passes to some particular place, retarded to the 5th, 6th or 7th hour, and even later.

The statements about the eastern Atlantic and Pacific coasts are in fact confirmed by modern measurements, excepting a stretch of the coast of West Africa, but they in no way hold consistently over the oceans as a whole, as now known. Nevertheless, it was no doubt the apparent agreement of the theory in question with currently known observations which deterred Newton from seeking refinements to his hypothesis of the sea's response to vertical forces.

In the three Propositions of Book III concerning tides, viz. Propositions 24, 36, 37, Newton explored the consequences of the Theorems and Corollaries of Book I, explaining the known features of ocean tides and evaluating the physical magnitudes of the tidal forces. After expressing confidence in the 3 hour delay discussed above, he went on to show that the combination of the tides raised by the sun and by the moon result quite simply in the cycle of spring and neap tides whose explanation had puzzled so many of his predecessors. Further, because of the inverse cube law of distance, he showed that the solar tides are greater in winter time (of the present era) because the earth is then near its *perihelion*; also, that spring tides have alternately greater and less amplitude than average when the moon's perigee (close approach to earth in the moon's elliptic orbit) is close to the syzygy. These properties are manifestly correct.

Continuing (in III, Prop. 24), Newton went on to show that as the declination (angular distance from the equator) increases, both luminaries have less propensity to generate regular (i.e. twice daily) tides, but greater propensity to produce a diurnal inequality in the tides. For example, in June, a place in the Northern hemisphere comes closer in angular distance to the sun in its daytime (upper) transit, but further in the the night-time (lower) transit. Accordingly, the daytime flood tends to be greater than the night-time flood in summer, assuming that the phase lag is small. The same applies to the lunar tide, since the moon's declination also increases in summer, and the effect is of course reversed for both luminaries in winter. These properties are manifest, and they admirably account for the main variations of the diurnal inequality at most places.

Newton noted that the diurnal inequalities at Plymouth and Bristol, where they had been most carefully observed, were considerably less in relation to the semidiurnal tidal amplitude than his theory suggested. He attempted to explain this in terms of inertia of local seas. The modern explanation, involving magnification of the semidiurnal and suppression of the diurnal responses of the North Atlantic, is much more complex, but Proudman(2) points out that Newton was essentially correct in every statement he made, except in his neglect of the effects of the earth's rotation.

Also in Proposition 24, Newton offered a famous explanation of a very large diurnal inequality at Tonquin (South China Sea) which had been reported to the Royal Society by F. Davenport with a commentary by Halley(8). It was, he said, owing to the regular tide arriving there by two different routes, from the Indian and Pacific Oceans, one producing 6 hours longer delay than the other, thus annulling the semidiurnal wave by phase cancellation but magnifying the diurnal wave. This again is an essentially correct, if greatly simplified, description of the local tidal regime, and it stands, incidentally, as the first cited example of interference between two wave trains.

At the conclusion of Book III, Proposition 24, he writes: 'Thus I have explained the causes of the motions of the moon and of the sea. Now it is fit to subjoin something concerning the amount of those motions.' Propositions 25–35 concern only the quantification of the moon's motion as disturbed by the sun. Quantification of the tidal forces on the sea is taken up in Proposition 36.

In Book III, Proposition 36 Newton relates the magnitude of the sun's tide-raising force to known terrestrial quantities. To do this, he ingeniously combined four relations already established in other parts of the *Principia*. In terms of the forces represented in Figure 5.1(c), which are equivalent to Newton's own expressions, the gravitational constant γ is simply related to terrestrial gravity g

(force per unit mass) by $\gamma T = ga^2$, where T is the mass of the earth. Therefore, the mean vertical component of the sun's tidal force,

$$\gamma S \frac{a}{d^3} = g \frac{S}{T} \frac{a^3}{d^3}.$$

In Proposition 25 he had shown that the ratio of the gravitational attraction between sun and earth to that between earth and moon, namely

$$\frac{S/d^2}{T/a'^2} \text{ is equal to } m^2 \frac{a'}{d},$$

where a' is the mean earth:moon distance and m is the ratio of their mean periods, 27.32:365.24. Accordingly,

$$\frac{Sa'^3}{Td^3} = m^2,$$

in agreement with Kepler's third law, namely that the square of a satellite's period is proportional to the cube of its mean distance. We therefore see that the mean vertical solar tidal force is

$$\frac{\gamma S a}{d^3} = gm^2 \frac{a^3}{a'^3}.$$

By employing an observational value for $a/a' = 1/60.5$, (discussed in detail in III, Prop. 4), the last expression was found by Newton to be

$$g/38{,}604{,}600 = g/K, \text{ say.}$$

Considering the roughness of the the astronomical quantities used, the quoted value for K compares plausibly well with the modern figure of $39{,}231 \times 10^3$.

The range (i.e. double amplitude) of the variable vertical tidal force is three times the value derived for g/K. In order to relate this to an equivalent range of water level, assuming an 'Equilibrium' theory, (that is, one which neglects inertial forces), Newton finally employed the ingenious device of comparing the vertical force with the ratio of the earth's equatorial bulge to its centripetal acceleration at the equator, divided by g. For this he used his own estimate of the excess radius,

85,472 'Paris feet' (26.49km),

which is roughly comparable with the modern value of 21.39km. His final result for the range of the solar equilibrium tide,

23 'Paris inches' (0.60m),

is much larger than the equivalent modern value of 0.25m, or 0.33m if one allows for self-attraction of the tidal bulge. Newton's still more dubious estimate of the range of the lunar tide in III, Proposition 37 is discussed in the following section.

The System of the World

Book III, which concludes the *Principia* is entitled *System of the World (in mathematical treatment)*. The English translations of the *Principia*, however, append a separate book, entitled *The System of the World*, in which Newton summarised his main findings with less mathematical rigor. Sections 50–53 of the appended book are of special interest to us, because therein Newton offered another set of arguments to quantify the range of the solar tide, which gives a quite different result from that given in the *Principia*. He considered two vertical wells penetrating the entire diameters from A to B and from C to D, intersecting at T (Figure 5.1(a)). When the sun is over A the downward vertical forces at A and at C are in the ratio

$$(1-2/K)\ /\ (1+1/K)=1-3/K$$

very approximately. Since gravity inside the earth as well as the tidal perturbations are proportional to distance from T (proved elsewhere in the *Principia*), the pressure at the bottom of each well is proportional to the square of its water depth. These pressures must balance at T, so the ratio of the water heights at A and C is the inverse square-root of the ratio of the forces, that is $(1+3/2K)$. The difference in the heights of the two tide levels is therefore $3a/2K=9$ inches (0.24 m). This is much less than the height of 23″ derived in *Principia*, Book III, Proposition 36, but it is close to the modern value derived from the *primary solar potential*, without adjustments for crustal deflection or for self-attraction.

Newton made no comment on the two different estimates, 23″ and 9″, but Proudman[2] states that the difference between them is entirely accounted for by the fact that Newton's theory for the equatorial bulge (*Principia*) does allow for the self-attraction of a deformed liquid sphere, whereas the lower value arose from considering only the direct attraction of a rigid sphere. Although neither figure is exactly correct for the natural configuration of the oceans, rigorously calculated values closely similar to the *lower* figure are adopted to the present day as a useful reference for comparison with observed tidal amplitudes.

In the absence of a reliable estimate for the mass ratio of moon:earth, M/T, Newton was unable to make a *direct* estimate of the tidal force or equilibrium displacement due to the moon. However, in III, Proposition 37 he had estimated these quantities indirectly by roughly comparing the relative contributions of the sun and moon to the mean spring and neap tides, as observed at Bristol by Sturmy (Chapter 6). If the sun:moon tide ratio is r, then the spring and neap ranges should by simple reasoning be approximately as $(1+\mathrm{r}):(1-\mathrm{r})$. There are various sources of inaccuracy in this simplified assumption, and Newton exacerbated them by equating r to the force ratio at the actual time of springs and neaps, about 36 hours later than the peak forces. Consequently, his estimate for r, namely 1:4.48 was much smaller than the true value, 1:2.15. (Newton's methods of selecting data have been strongly criticised by Westfall.[4]) Using his larger value for the solar force, 23″, he obtained 8.6 feet

(2.7 m) for the range of the lunar equilibrium tide, very much larger than the value used today. The same ratio applied to the 9″ figure (*System*, Section 53), gives a lower result, but still exaggerated by the small solar:lunar ratio which Newton assumed. Better results were made by Bernoulli[7], about 50 years later, using another method based on observations. Bernoulli's work will be described later in this Chapter.

Whatever its limitations, the use of tide observations to deduce the sun:moon mass ratio was for long considered by astronomers to be more reliable than any available astronomical method until well into the 19th century,[9] It was eventually superseded by observations of the minor planet *Eros* at its closest approaches to earth, and more recently by the precise methods of space geodesy.

Essays for the Académie Prize of 1740

To summarise the above account of Newton's revolutionary theory of tides, the basic theory was correct in principle, but, like many pioneering ideas, it was in need of refinement in quantitative detail and in application to real measurements. During the 18th century, natural philosophers in Britain paid scant attention to tides, but the *Académie Royale des Sciences* in Paris sponsored several important projects to study and improve tidal observation and prediction at French ports (Chapter 6). In the early years of the 18th century, the leading members of the Académie were slow to accept Newton's gravitational theory and still looked to Descartes' vortex theory to improve tidal predictions. Dissatisfied with progress, in 1738 the Académie offered one of its prestigious prizes for the best philosophical essay on *Le flux et le réflux de la mer* – the flood and ebb of the sea. In 1740 the prize was shared between four distinguished international contestants; Daniel Bernoulli, Professor of Anatomy and Botany at Basel; Antoine Cavalleri, Jesuit, Professor of Mathematics at Cahors; Leonhard Euler, Professor of Mathematics at St. Petersburg; and Colin Maclaurin, Professor of Geometry at Edinburgh. All four essays (the last two mentioned being in Latin) were printed by the Académie in their *Receuil* of 1752.[10] Those of Bernoulli, Euler and Maclaurin were later reproduced in the 'Jacquier & Le Seur' edition of Newton's *Principia*.[11]

Cavalleri's essay has the dubious reputation of being the last attempt in the history of science to justify the Cartesian theory of vortices. While he disagreed with Descartes' tidal theory on the grounds that it discounted the manifest direct influence of the sun, Cavalleri also objected to Newton's theory of remote gravitational attraction. He therefore attempted to prove that an inverse square law of attraction could result from Cartesian vortices. Apart from disregarding Newton's theoretical objections to the vortex theory, Cavalleri misinterpreted a modified vortex theory by Villemot[12] which attempted to reconcile it with Newtonian attraction. Cavalleri's essay (in French) may be seen in ref. 10, but being opposed to gravitational theory it was not included in ref. 11. Aiton's account of the work of the Cartesians[13] gives an adequate summary.

The other three authors all accepted Newton's theory of universal gravitational attraction as incontestably proven, and extended its implications for tides in significant ways. We hear no more of Newton's dubious hypothesis of a 3-hour delay in oceanic response, but these authors based their work on the force field described by Figure 5.1, with maximum upward force at the nadir A of the attracting body and at its opposite pole B.

Colin Maclaurin and Leonhard Euler

Maclaurin's essay *De Causa Physica Fluxus et Refluxus Maris*, (On the physical cause of the flux and reflux of the sea) is a purely geometrical exercise in the theory of *Fluxions*, of which he was a leading authority. By the fluxional calculus he was able to prove what Newton had merely assumed, that the shape of an otherwise spherical ocean in static equilibrium with the tidal force of a disturbing body is a prolate spheroid whose major axis points towards that body. Harris[15] has pointed out that Maclaurin also appears to be the first scientist to appreciate the deflecting action due to the earth's rotation, now known as the *Coriolis* effect. The important role of this effect in tidal dynamics was later properly analysed by Laplace (Chapter 7); Maclaurin did not go further into dynamical theory.

The most important contribution of Euler in his essay *Inquisitio Physica in Causam Fluxus et Refluxus Maris* (Physical enquiry into the cause ...) was to show that it is the *horizontal*, not the vertical component of the force field, which determines the tidal motion. The vertical component is balanced by pressure on the sea bed, but the ratio of the horizontal force per unit mass to vertical gravity g has to be balanced by an opposing slope of the sea surface, as well as by possible changes in current momentum. From the force components of Figure 5.1(c), the horizontal component is easily seen to be proportional to sin 2θ, where θ is the angle PTS. Euler also took into account the second term in the expansion of (a/r), and using the relation $\gamma T = ga^2$, showed that the elevation of the sea surface above its undisturbed level has the form:

$$\frac{a}{2}\frac{S}{T}\rho^3[(3\cos^2\theta - 1) + \rho\,(5\cos^3\theta - 3\cos\theta)]$$

where S/T is the mass ratio sun:earth, and ρ is the sun's equatorial parallax, a/d. The inner bracketed expressions are recognised today as simple multiples of *Legendre functions*. The first one is identical to the first approximation to a prolate spheroid, providing another confirmation of the form inferred by Newton from the vertical force component. Its range, $1.5a(S/T)\rho^3$, is about 0.25 meters, corresponding to Newton's *lower* estimate, 9 inches. The second term, proportional to the 4th power of the parallax, is quite negligible for the solar tides, but its leading lunar components have been detected in recent years by high resolution spectroscopy of long records of observations, on account of the moon's much larger value of $\rho = 1/60$.

Euler did not analyse the dynamics of horizontal motion, but rather incongruously, and unsuccessfully, attempted to develop a dynamical theory based on *vertical* water motion. The correct theory, as propounded by Laplace, (Chapter 7), develops the *horizontal* motion and invokes the vertical velocity through the continuity of water mass.

Daniel Bernoulli

Bernoulli's *Traité sur le flux et le réflux de la mer*,[7] though uneven in quality, had direct application to the analysis and prediction of the natural tides, and so it had a longer lasting reputation than the other three Prize Essays. Indeed, the so-called *Equilibrium Theory* is usually associated with Bernoulli rather than with Newton. English summaries of his essay were published much later by J.W. Lubbock[14] and in the historical survey by Harris.[15] See also Aiton.[3]

Bernoulli assumed, justifiably, that for purposes of tidal calculation the undisturbed figure of the earth may be taken as a perfect sphere, and that the form of the static tidal deformation of the surface is a prolate spheroid. At that time he was, of course, unaware of Maclaurin's rigorous justification of the prolate spheroid. In his first few chapters Bernoulli derived expressions for the gravitational attraction of spherical shells of uniform density, of a complete spheroid, and of a meniscus whose outer surface is spheroidal and inner surface spherical. Most of his results confirm those of Newton, but in applying them to the (solar) tide he noted that the amplitude depended on the density ratio ocean:earth. Unfortunately, a flaw in Bernoulli's arguments, first pointed out by D'Alembert,[16] led him to false conclusions in this area – see Lubbock.[14]

Bernoulli's most productive line of argument was to analyse the geometrical structure of the combined Equilibrium Tide generated by the sun and the moon. Uncertainty about the relative amplitudes was no handicap, because his practical objectives were to fit the characteristics of any observed tide regime to a canonical form determined from astronomy, from which predictions could be made. In this he was remarkably successful within the limited reality of purely semidiurnal tides. His computed tide-tables (Figure 5.4) were the first practically reliable advance in prediction technique since the crude rules of thumb of the 13th century.

In Chapter 5 of his essay, Bernoulli proved first that to conserve volume the tidal spheroid due to either luminary must have an elevation at nadir equal to twice the depression at the quadratures. That is, Aa = 2Cc in Figure 5.3. This result is the same as Newton's, except that it was here derived from analytic geometry instead of from force vectors. More generally, the elevation Xx at an arbitrary angle θ from CS is

$$h(\cos^2\theta - \tfrac{1}{3})$$

where h is the range (Aa + Cc) of the solar tide. This accords with the leading term of Euler's result, of which Bernoulli was of course unaware.

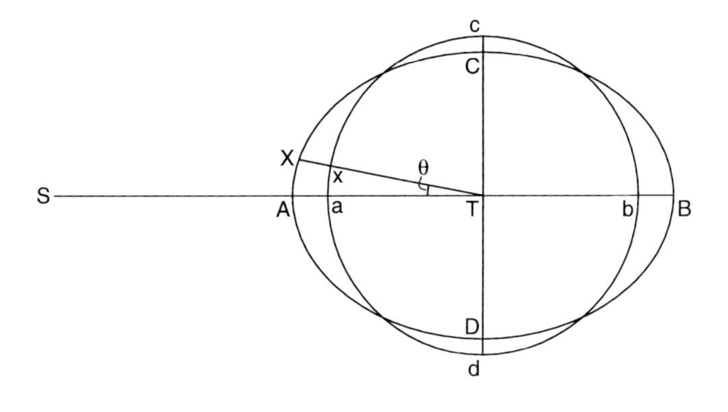

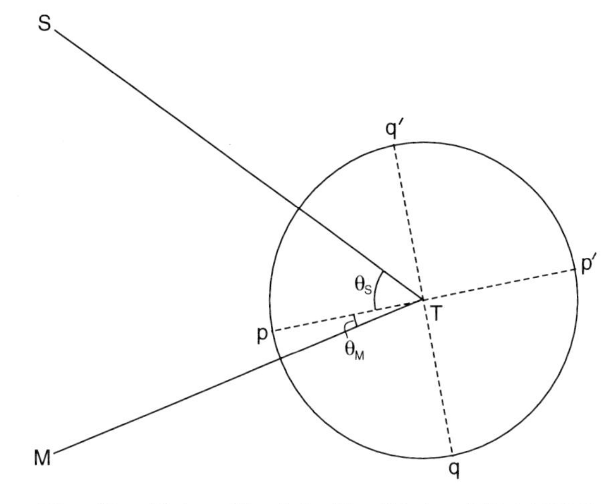

Figure 5.3. Above: The ellipsoidal profile of the 'Equilibrium Tide' ADBC compared in exaggerated scale with a spherical earth adbc. S is the sun (or moon) as in Newton's diagrams, and θ is the angular parameter for defining the height difference Xx, (see text). Below: Bernoulli's diagram for the parameters of the combined Equilibrium Tide due to the sun S and the moon M. pp′ is the line of syzygies.

Denoting by H the corresponding range of the lunar equilibrium tide, Bernoulli then considered the combined profile:

$$h(\cos^2\theta_S - \tfrac{1}{3}) + H(\cos^2\theta_M - \tfrac{1}{3})$$

where θ_S, θ_M are the angles relative to the sun and moon, shown in Figure 5.3 (lower). Such a profile has maxima at the longitudes of p, p′ and minima at q, q′. By differentiation, he showed that the angles θ_S at p, p′, q, q′ are given by

$$\sin^2\theta_S = \tfrac{1}{2}[1 \pm 1/\sqrt{1+R^2}], \quad \text{where } R = -\sin 2\alpha \bigg/ \left(\frac{h}{H} + \cos 2\alpha\right)$$

and $\alpha = \theta_S + \theta_M$ is the angle STM. (Bernoulli used the notation: $A = 2/R$, $m = \sin\alpha$, $n = \cos\alpha$). Similar expressions with h and H interchanged give θ_M.

Notes and references

1. Newton, Sir Isaac. *Philosophae Naturalis Principia Mathematica.* London, The Royal Society, 1687. English transl. by A. Motte (1729), revised by F. Cajori, 2 vols, Univ. California Press, Berkeley, 1962. (See also ref.10.)
2. Proudman, J. Newton's work on the theory of the tides. Pp. 87–95 in W.J. Greenstreet (ed.), *Isaac Newton 1642–1727, Memorial Volume*, G.Bell & Sons, London, 1927.
3. Aiton, E.J. The contributions of Newton, Bernoulli and Euler to the theory of the tides. *Ann. Sci.*, **11**, 206–223, 1955.
4. From close examination of the texts of all three editions of *Principia*, Professor R.S. Westfall, (Newton and the Fudge-Factor, *Science*, **179**, 751–758, 1973), has shown that in collusion with his editor, Roger Cotes, Newton was over-zealous in choosing and manipulating data in order to confirm his theories to unwarranted precision. In particular, Newton's figure 4.48 for the moon:sun tidal force ratio was devised and selected by him because it led to impressive agreement of his theory with the known rate of precession of the equinoxes. That figure's implication for the tidal forces themselves was not so easily verifiable.
5. Euler, L. *Inquisitio physica in causam fluxus ac refluxus maris* (1740). Original (Latin) in refs. 9, 10. English summary in ref. 3.
6. Laplace, P.S. Recherches sur plusieurs points du Système du Monde. *Mém. Acad. roy. des Sciences*, **88**, 75–182, 1775 and **89**, 177–264, 1776.
7. Bernoulli, D. *Traité sur le flux et réflux de la mer*, Original (French) in refs. 10, 11. English summary in ref. 14.
8. Halley, E. An account of the course of the tides at Tonqueen (sic) in a letter from Mr. Francis Davenport, July 15, 1678, with the theory of them at the Barr of Tonqueen (later: 'Bar of Tunking') by the learned Edmund Halley, FRS. *Phil. Trans. R. Soc. London*, **14**, 677–688, 1684.
9. De Pontécoulant, G. *Théorie analytique du Système du Monde*, Tome IV, *Théorie de la Lune*. Paris, xxviii + 664pp., 1846.
10. *Receuil des pièces qui ont remporté le prix de l'Académie Royale des Sciences*, **4**, containing the original essays of A. Cavalleri (pp. 2–51), D. Bernoulli (pp. 55–191), C. Maclaurin (pp. 195–234), and L. Euler (pp. 237–350), Paris, 1752.
11. Newton, I. *Principia* ... ed. F. Jacquier & T. Le Seur, 3 vols, Geneva, 1742, reprinted 1760. (The widely circulated so-called 'Jesuit' edition – the editors were actually members of a Franciscan, not a Jesuit Order.) Vol. 3 includes copies of the Prize Essays by Bernoulli, Euler and Maclaurin, (ref. 10).
12. Villemot, P. *Nouveau système, ou explication nouvelle des mouvements des planètes*, Lyon, 1707.
13. Aiton, E.J. Descartes' theory of the tides. *Ann. Sci.*, **11**, 337–348, 1955. (pp. 346–348 contain a commentary on Cavalleri's essay – ref. 10.)
14. Lubbock, J.W. Account of the *Traité* ... of Daniel Bernoulli and Treatise on the attraction of ellipsoids, London, 1830.
15. Harris, R.A. *Manual of Tides*, Book 1, Govt. Printing Office, Washington, D.C., 1897.
16. D'Alembert, J. Reflexions sur la cause générale des vents – (pièce qui a remporté le prix proposé par *l'Académie Royale de Berlin* pour l'an 1746), Paris, 1747. (Both the original paper in Latin and the author's translation to French are included in this publication.)

6

Measurements and empirical studies, 1650–1825

The work of Newton and his followers has led us into the theoretical world of classical mechanics. We must not lose sight of the important fact that advances in theory are dependent on careful measurements of the phenomena under study. In this Chapter we retrace our steps back to the mid 17th century in order to consider the earliest serious efforts to measure and compile tidal data, providing the foundations on which theoretical ideas could be reliably built. Some of the data stimulated ideas for methods of local prediction, but with the exception of the work of Bernoulli (Chapter 5) these were largely empirical and particular to the place of observation. Eventually, advances in theory led not only to improved methods of prediction but also to longer and more accurate series of observations, so theory and measurement proceeded hand-in-hand.

High Waters, Low Waters, times and heights

The measurements of this period invariably consisted of visual observations of the times (HWT) and heights (HWH) of High Water and, later, the times and heights of Low Water (LWT, LWH) – see Appendix B for definitions. These provided a maximum of eight statistics per 24-hour day, but it was rare to record all eight except in the later and best organised campaigns. In the early years, two values per day of HWH or of HWT had to suffice. At best, HWT/H and LWT/H were considered adequate to define the whole oscillation. They were in any case the quantities which mariners traditionally were used to. However, it is worth remarking that the insistence on the *extrema* of a continuous motion in the vertical, invariably used in tide-tables to the present day, has introduced tiresome complications to all measurements, theory, and prediction techniques.

Measurement of the *heights* of extrema is probably the least difficult, provided one has a graduated vertical scale related to some fixed point on land, and one arranges to be there just before the right time. Some degree of visual averaging over wave motion is inevitable, including possibly *seiche-like* motion of several minutes' period. Accurate determination of the *times* is much more difficult because, apart from the need of a reliable clock, there is by definition little change to indicate the moment of extreme height. Estimates of time would tend to have limited accuracy and a personal bias towards lateness. The commonest reliable method (for HWT) was to take the average of the two times when water level rises and falls past a recognisable level not far below the expected extreme – see caption below Figure 6.1. This demanded a fair amount of time and patience as well as good judgement of an appropriate level. If too high, one missed the event altogether; if too low, one had a long time to wait for the time of fall, and the fact that the tide usually rises slightly faster than it falls introduces a bias. On a cold blustery night, one may well imagine the readings being omitted or replaced by 'guesstimates'.

The very concept of time measurement would have been difficult for the humbler type of waterman. What is now called *Local Apparent Time* (or *Temps Vrai*), based on observed transits of the sun across the local meridian, was assumed. Any mechanical timepiece would require frequent adjustment to accord with transit measurements (clear sky provided) by telescope, or at worst a sundial. It was really a task for observers trained in astronomical practice, but such people could not often be persuaded to spend months living by the waterside. It is not surprising, therefore, that the first reliable sets of HW timings were made under the surveillance of an astronomer or a geodesist.

Observations for the Royal Society of London

The early Royal Society of London was strongly inclined to experimentation, and was the first institution to encourage careful series of tide measurements, as distinct from casual reports from mariners. The need arose from the controversy stimulated in 1666 by the primitive tidal theory of John Wallis(1) (Chapter 4 – see Ch. 5 of Margaret's Deacon's book(2) for details and extracts). Wallis's theory, it may be recalled, was basically an adaptation of Galileo's solar tidal theory (Chapter 4) to include the moon, in virtue of Wallis's hypothesis that the earth has a rotatory component about the common center of gravity. Like Galileo, he had difficulty accounting for two tides a day (solar or lunar), but the theory seemed to accommodate the twice-monthly inequality. Wallis, however, failed to see why spring tides should be greater at the equinoxes. He clung to reports that high spring tides had been observed in February and November, partly supported by his own observations at London in November 1660 and 1665.

As a result of controversy, Wallis suggested that the Royal Society should sponsor observations of spring HWH and LWH over several months at as many sites as possible, in order to settle dispute as to when the highest tides occurred and whether or not high HW was invariably accompanied by low LW. Like many a theoretician, he belittled the difficulties involved in making such observations:

> ... And it were not amiss that the Low Waters were observed too. Which may be easily done by a mark made upon any standing Post in the Water by any Waterman, or understanding Person, who dwells by the Waterside.[3]

Sir Robert Moray

A formal set of recommendations was assigned by the Society to Sir Robert Moray, who took a much more serious view of matters of practical measurement. Moray, like Wallis one of the Founder Members of the Society, was a man of wide scientific interests. He had already communicated a paper on tides to the first volume of the *Transactions*,[4] reporting extraordinary behavior of the currents between some of the Western Isles of Scotland, as observed by him and by local fishermen. Incidentally, the feature in question, namely currents of once-daily periodicity in an otherwise twice-daily tide regime, remained unaccountable theoretically for over three centuries – see Chapter 13. Moray's *Considerations and Enquiries concerning Tides*[5] were also ahead of his time:

> In some convenient place upon a Wall, Rock or Bridge &c let there be an *Observatory* standing, as neer as may be to the brink of the Sea ... and if it cannot be well placed just where the Low water is there may be a Channel cut from the Low water to the bottom of the Wall, Rock &c. The Observatory is to be raised above the High water 18 or 20 foot; and a *Pump* of any reasonable dimension placed perpendicularly by the Wall, reaching above the High water as high as conveniently may be. Upon the top of the Pump a Pulley is to be fastned, for letting down into the Pump a piece of floating wood, which as the water comes in, may rise and fall with it. ... And because if the Hole by which the water is let into the Pump be as large as the Bore of the Pump itself, the weight that is raised by the water will rise and fall with an Undulation according to the inequality of the Sea's Surface, 'twill therefore be fit that the hole by which the water enters be less than half as bigg as the bore of the Pump; any inconvenience that may follow thereupon, as to the Periods and Stations of the Flood and Ebb, not being considerable.

The 'pump' with small-bore inlet, now called a *stilling-well*, is an essential feature of most harbor-gages to the present day.

Bristol or 'Cheap-Stow' (Chepstow) were especially recommended as observing sites because of the large vertical range in the Severn Estuary. Measurements were to include:

> 1. The degrees of the Rising and Falling of the water every quarter of an hour (or as often as conveniently may be) from the Periods of the Tides and Ebbs; to be observed night and day for 2 or 3 months.
> 2. The degrees of the velocity of the Motion of the Water every quarter of an hour for some whole Tides together; to be observed by a second Pendul-Watch; and a logg fastened to a line of some 50 fathoms, wound about a wheel.

3. The exact measures of the Heights of every utmost High-Water and Low-Water, from one Spring-tide to another, for some Months or rather Years.
4. The exact heights of Spring-tides and Spring-Ebbs for some years together.

Additionally, each observation was to be accompanied by a log of several weather parameters and the azimuths of the moon and sun.

In practice, no such elaborate programme of observations was ever carried out until the age of automatic recording instruments, but Moray's instructions set standards to which some later campaigns aspired. For the time being, a few amateurs sent summary reports of collections of tide observations made by experienced watermen. Some contradictions apart, they mostly agreed as to certain qualitative facts, such as that the highest floods are accompanied by the lowest ebbs and generally follow the syzygies by about three tides, that is 37 hours, and that the evening tides are somewhat higher than the morning tides in summer, *vice versa* in winter. (This last property, which depends on the phase of the harmonic component K_1, does not apply everywhere, but it holds in most of the seas surrounding southwest Britain and Brittany where the early measurements were taken.)

Henry Philips

One Henry Philips made an original submission about High Water *times* rather than heights.[6] Philips seems to have been the first to call attention to the fact that at London, HWT at the moon's quadratures is about 90 minutes earlier than at the syzygies in relation to the moon's southing. He remarked: 'Now this being a very considerable difference of time, which might very well make many Seamen and Passengers to lose their Tides, I set myself to watch this difference of the time of the Tides, and to find out some *Rule*, how to proportion [it] between Spring-tides and Neap-tides ...'. His *Rule* consisted of a correction to the simple progression of 48 minutes per day, then still in common use, which he expressed in terms of a geometrical construction. In modern terms, this correction amounted to the addition of a cosine variation of amplitude 45 minutes about the *mean* luni-tidal variation, with a period of 15 days. Philips incorporated it in a revised tide-table for 1668. We shall see that his scheme was further improved by Flamsteed, 15 years later; both Philips' and Flamsteed's rules were in turn superseded in 1740 by Bernoulli's more rational formulation (Chapter 5).

Observations of the *dates* of the annual maxima were various and inconclusive. Henry Powle, reporting from the River Severn,[2,7] gave dates near the equinoxes, and these were supported by Captain Samuel Sturmy from Bristol,[2,8] as well as by the general opinion of mariners. (We may recall from Chapter 5 that the figures quoted by Sturmy for the mean heights of springs and neaps were used by Isaac Newton in an attempt to deduce the moon:sun tide ratio.) In contrast, Samuel Colepresse from Plymouth reported an annual maximum in July, and later one in February 1667,[2] more in keeping with Wallis's preferred hypothesis. All were agreed that strong winds and river run-off could produce anomalous features, especially in estuaries like Severn and Thames.

Joshua Childrey

The most interesting of these early amateur observers and collectors of data was the Reverend Joshua Childrey (1623–1670), rector of Upwey near Weymouth on the Dorset coast. Childrey had much enthusiasm for the ancient lore of Astrology and for all natural phenomena, with books to his credit in both areas of knowledge. His interest in tides and weather started in boyhood, and he amassed several volumes of records, observed by himself and by '*a very able ancient Seaman*' at Weymouth. To express his interest in the Royal Society's correspondence, Childrey contributed a list of *Animadversions*(9) to Wallis's supposition that the highest tides occur in February and November. He ascribed the instances cited by Wallis and Colepresse largely to the strength and direction of the wind, quoting many examples of such effects from the Kent coast where Childrey once lived, and from historical records going back to the 13th century.

Childrey's most original contribution, however, was to notice that many instances of unusually high spring tides coincided with the moon's being close to *perigee*. Since the time intervals between passage of the moon through perigee (the *anomalistic month*) is incommensurate with the synodic month or the tropical year, such *perigean spring tides* (see Appendix A), could occur at any season. Childrey cited many examples from his own and from historical observations which proved the importance of this influence with little room for doubt. He further suggested that it would be interesting to observe whether neap tides were also higher than normal at perigee, and he promised to continue with his records from Weymouth to ascertain this.

Wallis's reply, printed in the *Transactions* together with Childrey's *Animadversions*,(9) was patronising and probably, as Deacon says,(2) disingenuous. There were no thanks to Childrey for his zealous collection of facts; Wallis agreed with the dependence on weather; he was aware that the passage of perigee could be significant, but had not insisted on it because it lacked annual periodicity. We now know, of course, that the closeness of the moon to perigee or apogee has a stonger effect on the amplitude of spring tides than the passage of the equinoxes, whereas any theory involving hours of daylight, which Wallis was hoping to prove, was doomed to failure. In practically all respects, Childrey's empirical findings were closer to truth than any of Wallis's theorising about tides.

Flamsteed and Halley

According to the article on Joshua Childrey in the *Dictionary of National Biography*, (1887 edition), Childrey, who died shortly after his correspondence with Wallis, had expressed intention to bequeath his volumes of tide and weather records to the Royal Society – *'but they seem to have been lost'*. Such was no doubt the fate of most 17th century tidal data collections. But not so the semi-empirical tide-tables for London Bridge compiled by John Flamsteed (1646–1719), the first *Astronomer Royal*. Flamsteed does not appear to have taken part in the discussions raised by Wallis, but he became interested in tides when, in his frequent river crossings between London and Greenwich, he noticed that the times of High Water were often earlier than those predicted by Philips' method[6].

Flamsteed was also conscious of the fact that Greenwich Observatory had been set up primarily to aid navigation.[10] Accordingly, he had careful measurements of HWT made in late 1681 'by an ingenious friend at Tower-Wharfe', and, mindful of possible disturbance by winter weather, carried out 'above 80 High Waters at Tower-Wharfe and Greenwich' in summer 1682 'with the help of my Friends and Servants'. From these observations and his own ephemerides of the moon and sun, he produced a predicted tide-table of all HWT for the year 1683, and for subsequent years to 1688, all of which were published in advance by the Royal Society.[11] These tables (of times only) proved to be better than any previous predictions for that locality, at least to about 15 minutes' accuracy.

Flamsteed gives no account of his technique of measurement, but he specifically mentions 'how difficult it is to determine the Time of an High-Water exactly'. The problem is similar to the astronomer's problem of determining an accurate meridian, and the Astronomer Royal would doubtless have provided impeccable standards of time-keeping. I have therefore used one of his tables – for May 1687 – in Figure 3.3, as proxy for contemporary data in comparison with other predictions. The method devised by Flamsteed to relate HWT to the lunar ephemeris is, however, unknown. We can only assume that he employed an empirical relation to the times of lunar transit, depending on the moon's phase, somewhat more sophisticated than Philips's cosine rule.

In his account of the table for 1683, Flamsteed suggested that the same figures might be converted to a tide-table for any other port, by simply adding or subtracting an appropriate constant time interval. With the table for 1686 (pub. 1685), he listed such time intervals for several ports in Great Britain and continental Europe[11] – as later became common practice. Such adjustments were soon known to be approximately correct only within a few miles of the primary station.

Edmond Halley, as Editor of the *Transactions* at that time, already knew this to be so and added a critical editorial comment in the volume containing Flamsteed's table for 1687, citing an example of dissimilarity between HWT at London and Dublin described by William Molyneux.[12] Flamsteed did not take kindly to criticism, and he had a personal dislike for Halley (who eventually succeeded him as Astronomer Royal). The incident of 1686 sparked off a series of bitter attacks on Halley by Flamsteed in his letters to Halley, Newton and others;[10] their details do not concern us here.

Edmond Halley (1656–1742), who is of course otherwise famous as an astronomer, was mentioned in Chapter 5 as the patron and editor of Newton's *Principia* and for publicising the extraordinary properties of the tides at Tonquin. Halley was the archetypal polymath-scientist, with a 'hands-on' approach to all natural phenomena which he perceived to require direct measurement. Apart from his disputes with Flamsteed over tide tables, Halley contributed little to tidal theory himself, considering Newton to have solved all the essential problems,[13] but he made some original measurements of the tides in the English Channel.

Halley's expedition of 1701 in command of the naval pink, '*Paramore*', to survey the English Channel, is overshadowed in scientific history by his voyages of 1698–1700 in the same vessel to map the magnetic variation in the entire Atlantic Ocean. The true motive of the Navy in commissioning the 1701 expedition is obscure. Certainly, Halley had expressed a desire to undertake it, but it could also have been a cover for keeping an eye on French shipping and naval resources. Whatever the motives, the *Paramore* occupied many anchor-stations on both sides of the Channel from North Foreland to the Lizard, observing the directions, speeds and times of the tidal streams and rough estimates of the rise and fall through repeated soundings. Halley's navigational chart of the English Channel, published in 1702,[14] was the first ever to show isolines of tide times (relative to lunar transit) and currents across open sea, and to indicate such features as the area east of Kent where the up-channel progression of the tide merges with the southward progression down the western side of the North Sea. Unfortunately, this classic chart is too large and detailed to reduce to a legible illustration here. Part of it is reproduced in Margaret Deacon's book.[2]

Nevil Maskelyne

The next, and indeed the last occasion when an astronomer was known to 'get his feet wet' in the cause of tidal observation was in 1761, when the Reverend Nevil Maskelyne (later to become the 5th Astronomer Royal) sojourned on the oceanic island of St. Helena[15] to record a Transit of Venus across the sun's disc – an idea previously suggested by Halley to estimate the sun's distance in terms of the earth's radius. The transit was obscured by clouds at St. Helena, and, perhaps to add justification for his long journey, Maskelyne set about observing the tides there, of rather small amplitude, with the aid of a graduated vertical staff erected offshore and his own very precisely calibrated pendulum clocks. This was no easy task in conditions of continual swell, which obliged him and his assistant to average over many recorded heights for some hours before and after HW and LW, but he managed to keep the records going for some 40 days, and all were reproduced in the *Transactions* of 1762.[16] They are the first example of tide heights recorded at arbitrary instants of time rather than as explicit maxima and minima. Many, perhaps most, of the readings were actually taken by Charles Mason, an Assistant of the Royal Society, who had successfully recorded the Venus Transit at the Cape, and who was later to

388 OBSERVATIONS ASTRONOMIQUES

Septembre.		H. M. S. du Soleil.	H. M. S. de la Lune.	
Vent d'Oûëſt,	18	2.25.30. du ſoir.	3.51.10. Orient.	Haute mer.
Vent d'Oûëſt,	19	3.13.30. du ſoir.	3.43.30. Orient.	Haute mer.
Vent de Nord,	21	10.29.30. du matin.	9.17.10. Occid.	Baſſe mer.
Vent de Nord,	22	11.41.45. du ſoir.	9. 8. 0. Orient.	Baſſe mer.
Calme,	24	0.25.30. du matin.	8.52.30. Orient.	Baſſe mer.
		0.46.30. du ſoir.	8.46.30. Occid.	Baſſe mer.
	25	1.12.30. du matin.	8.43. 0. Orient.	Baſſe mer.
		1.34 30. du ſoir.	8.36.10. Occid.	Baſſe mer.
Vent d'Oûëſt,	26	1.56.40. du matin.	8.31. 0. Orient.	Baſſe mer.
		8. 6.45. du matin.	2.28.30. Occid.	Baſſe mer.
	27	3.38.30. du matin.	9.18.30. Orient.	Baſſe mer.
		9.16.30. du matin.	2.45. 0. Occid.	Haute mer.
		10. 9.30. du ſoir.	3. 6.30. Orient.	Haute mer.
Calme,	28	10.47. 0. du matin.	3.23. 0. Occid.	Haute mer.

Pour ces Obſervations, on n'attendoit pas que la Mer fut tout-à-fait haute ou tout-à-fait baſſe, parce qu'alors elle demeure trop long-temps en état ; mais on marquoit deux temps éloignez devant & après, auſquels elle ſe trouvoit à certaine hauteur préciſe qui duroit ſi peu que nous n'avons point fait de difficulté de marquer juſques aux ſecondes ; puis on prenoit le milieu du temps qui s'étoit écoulé entre les Obſervations. La colomne qui contient les heures de la Lune fait voir dans quel cercle horaire la Lune ſe trouvoit ſoit vers l'Orient, ſoit vers l'Occident, au moment que la Mer étoit haute ou baſſe ; ce qui n'a

FAITES A BREST ET A NANTES. 389

pas été ſans une variation conſidérable, laquelle pourroit bien avoir été cauſée par les fréquentes tempêtes dont l'Océan fut agité durant ce temps-là.

Nous laiſſâmes un Barometre ſimple entre les mains de M. Olivier Medecin de la Marine, très-habile & très-curieux, qui après environ ſix mois d'Obſervations, nous fit rapport qu'à Breſt la hauteur du Vif-argent avoit varié entre 27 pouces 8 lignes, & 26 pouces 1 ligne ; ce qui eſt fort different de ce qu'on obſerve à Paris & à Montpellier, comme on peut voir ci-deſſus.

A NANTES.

Hauteurs Méridiennes obſervées au mois de Décembre 1679.

La Luiſante d'Aries.	64°	42'	35"	à Nantes.
	64	13	55	à la Fléche.
Difference	0	28	40	
Menkar	45	36	30	à Nantes.
	45	7	45	à la Fléche.
Difference	0	28	45	
La Polaire	50	7	25	à la Fléche.
	49	38	45	à Nantes.
Difference	0	28	40	

Les Obſervations furent faites à Nantes proche le Château. On les a miſes en comparaiſon avec d'autres qui furent faites à la Fléche peu de jours après, parce qu'on n'en avoit point de Paris.

Hauteur du Pole de la Fléche	47	41	50
Difference à ôter.	0	28	40
Donc à Nantes hauteur du Pole	47	13	10

Ccc iij

Figure 6.1. The first accurately timed records of High and Low Water (Basse mer) – made in the harbour of Brest in September 1679 by the geodesists Picard and La Hire.[17] The times are expressed in both (Apparent) solar and lunar times to about 10s precision. The text beneath the table is probably the first written specification of how to judge HWT and LWT – by taking two times at the same level a little short of the extremum – but warns of variations due to frequent storms. Barometric readings were also taken at Brest by a naval doctor, as described below, in inches (pouces) and twelfths (lignes) of mercury (vif-argent). The last half page concerns astronomical measurements at Nantes by the same authors.

achieve popular fame as the principal surveyor of the *Mason-Dixon Line* of Pennsylvania and of the first meridional arc in North America.

Picard and La Hire at Brest

As we saw in Chapter 5, after about 1700 the initiative for tidal investigations passed from the Royal Society of London to the Académie Royale des Sciences of Paris. To that Academy we owe not only the Essay Prize of 1740 but also the first accurate measurements of tide heights and times recorded over usefully long periods. Times of HW and LW recorded at Brest by the geodesists Abbé Jean Picard (1620–1682) and Phillipe de la Hire (1640–1718) in 1679[17] to an apparent precision of about 10 seconds in solar and lunar time serve as a precedent for more scientific standards of observation of tidal phenomena, as well as being the start of a long history of recording at Brest. Their recorded times are few in number and are not accompanied by vertical heights, but the authors exercised proper astronomical control of time. The text below their Table, reproduced in Figure 6.1, shows the care they took to define the extrema. Picard and La Hire took these observations in the course of their geodetic survey of the whole of France; their intended use is not quite clear. But they predate by two years Flamsteed's accurate measurements in the Thames,[11] and the publication of the actual data also sets a precedent.

Observations for the Académie Royale des Sciences

Early in 1701, appreciating the need for more exact information of the tides at all ports of France, the Académie issued a *Mémoire de la Manière d'observer dans les ports le Flux et le Reflux de la Mer*, addressed to qualified and capable people at those places. The anonymous Introduction begins:

> Quoique le Flux & le Reflux ait passé pour une merveille impénétrable à l'esprit humain, peut-être la cause est-elle découverte, & tout l'honneur en seroit dû à M. Descartes. Mais ce qui pourra paroître surprenant, on peut plutôt se flatter d'avoir le Systême, que s'assûrer d'avoir les Phénoménes avec assez d'exactitude. L'Académie songea donc à tirer de différens endroits des Observations de la Flux & le Reflux, faites par des gens habiles, & à profiter d'un avantage qu'elle avoit pour cela, le plus grand qu'elle pût jamais souhaiter.
>
> (*Memorandum on how the tides should be observed. (Introduction)*: Although the [tides] had been regarded as an impenetrable wonder, perhaps their cause has now been discovered, to the honor of M. Descartes. But surprisingly, we have the *System* without assurance that the *Phenomenon* itself is known with sufficient exactitude. The Académie therefore decided to obtain direct observations of the tides by able persons from various locations, and to profit from the greatest advantage she could ever hope to have in [the availability of such persons].)

The passage incidentally shows that the Academiciens were still attached to Cartesian theory while realising the dearth of exact data with which the theory might be tested in a predictive mode for the benefit of Navigation.(18)

The instructions themselves, drawn up partly on the advice of La Hire, covered nine points to be observed for good measurements, on similar lines to the earlier recommendations of Moray(5) but without specific advice on instrumental aids. The ninth point, like Moray's, recommended measurements of *current* on calm days, but if any such were made they were not further discussed in the literature of that time. Overall authority for the observations was put in the hands of M. le Comte de Pontchartrain, who was also responsible for the Navy. Within ten months of issuing the *Mémoire*, a long series of excellent records of HW times and heights for the greater part of 1701 was received from Dunkerque, soon followed by data of similar duration and quality from 'Havre de Grace' (Le Havre). Records from all the major oceanic ports of France were achieved within the next 15 years.

The operations themselves were undertaken by the *Professeurs d'Hydrographie* attached to all ports, professional men with a fair degree of scientific understanding of the physical behavior of the sea and of measuring technique. All were proficient at and suitably equipped for accurate time-keeping. The Professeur who so promptly sent data from Dunkerque, a Monsieur Baert, constructed a novel apparatus to facilitate height measurement. A rough sketch of his apparatus was reproduced in the first paper to present an analysis of the data, by Cassini.(19) It consisted of a free float inside a long vertical pipe of square section fixed to the side of the wharf; a light vertical rod attached above the float allowed its top position to be easily measured against a graduated scale at eye level. It is not clear whether it had an inlet of reduced diameter, or whether similar devices were used at the other ports, but all observers used a vertical scale graduated in *pieds* (feet) and *pouces* (inches), newly erected for the purpose. Even *lignes* (twelfth parts of an inch) were occasionally mentioned, (Figure 6.2).

Jacques Cassini

All observations were transmitted to the Académie in Paris, where they were analysed in due course by Jacques Cassini (1677–1756), son of Jean Dominique Cassini (1625–1712), the first 'Head' of the Observatoire Royale. (The title *Directeur de l'Observatoire* was not officially appointed until 1771.) Jacques Cassini is chiefly remembered as a geodesist who organised the survey of a long meridional arc across France. His frequent papers in the *Mémoires de l'Académie* show him to have been continually busy recording eclipses, sunspots and other astronomical features, as well as discussing the tidal data in his care. He shared many attitudes with his father, (whom he succeeded as Head of the Observatory), including the now-curious notion that the earth is elongated instead of depressed at the poles, and a rejection of Newton's gravitational theory and all its consequences. Not surprisingly, Cassini was a firm believer in Descartes' vortex theory of tides (Chapter 4); he was even able to reconcile the unique collection of facts at his disposal with Cartesian notions.

As late as 1720, some 32 years after publication of Newton's *Principia*, Cassini was able to write apropos of his results from many years of observations at Brest and L'Orient:

> Nous avons déja remarqué que le Systême de Descartes ...est favorable à notre opinion, que le Soleil contribue à la hauteur des Marées, & il n'auroit pas manqué d'en faire application, s'il avoit eu comme nous les connaissances des diamètres apparents de la Lune; car de dire que le Tourbillon de la Terre est toujours applati du côté qu'elle regarde le Soleil, ou de soutenir que le Soleil fait une pression sur le Tourbillon de la Terre qui peut se communiquer jusques sur les eaux de la Mer, cela revient à peu-près au même, puisque l'un résulte de l'autre ...
>
> (We have already remarked that the System of Descartes supports our opinion that the sun contributes to [the tides], and he would certainly have applied this fact if he had had, like us, data for the apparent lunar diameter; for to say that the earth's vortex is always flattened on the side facing the sun, or that the sun exerts a pressure on the earth's vortex which can communicate itself to the sea surface, comes to the same thing, since one is the result of the other ...)

After this from the Head of the Observatoire Royale of Paris, it is easier to understand how Cavalleri could write his essay of 1740 in anachronistic defence of the vortex theory, and indeed how the Académie could see fit to award Cavalleri one of its prizes.

The series from Brest, though late in starting, proved to be the most valuable in terms of fulness and accuracy. The first page from June 1711 is reproduced in Figure 6.2 from the later complete publication by Lalande.[21] HW and LW were recorded day and night for the most part, with times estimated to the half-minute and heights in pieds, pouces and lignes. The most relevant lunar ephemerides are listed on the right. The *Professeur d'Hydrographie* was a certain Sieur Coubard, who was evidently quite conscientious in recording at least three events per day, no doubt with an assistant for some of the night watches. A distinguished local ship-designer named Sieur Charles Montier des Longchamps is also mentioned by Cassini as having taken an active part in the work. Cassini soon perceived the importance of continuing this series for a number of years, and managed to have observations at Brest maintained to the end of September 1716. It was supplemented by another series, from L'Orient, 100 km southwest of Brest, which extended from June 1716 to June 1719 as well as a year in 1711 to 1712 overlapping the Brest observations.[20]

Cassini's several analyses of these data, published from 1710 to 1720,[19,20] are somewhat verbose and repetitive, but they brought out some interesting facts, despite his adherence to an obsolete physical theory. He noted that the spring tides followed the Full or New Moons by about two tides ('Age' 1 day), whereas

ET DU REFLUX DE LA MER. 161

Obſervations des marées faites à Breſt, depuis 1711 *juſqu'à* 1716.

Jours du mois.	Temps de la haute mer.*	Hauteur de la pleine mer, au-deſſus du point fixe.	Temps de la baſſe mer.*	Hauteur de la baſſe mer, au-deſſ. ou au-deſſ. du point fixe.	Circonſtances des Obſervations.
1711.	H. M.	pi. po. li.	H. M.	pi. po. li.	
Juin 10	11 34 m.	12 7			D. Q. le 8 à 2 h. 35 m. du matin.
11	0 29½ ſ.	12 11 6			
12	1 28	13 7 4	7 18½ m.	3 6 8	Apog. moyen
			7 50 ſ.	3 6	7 h. ſoir.
13	2 9½	14 3 6	8 10 m.	3 2	
14	2 46½	14 10 6	8 12	2 9	
15	3 25½	15 5 6	9 31	2 4 6	Luniſt. Bor.
16	3 47. m.	15 3			décl. 26° ½
	3 59	15 10 8	10 2	2 0 6	N. L. 6^{h} 18′ m.
17	4 23	15 5 6			
	4 38½	16 2 6	10 37½	1 9 6	
18	4 56½	15 9			
	5 9	15 10 6	11 14½	1 4 6	
19	5 36	15 2 6			
	5 47	15 6 4	11 49	1 0 8	
20	6 11	14 8 6			
	6 36	15 1 6	0 33½ ſ.	1 3 6	
21	6 56	14 4 6			
	7 18	14 9 4	1 8½	1 6 8	
22	7 34	14 0			
	8 7½	14 5	2 1½	1 11 4	L. dans l'éq.
23	8 41	13 4 8			
	9 0	13 7	2 56½	2 5	P. Q. 5^{h} 58′ ſ.
24	9 37	13 7 8			
	10 0	13 11	4 10	2 8	
25	10 40	13 9			
	11 10	14 1	5 1½	2 7 8	
26	11 44 m.	14 2	6 16	2 4 4	Périgée 2^{h} ſ.
27	0 27	14 7			
	0 59	15 3	7 14	1 9	
28	1 23	15 5			
	2 0	16 1	8 15	0 8 4	
29	2 29	16 1			
	2 54½	16 11 4	8 48½ m.	0 8	Luniſt. mér.
30	3 43 ſ.	17 6	9 40½	0 0 6	P. L. 9^{h} 7′ m.

* Il faut ôter 17 min. des temps de ces Obſervations en 1711 & 1712, ſuivant M. Caſſini. (*Mém. de l'Acad.* 1714, *p.* 248); mais j'ai cru devoir les rapporter ici telles qu'on les trouve dans les manuſcrits de M. Caſſini. Les ſignes – indiquent l'abaiſſement au-deſſous du point fixe.

Tome IV. X

Figure 6.2. The first page of tidal records from Brest observed by M. Coubard, *Professeur d'Hydrographie* at the port, and his assistants, as published by Lalande.[21] Apparent Times (cols. 2, 4) are given in a 12-hour system with occasional indications of a.m. (matin) and p.m. (soir). HW heights (pleine mer) recorded above a fixed datum in feet (pieds), inches (pouces), and twelfths (lignes), are listed in col. 3. LW heights (basse mer) are in col. 5. Col. 6 gives times of lunar phase, apogee, perigee, declination, and equatorial crossings. The footnote warns that 17m should be subtracted from the recorded times in order to allow for a meridian error in the transit telescope, discovered by Coubard and noted by Cassini in 1714. Similar records continue to 31 August 1716, with the exception of 1713 and a few other months which Lalande could not find among Cassini's papers.

at Dunkerque and Le Havre he had found the interval to be 3 or 4 tides ('Age' 2 days). He attributed the difference in tidal age to the effects of propagation up the Channel, as justified later theoretically by Laplace. He noticed also that HWT on days of syzygy varied from its mean time by up to ±30 minutes, and that the discrepancies could be accommodated by the empirical rule of adding (subtracting) to the mean time 2 minutes for each hour that the time of Full or New Moon precedes (follows) the mean HWT. This rule accords with the old linear rule of 48 minutes/day, but over a very limited period. On days of quadrature he found a similar rule with 2.5 minutes per hour, in keeping with the longer interval between successive HWT. Cassini had little to say about behavior *between* springs and neaps, but he implied a smooth transition from 2 to 2.5 minutes per hour, depending on the luni-solar angle. In brief, these rules allow a refinement to Philips's empirical rule,[6] to which Cassini referred, or possibly to Flamsteed's improvement,[11] but with parameters more suitable to the ports studied by Cassini.

Regarding amplitudes, Cassini found that the diurnal inequality at Brest and L'Orient showed the same property as observed at Plymouth and Bristol,[2] namely higher levels of HWH and LWH in summer afternoons and winter mornings. This was for a time thought to be a general rule, until good data became available from a wider variety of ports. He also showed categorical support for Childrey's[9] discovery of the enhancing effect of lunar perigee, as the last extract suggests.

In general, the great advantage in having such a long series of observations is the ability to fix one astronomical variable and study the variation due to other variables. Additionally, one may take averages over random effects of weather. In working mainly with the amplitudes at syzygies and quadratures, Cassini was fixing the luni-solar angle, and could additionally fix either lunar declination or parallax and vary the other. On choosing spring tides at summer and winter solstices, for example, and selecting cases of nearly equal parallax, he demonstrated that solsticial spring amplitudes were greater in winter than in summer. This was the first demonstration that the parallax of the sun, which was least at the end of December at that epoch, has a measurable effect on tides.

Figure 6.3 epitomises Cassini's effort in his paper of 1713[20] to quantify from the Brest records the variation of the HW heights at syzygy, above the chosen datum, as a function of both declination and earth–moon distance (inverse parallax). Such heights are slightly lower than peak spring heights. Though based on only the first 13 months of data, these sufficed to determine the parameters of a simplified empirical linear model with ability to smooth over random effects. The only variations not accounted for are the sun's distance (parallax) and the position of the moon's node, but these are relatively minor. The decrease of amplitude with increasing declination and distance are clearly seen. A similar diagram was constructed for the tides at quadrature. These were the most complete analyses of a tidal regime yet achieved by purely empirical means, and a precedent for the more thorough treatment by J.W. Lubbock (Chapter 8).

TABLE de la Hauteur des Grandes Marées dans les Nouvelles & Pleines Lunes.

Déclinaison de la Lune.		*Distance de la Lune à la Terre.* 94		95		96		97		98		99		100		101		102		103		104		105		106	
		Hauteur de la Pleine Mer à Brest dans les Nouvelles & Pleines Lunes.																									
Deg.	Min.	Pieds.	Pouc.	Pieds.	Pouc.	Pieds.	Pouc.	Pieds.	Pouc.	Pieds.	Pouc.	Pieds.	Pouc.	Pieds.	Pouc.	Pieds.	Pouc.	Pieds.	Pouc.	Pieds.	Pouc.	Pieds.	Pouc.	Pieds.	Pouc.	Pieds.	Pouc.
0	0	20	0	19	9	19	6	19	3	19	0	18	9	18	6	18	3	18	0	17	9	17	6	17	3	17	0
6	56	19	11	19	8	19	5	19	2	18	11	18	8	18	5	18	2	17	11	17	8	17	5	17	2	16	11
9	48	19	10	19	7	19	4	19	1	18	10	18	7	18	4	18	1	17	10	17	7	17	4	17	1	16	10
12	0	19	9	19	6	19	3	19	0	18	9	18	6	18	3	18	0	17	9	17	6	17	3	17	0	16	9
13	50	19	8	19	5	19	2	18	11	18	8	18	5	18	2	17	11	17	8	17	5	17	2	16	11	16	8
15	26	19	7	19	4	19	1	18	10	18	7	18	4	18	1	17	10	17	7	17	4	17	1	16	10	16	7
16	54	19	6	19	3	19	0	18	9	18	6	18	3	18	0	17	9	17	6	17	3	17	0	16	9	16	6
18	14	19	5	19	2	18	11	18	8	18	5	18	2	17	11	17	8	17	5	17	2	16	11	16	8	16	5
19	28	19	4	19	1	18	10	18	7	18	4	18	1	17	10	17	7	17	4	17	1	16	10	16	7	16	4
20	37	19	3	19	0	18	9	18	6	18	3	18	0	17	9	17	6	17	3	17	0	16	9	16	6	16	3
21	42	19	2	18	11	18	8	18	5	18	2	17	11	17	8	17	5	17	2	16	11	16	8	16	5	16	2
22	44	19	1	18	10	18	7	18	4	18	1	17	10	17	7	17	4	17	1	16	10	16	7	16	4	16	1
23	43	19	0	18	9	18	6	18	3	18	0	17	9	17	6	17	3	17	0	16	9	16	6	16	3	16	0
24	39	18	11	18	8	18	5	18	2	17	11	17	8	17	5	17	2	16	11	16	8	16	5	16	2	15	11
25	33	18	10	18	7	18	4	18	1	17	10	17	7	17	4	17	1	16	10	16	7	16	4	16	1	15	10
26	25	18	9	18	6	18	3	18	0	17	9	17	6	17	3	17	0	16	9	16	6	16	3	16	0	15	9
27	15	18	8	18	5	18	2	17	11	17	8	17	5	17	2	16	11	16	8	16	5	16	2	15	11	15	8
28	3	18	7	18	4	18	1	17	10	17	7	17	4	17	1	16	10	16	7	16	4	16	1	15	10	15	7
28	50	18	6	18	3	18	0	17	9	17	6	17	3	17	0	16	9	16	6	16	3	16	0	15	9	15	6

Figure 6.3. Cassini's digest of the dependence of HW heights (feet, inches) at Brest at the times of syzygy (Full and Change of Moon), on the two variables: lunar Declination (col. 1) and earth–moon distance as percentage of mean distance (top row). A similar Table on the following page describes the variation of HWH at quadratures (First and Last Quarters). This was the first attempt to define such relationships in bivariate form, and is a predecessor of J.W. Lubbock's 'synthetic' method of tide prediction, (Chapter 8).

J.J. de Lalande

There is no evidence that Jacques Cassini or his son – also a noted geodesist – took any further active part in the study of tides, but it is probable that he was a member of the jury for the Prize Essays of 1740 (Chapter 5). The next French astronomer to show serious interest in the subject was Joseph Jérôme Lefrancois de Lalande (1732–1807), who devoted a substantial part of Tome 4 of his treatise *Astronomie* (1781) to tides.[21] Lalande's book was the first to include a comprehensive history of the subject from classical times to mid 18th century. He was pro-Newton, and rather contemptuous of the die-hard Cartesians of the early part of the century, while setting great store by the Academy's campaign of direct measurement.

Having particular interest in distinguishing the equinoctial/solsticial effect on amplitude from weather disturbances, Lalande recognised the importance of having access to all the observations rather than just the features selected for illustration by Cassini. He had to seek out the original records, which he eventually found among Jacques Cassini's papers through the help of Cassini's grandson. As well as making use of these himself, Lalande had the munificence to publish the entire extant series from Brest, June 1711 to August 1716, in his book.[21] Only the year 1713 and a small part of 1714 were not found; the last month, September 1716, was discovered too late for printing. They occupy 71 pages; the first page is reproduced in Figure 6.3.

Lalande also had substantial synoptic tidal records made in the years 1773–75 for his own investigations, through the assistance of a certain M. Blondeau, Professor of Mathematics at Brest. He mentions results from them but did not reproduce the records themselves, possibly because of insufficient accuracy. He did however reproduce less comprehensive data sets from Dunkerque, Le Havre, Rochefort, St. Malo, and the Mediterranean port of Toulon. The last three were recorded in the 1770's, showing that the tradition of keeping tide records had persisted. He concluded his treatise with an exhaustive list of known *Etablissements* (HWT at Full and Change) at worldwide ports.

The results of Lalande's experiments to determine whether equinoctial tidal range is systematically greater than the solstitial range were somewhat inconclusive,[21] partly owing to the presence of many other sources of variation and partly because his analytical methods lacked the sophistication which Laplace was to apply to his own data from Brest in the following century. Lalande made some significant contributions to astronomy, but his involvement with tides is chiefly remembered for his expository writings and for his assiduous collection and publication of tidal data.

Observations at Brest for Marquis de Laplace

Carrying this survey of pioneering measurement campaigns into the early 19th century, the Brest series were extended still further under the instigation of the famous Marquis de Laplace. An account of Laplace's contributions to the theory of tides will be given in the next Chapter. Here, it suffices to say that he needed even longer series than were available from Lalande's treatise, in order to separate lunar from solar effects and diurnal from semidiurnal and other species. He recommended 18–19 years in order to sample all positions of the moon's node. Laplace managed to instigate such a renewed campaign in 1806.

New vertical scales were set up and HW and LW times and heights were recorded day and night in a manner similar to the 1711 exercise. Despite great care in planning, standards were variable. The readings from 1806 were deemed sub-standard and were not preserved. During 1807–11 the timing proved erratic. A new observer, noted for his zeal and exactitude, was appointed at a better site and he maintained the effort alone from 1812 to 1822, with markedly improved results. Even so, Laplace was able to detect subtle differences from the 1711–16 series; these differences vanished when yet another observer was appointed in 1823. One thus sees the inevitable personal biases in such observations as well as the excellence of the standards of 1711, which Laplace recognised.

While hoping for 18 years of records, Laplace made do with those from 1807 to 1822. The results of his analysis were published in his monumental *Traité de Mécanique Celeste* – see Chapter 7. Visual observations were continued at Brest at least into 1843 and possibly until they were replaced by automatic instrumental recording in the 1850's; the latter has been continued with only minor interruptions up to the present day. In 1843 the Bureau des Longitudes published all the observations of 1807–35 in a dedicated volume.[22] The Introductory section of this volume gives a detailed account of where and how they were recorded.

Observations at Liverpool and London Docks

Finally, within the subject of this Chapter mention should be made of the 19-years series of observations of High Waters at Liverpool (1774–1792) and London Docks (1808–1826), later obtained and used for his own analyses by J.W. Lubbock in the 1830's. Lubbock's work will be discussed in Chapter 8. There appears to have been no scientific incentive for the observations themselves; they were made by the respective Dockmasters as part of their daily duties. The observations would have been still more valuable if Low Waters had been included, but it is doubtful whether these were accessible to observation; in any case they were presumably of little interest to those concerned with shipping.[23] The statistics from both sets of observations are preserved in the Archives of the Royal Society.

Notes and references

1. Wallis, J. Hypothesis on the flux and reflux of the sea. *Phil. Trans. R. Soc. London*, **1**, 263–298, 1666.
2. Deacon, Margaret. *Scientists and the Sea, 1650–1900*. Academic Press, London, 445pp., 1971 – reprinted 1997.
3. Wallis, J. Some Inquiries and Directions concerning Tides proposed by Dr. Wallis for the proving or disproving of his lately published discourse. *Phil. Trans. R. Soc. London*, **1**, 297–298, 1666. (Followed by ref. 5)
4. Moray, R. A relation of some extraordinary Tydes in the West Isles of Scotland. *Phil. Trans. R. Soc. London*, **1**, 53–55, 1665.
5. Moray, R. Considerations and Enquiries concerning Tides; likewise for a further search into Dr. Wallis's newly publish't Hypothesis. *Phil. Trans. R. Soc. London*, **1**, 298–301, 1666.
6. Philips, H. A letter written to Dr. John Wallis containing observations about the true time of the tides. *Phil. Trans. R. Soc. London*, **2**, 656–659, 1668.
7. Powle, H. Observations concerning tydes. *Royal Soc. Published Papers*, **6**, (18), 1666.
8. Sturmy, S. An account of some observations, made this present year . . . within four miles of Bristol, in answer to some of the Quaeries concerning the Tydes. *Phil. Trans. R. Soc. London*, **3**, 813–817, 1668.
9. Childrey, J. (first name Joshua, wrongly printed 'Joseph'). Animadversions on Wallis's hypothesis about the flux and reflux of the sea, *Phil. Trans. R. Soc. London*, **5**, 2061–2068, 1670. (Followed by Wallis's reply, 2068–2074).
10. Forbes, E.G. *Greenwich Observatory*, Vol.1 – *Origins and early History*, 1675–1835. Taylor & Francis, London, 204pp., 1975.
11. Flamstead, J. (sic). A correct Tide-Table shewing the true Times of High-Water at London-Bridge, to every day in the year 1683 – followed by: 'An account of the foregoing

Tide-Table, by the same Hand'. *Phil. Trans. R. Soc. London*, **13**, 10–15, 1683. (Similar Tables were printed annually by Flamsteed to 1688, without further explanation.)
12. Molyneux, W. An account of the course of the tides in the port of Dublin, *Phil. Trans. R. Soc. London*, **16**, 192–193, 1686.
13. Halley, E. The true theory of the tides, extracted from that admired treatise of Mr. Isaac Newton ..., *Phil. Trans. R. Soc. London*, **19**, 445–457, 1696.
14. Halley, E. A new and correct chart of the Channel between England and France ... with the flowing of the tydes and setting of the currents, London, 1701. (Copy in British Museum, map collection)
15. Tatham, W.G. and K.A. Harwood. Astronomers and other scientists on St. Helena. *Ann. Sci.*, **31**, 6, 489–510, 1973.
16. Maskelyne, N. Observations of the tides in the island of Saint Helena, *Phil. Trans. R. Soc. London*, **52**, 586–606, 1762.
17. Picard, J. and P. De la Hire. Observations astronomiques faite à Brest et à Nantes, *Mém. Acad. royale des Sciences*, 1666–1699, 7, 379–390, 1680.
18. Anon. Mémoire de la Manière d'observer dans les ports le Flux et le Reflux de la Mer. *Histoire Acad. royale des Sciences*, 11–13, 1701. (Page 11 is introductory, and simply entitled 'Sur le Flux et le Reflux'.)
19. Cassini le fils (Jacques). Réflexions sur les Observations du Flux et du Reflux de la Mer faite à Dunquerque par M. Baert, 1701–1702. *Hist. et Mém. Acad. roy. Sciences; (Mém.)*, 318–341, 1710.
20. Cassini, Jacques. Papers on 'le Flux et le Reflux' at a series of French ports in *Mém. Acad. roy. Sciences*, 1710–1714 and 1720.
21. Lalande, J.J.L. *Astronomie*, (4 vols.), 'Traité du flux et du reflux de la mer' in **4**, 1–348, 1781.
22. Anon. Observations des marées dans le port de Brest, 1807–1835. Bureau des Longitudes, Paris, 1843.
23. Dr. Philip Woodworth of the Proudman Oceanographic Laboratory informs me that the Old Dock at Liverpool would have dried out at most Low Waters, if a minimum water level were not maintained by a tidal lock. The 18th century observer, a Mr. Hutchinson, was known to be of a calibre to have recorded all Low Water times and heights if it had been possible. Hutchinson's original observations for 25 years have recently been re-discovered in a Liverpool Archive.

7

Laplace and 19th century hydrodynamics

The development of tidal theory by Pierre Simon, Marquis de Laplace (1749–1827) was no less radical than that of Isaac Newton. Newton had defined the forces which drive the tides resulting from his theory of gravitation, but he and the Essayists of 1740 (Chapter 5) had assumed the ocean's response to these forces to be quasi-static. Laplace defined the *dynamic* relations which determine the ocean's response to tidal forces, and, having identified three distinct nearly-periodic components of the force field, showed that the response is far from static and different for each periodic component. Although at first considered too unwieldy for practical application, Laplace's theory eventually provided the basis for the solution of many important practical problems.

Like Newton, Laplace approached the theory of tides as just one of many facets of the mechanics of the solar system, treated as a deterministic whole. Finding their solution to be much more difficult than had been supposed, he returned to the tides of ocean and atmosphere several times over a span of fifty years with publication dates from 1775 to 1825. His appreciation of the difficulty is apparent in the following sentence from the opening paragraph of his paper of 1790 to the Académie:[(2)] 'Aidé des déscouverts que l'on a faites depuis sur ces deux objets, j'ai repris, dans nos Mémoires pour les années 1775 et 1776, ce problème, le plus épineux de toute la Mécanique Céleste.'

The 18th century saw great advances in the theory of the dynamics of fluids due to Continental writers such as Bernoulli, Euler, d'Alembert and Lagrange, as well as Laplace. The influence of his contemporaries is not always evident in Laplace's writings. 18th century mathematicians were less conscientious in acknowledging the sources of their inspiration than they are today; Laplace has even been accused of occasional plagiarism. This does not detract from the excellence of his work as a whole, but it should be seen as the epitome of a new school of scientific thought, to which others also contributed, and which formed the basis of the field of *geophysical fluid dynamics* to be developed in the 19th and 20th centuries.

Consideration of the works of Laplace and most later tidal researchers demands recourse to some mathematical discussion. Mathematical equations have hardly been necessary in previous chapters because natural philosophers of the 17th and early 18th centuries expressed their arguments in words aided by geometric diagrams and a little arithmetic. The infinitesimal calculus was invented independently by Newton and by Leibnitz around 1670, but Newton did not employ differential notation explicitly in the *Principia*. By the time of Laplace, the calculus of variations, as well as the earlier notation and properties of the trigonometric functions, were accepted as necessary tools for all theoretical mechanics. We therefore enter a world of more sophisticated argument in which mathematics plays an increasing part. Comparison of Figure 5.2 from Newton's *Principia* of 1687 with Figure 7.1 from Laplace in 1776, well illustrates the changing mode of expression.

As explained in my Preface, the emphasis of this book is on the historical sequence of ideas, rather than on elementary pedagogy or the detailed examination of every explored path. Accordingly, the work of Laplace and his followers will be presented in summary, with the minimum of mathematical detail necessary to make it comprehensible. Some of the algebraic derivation is given in Appendix C for those who require it.

The *Mécanique Céleste*

Laplace's most important work on tidal theory is contained in three long papers presented to the Académie des Sciences.[1,2] (By 1790, the epithet 'Royale' had been removed by the Revolutionary authorities.) The substance of these papers was repeated in Book IV of his great *Traité de Mécanique Céleste*, ('Treatise on Celestial Mechanics'), published in 1799.[3] Some aspects had to await the completion and analysis of many years' observations of the tide at Brest (see Chapter 6). The results of this analysis are contained in Book XIII, published in 1825, which also contains some conclusions on air tides.[4] Book IV is included in the second volume of the abundantly annotated English translation of the *Mécanique Céleste* by the American mathematician Nathaniel Bowditch.[5] (Bowditch, who also authored a famous treatise on navigation, did not live to translate Laplace's fifth volume, containing Book XIII.)

Another notorious feature of Laplace's writing is his frequent omission of derivatory exposition obvious to him, leaving the reader to fill the gaps. In the Preface to his translation of *Mécanique Celeste* Bowditch writes: 'Whenever I meet in LA PLACE (sic) with the words "Thus it plainly appears ... [Ainsi il est clair que ...]", I am sure that hours, perhaps days of hard study will alone enable to discover *how* it plainly appears.' Fortunately for the present discussion, these lacunae do not often occur in the sections dealing with tides.

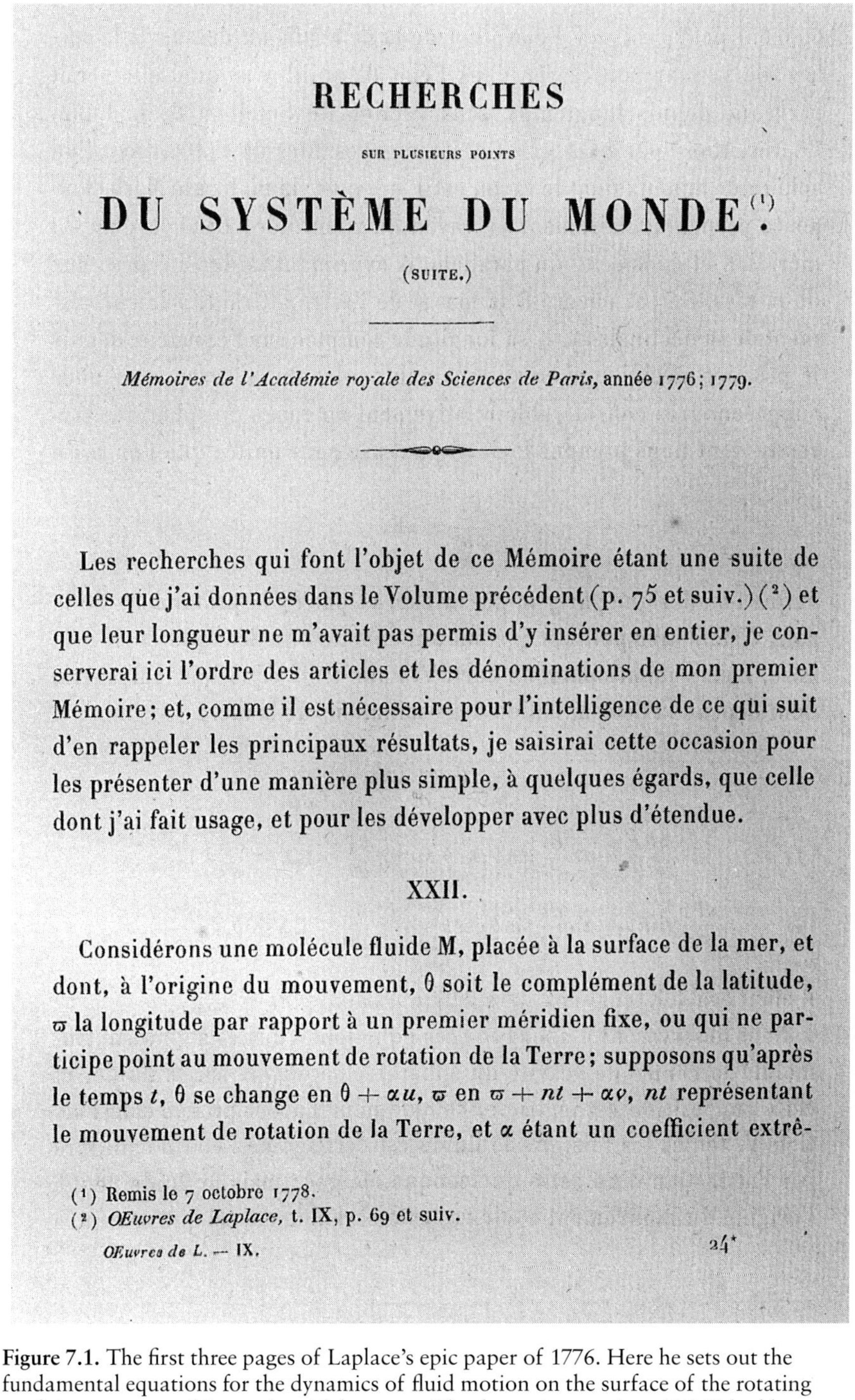

RECHERCHES

SUR PLUSIEURS POINTS

DU SYSTÈME DU MONDE[1]

(SUITE.)

Mémoires de l'Académie royale des Sciences de Paris, année 1776; 1779.

Les recherches qui font l'objet de ce Mémoire étant une suite de celles que j'ai données dans le Volume précédent (p. 75 et suiv.) [2] et que leur longueur ne m'avait pas permis d'y insérer en entier, je conserverai ici l'ordre des articles et les dénominations de mon premier Mémoire; et, comme il est nécessaire pour l'intelligence de ce qui suit d'en rappeler les principaux résultats, je saisirai cette occasion pour les présenter d'une manière plus simple, à quelques égards, que celle dont j'ai fait usage, et pour les développer avec plus d'étendue.

XXII.

Considérons une molécule fluide M, placée à la surface de la mer, et dont, à l'origine du mouvement, θ soit le complément de la latitude, ϖ la longitude par rapport à un premier méridien fixe, ou qui ne participe point au mouvement de rotation de la Terre; supposons qu'après le temps t, θ se change en $\theta + \alpha u$, ϖ en $\varpi + nt + \alpha v$, nt représentant le mouvement de rotation de la Terre, et α étant un coefficient extrê-

(1) Remis le 7 octobre 1778.

(2) *OEuvres de Laplace*, t. IX, p. 69 et suiv.

OEuvres de L. — IX. 24*

Figure 7.1. The first three pages of Laplace's epic paper of 1776. Here he sets out the fundamental equations for the dynamics of fluid motion on the surface of the rotating earth in response to Newtonian tidal stresses. The dynamic equations (6,7,9) had been derived in the preliminary paper of 1775, containing Sections I to XXI. See text and Appendix C for explanantion. (Photographs courtesy of the Royal Society.)

188 RECHERCHES SUR PLUSIEURS POINTS

mement petit; soit αy l'élévation de la molécule au-dessus de la surface de la mer considérée dans l'état d'équilibre auquel elle serait parvenue depuis longtemps, sans l'action du Soleil et de la Lune. Représentons par $\alpha B\Delta$ et $\alpha C\Delta$ les composantes de l'attraction d'un sphéroïde aqueux dont le rayon est $1 + \alpha y$ sur la molécule M, décomposée perpendiculairement au rayon du sphéroïde, dans le plan du méridien et dans celui du parallèle, Δ exprimant la densité des eaux de la mer. Soient encore S la masse de l'astre attirant, ν le complément de sa déclinaison, φ sa longitude comptée sur l'équateur depuis le premier méridien, h sa distance au centre de la Terre, que nous supposons très considérable relativement au rayon du sphéroïde terrestre dont nous prenons le demi petit axe pour unité; que l'on fasse

$$\frac{3S}{2h^3} = \alpha K,$$

et que l'on désigne par g la pesanteur, et par $l\gamma$ la profondeur de la mer, l étant très petit, et γ étant une fonction quelconque de θ; cela posé, nous sommes parvenu (art. VI) aux trois équations suivantes, dont dépend la détermination des oscillations de la mer,

$$y = -\frac{l}{\sin\theta}\frac{\partial . u\gamma\sin\theta}{\partial\theta} - l\gamma\frac{\partial v}{\partial\varpi}, \tag{6}$$

$$\frac{d^2u}{dt^2} - 2n\frac{dv}{dt}\sin\theta\cos\theta = -g\frac{\partial y}{\partial\theta} + B\Delta + \frac{\partial R}{\partial\theta}, \tag{7}$$

$$\frac{d^2v}{dt^2}\sin^2\theta + 2n\frac{du}{dt}\sin\theta\cos\theta = -g\frac{\partial y}{\partial\varpi} + C\Delta\sin\theta + \frac{\partial R}{\partial\varpi}, \tag{9}$$

R étant égal à $K[\cos\theta\cos\nu + \sin\theta\sin\nu\cos(\varphi - nt - \varpi)]^2$.

Nous observerons d'abord sur ces équations qu'elles supposent immobile le centre de gravité du sphéroïde recouvert par le fluide, et cette supposition est légitime, comme nous l'avons prouvé dans l'article V, toutes les fois que le fluide est dérangé de l'état d'équilibre par l'attraction d'un astre quelconque éloigné; mais le fluide peut à l'origine du mouvement avoir reçu un ébranlement tel que ce centre

Figure 7.1 (*cont.*)

ne reste pas immobile, et qu'il fasse des oscillations autour du centre de gravité du système entier du sphéroïde et du fluide, que l'on peut toujours regarder comme immobile. Pour être en droit de considérer alors le centre de gravité du sphéroïde comme étant en repos, il faut transporter continuellement en sens contraire aux molécules fluides les forces qui l'agitent. Maintenant il est clair que ce centre ne peut faire que des oscillations de l'ordre αy; d'où il suit que la force qui l'anime à chaque instant ne peut être que de l'ordre de $\alpha \frac{d^2 y}{dt^2}$; en transportant en sens contraire cette force à la molécule M, il en résultera, dans les équations précédentes, des termes de l'ordre de $\alpha \frac{d^2 y}{dt^2}$, que l'on peut rejeter comme étant de l'ordre de $\alpha l \frac{d^2 u}{dt^2}$. Ces équations expriment donc généralement les oscillations d'un fluide qui recouvre un sphéroïde dont le centre est supposé immobile, quelle qu'ait été d'ailleurs la nature de l'ébranlement primitif, pourvu qu'on le suppose de l'ordre α.

Nous observerons ensuite que l'on a par l'article I, en y changeant μ en y et en y supposant $a = 1$,

$$B = 2\frac{\partial A}{\partial \theta} + \frac{4}{3}\alpha\pi\frac{\partial y}{\partial \theta},$$

$$C \sin\theta = 2\frac{\partial A}{\partial \varpi} + \frac{4}{3}\alpha\pi\frac{\partial y}{\partial \varpi},$$

π exprimant le rapport de la demi-circonférence au rayon; donc

$$B\,d\theta + C\,d\varpi \sin\theta = 2\left(\frac{\partial A}{\partial \theta}d\theta + \frac{\partial A}{\partial \varpi}d\varpi\right) + \frac{4}{3}\alpha\pi\left(\frac{\partial y}{\partial \theta}d\theta + \frac{\partial y}{\partial \varpi}d\varpi\right).$$

Soit

$$D = 2A - \tfrac{8}{3}\pi + \tfrac{4}{3}\alpha\pi y,$$

et l'on aura

$$B\Delta = \frac{\partial D}{\partial \theta}\Delta, \qquad C\Delta \sin\theta = \frac{\partial D}{\partial \varpi}\Delta.$$

Figure 7.1 (*cont.*)

Laplace's Tidal Equations

The opening pages of the famous paper presented in 1776,[1] reproduced in Figure 7.1, provide a good starting point for discussion. Laplace here recapitulates the now famous *Laplace Tidal Equations* (LTE), which had been derived in the context of the earth's nutation in the previous part of the paper (1775), to which the present excerpt is a direct sequel. In this derivation, Laplace acknowledged a debt to the French hydrodynamicist Jean le Rond d'Alembert (1717–1783), who in 1747 had developed a similar formalism for the response of the atmosphere to tidal forces in an attempt to account for the Trade Winds.[6] However, d'Alembert had omitted the deflective acceleration due to the earth's rotation, whereas Laplace showed it to be an essential ingredient of LTE.

In his introductory paragraph of Figure 7.1 Laplace announces his intention of presenting a simplified account of his main findings of 1775,[1] followed by an extended development. Article XXII, shown almost in entirety, explains the basic equations of fluid motion on the earth's surface when subject to tidal forces. Some of Laplace's notation is now archaic; I shall paraphrase his exposition in a more modern notation for easier reference later.

Let (θ, ϕ) stand for co-latitude (i.e. the complement of north latitude), and east longitude, respectively, of a fluid element on a spherical earth of radius a, angular velocity of rotation Ω, and let (u,v) be the components of horizontal *velocity* relative to the earth in the directions of increasing (θ, ϕ). u and v are supposed uniform over the depth D of the fluid. (Laplace uses $\varpi + nt$ for ϕ, $l\gamma$ for D, takes a as the unit of length, and defines (u,v) as horizontal angular *displacements*.) Continuity of fluid mass requires the surface displacement ζ to be related to u and v by an equation:

$$\frac{\partial}{\partial \theta}(vD\sin\theta) + \frac{\partial}{\partial \phi}(uD) + a\sin\theta\frac{\partial \zeta}{\partial t} = 0,$$

which corresponds to Laplace's equation (6).

If $U(\theta, \phi, t)$ denote the scalar *potential* corresponding to the tidal forces specified by Newton (Appendix C), and δU the secondary potential due to the self-attraction of the global fluid deformation, then balancing the rates of change of the two components of horizontal momentum relative to the earth with the applied force per unit mass yields two further equations:

$$\frac{\partial u}{\partial t} - 2\,\Omega\cos\theta . v = -\frac{g}{a}\frac{\partial}{\partial \theta}[\zeta - U - \delta U]$$

$$\frac{\partial v}{\partial t} + 2\,\Omega\cos\theta . u = -\frac{g}{a\sin\theta}\frac{\partial}{\partial \phi}[\zeta - U - \delta U],$$

corresponding to Laplace's equations (7,9).

Laplace's introduction of the terms proportional to $2\Omega\cos\theta$ in these equations was a most important innovation, allowing for the relatively large deflective force caused by the rotation of the earthbound coordinate system. Such terms had been ignored by D'Alembert in his 1747 study of tidal dynamics.[6] The coefficient $2\Omega\cos\theta$, often denoted in later literature by f, is called the *Coriolis frequency* after the French mathematician Gaspard Coriolis (1792–1843), who in 1835 wrote a paper about motion relative to rotating axes.[7] The term 'Coriolis acceleration' $(fv, -fu)$ is frequently used by oceanographers and meteorologists without appreciation that it was first introduced by Laplace before Coriolis was born.

$U(\theta,\phi,t)$ depends also on the co-declination, right ascension, and parallax (Θ, Φ, ϖ) of the luminary causing the tide (i.e. its position relative to the earth in space) – see Appendix A for definitions of these coordinates. Laplace gives an expression for U (his R, with ν, ϕ in place of Θ, Φ) below his equation (9). The derivation of this expression, which is quite straightforward, was given in the preceding paper of 1775. A derivation in the present choice of notation is given in Appendix C for the sake of completeness, and because of its importance in Laplace's definition of tidal *species*.

On the third page included in Figure 7.1, Laplace first argues that the slight changes in the center of gravity of the earth ('spheroide') due to the tidal displacement of the fluid are entirely negligible, as one would expect from the very small ratio ζ/a. The fact that he considers such a minor perturbation at all testifies to his careful attention to rigor. Finally, at the end of Section XXII, Laplace gives expressions for the components of the gradients of the secondary potential δU, denoted in his text by $B\Delta$ and $C\Delta\sin\theta$, where Δ is the density of sea water. These depend on an integral of the surface displacement ζ (his y), which can be expressed quite simply for the forms of solution later considered.

Laplace's tidal equations (6,7,9), translated above into a modern notation, (as used for example by Lamb[14]), were the first to define the dynamic relations which the three unknown variables (ζ, u, v) must satisfy in order to represent the motions generated by the Newtonian tidal forces. Apart from slight modifications they have stood the test of time, and provide the basis of practically all modern tidal dynamics. However, although simple in appearance, LTE are extremely difficult to solve analytically except in highly simplified and artificial sea geometry. The search for more general, or at least approximate, solutions continued for nearly two centuries after the first enunciation of LTE, and became one of the outstanding mathematical problems in oceanography for those sufficiently motivated and competent to tackle them. Finally, realistic solutions for the tides have become possible only in the last few decades of the 20th century, by use of large digital computers. We shall reach that stage in Chapters 12 and 14, after considering many other diverse topics.

For the sake of brevity, I will eschew much mathematical detail and will try to bring out the essential results of Laplace's researches as clearly as possible. Laplace worked out solutions to LTE only for the simple case of an ocean completely covering the globe, whose depth is either constant or has D as a slightly variable function of co-latitude θ only. This geometry would appear to be more suitable to the atmosphere, but there the dynamics are complicated by density stratification, thermal heating and compressibility. Laplace discusses some of these complications in Chapter V of Book IV of the *Mécanique Celeste*; more recent researches on air tides will be described in Chapters 10 and 15 of this book.

Tides without earth rotation

Taking first the hypothetical case of no rotation ($\Omega = 0$) and constant depth, Laplace noted that equations 6,7,9 then easily reduce to the form

$$\partial^2 \zeta / \partial t^2 = gD\, \nabla^2 \zeta - D\, \nabla^2 U',$$

where ∇^2 denotes the *Laplacian operator*, equivalent to $(\partial^2/\partial X^2 + \partial^2/\partial Y^2)$ in plane Cartesian coordinates (X, Y), and U' is the *total* potential $(U + \delta U)$. A solution of this equation can generally be expanded in time-dependent surface harmonics

$$\zeta = W^{(0)} + W^{(1)} + W^{(2)} + \cdots,$$

for which Laplace showed that the corresponding secondary potential takes the simple form

$$-\delta U = (3g/\rho)\ (W^{(0)} + W^{(1)}/3 + W^{(3)}/5\ \cdots),$$

where ρ is the ratio of mean densities of earth:ocean, about 5.3. A solution for the full potential $U + \delta U$ can thereby be easily derived from the simpler case in which δU is neglected.

Laplace was curious, as others (e.g. Newton, Bernoulli) had been before him, as to whether self-attraction could produce *instability*, whereby a small perturbation to the dynamical system would grow indefinitely large in time, or to a very large amplitude limited mainly by friction. If $U^{(i)}$ denotes the ith term in the harmonic development of U, then from the preceding equations, $W^{(i)}$ must satisfy

$$\partial^2 W^{(i)}/\partial t^2 + \lambda_i^2\, W^{(i)} = i(i+1) D U^{(i)},$$

$$\text{where } \lambda_i^2 = i(i+1) gD[1 - 3/(\rho(2i+1))].$$

The condition for stability is $\lambda_i^2 > 0$. One sees that this is satisfied for all i, if $\rho > 1$, so all scales of tidal motion are stable for the real ocean. Further on, Laplace showed that stability holds equally well for a *rotating* earth, $(\Omega > 0)$. The matter of stability is therefore irrelevant to terrestrial tides, but it is relevant to more general cases in astronomy, with which Laplace was also concerned.

He at first paid special attention to solutions which depended on assumed *initial* conditions, as d'Alembert had done. But by adding linear frictional terms, $\varepsilon \partial u/\partial t$, $\varepsilon \partial v/\partial t$, to the left sides of the equations 7 and 9 respectively,

Laplace realised that the effects of initialisation are always exponentially transient, so that the only solutions of practical interest are oscillatory with the same periods as occur in the forcing functions. Familiarity has made this conclusion seem obvious today, but it was novel at the time. Many years later, in Book XIII,[4] Laplace saw fit to emphasize it by re-statement in italics: '*L'état d'un système de corps dans lequel les conditions primitives du mouvement ont disparu par les résistances que ce mouvement éprouve est périodique comme les forces qui l'animent.*' (The state of a system of bodies whose initial conditions [of motion] have disappeared through friction is periodic like the forces which maintain the motion.)

Nonzero rotation, and the three principal species of tide

Turning now to the natural world with finite speed of rotation, $\Omega > 0$, Laplace observed that the forcing function $\nabla^2 U = -6U$ can be expanded as

$$-(K/4)\,(1+3\cos 2\theta)\,(1+3\cos 2\Theta)$$
$$-3K\sin 2\theta\,\sin 2\Theta\cos\,(\Omega t+\phi-\Phi)$$
$$-3K\sin^2\theta\sin^2\Theta\cos 2(\Omega t+\phi-\Phi),$$

where K is the astronomical constant $(3\gamma S\,\varpi^3/2a)$ – see Appendix C, where the three terms are equivalent to $-6(f_1+f_2+f_3)$. This implies three *species*, (Laplace's word *espèces*), of tidal oscillation with distinct bands of period or frequency. The first species is purely zonal, being independent of ϕ or Φ and varying slowly with the co-declination Θ (and parallax) of the luminary. The second species also depends on co-declination and parallax but the presence of the term Ωt produces periods close to a solar or lunar day, that is the *diurnal* tides. The third species, containing $2\Omega t$, similarly varies with periods close to half a lunar or solar day – the *semidiurnal* tides. (A fourth species, which arises when a higher approximation to U is taken, including terms of order ϖ^4, was also recognised by Laplace. Its *terdiurnal* tides, as well as small additional contributions to the first three species, are detectable only in lunar tides and are ignored in most discussions.) Laplace was the first to make such a classification. Some modern authors prefer to label Laplace's first four species 0,1,2,3, reflecting the coefficient of Ω involved.

The equations being linear in ζ, u, v, there is no restriction to treating them as independent oscillations of the appropriate period. Small nonlinear dynamic terms are, however, introduced in shallow coastal seas. With the linear assumption, Laplace expressed the variables $\zeta, \zeta' = \zeta - U'/g, u, v$, in the general forms

$$(\zeta,\zeta',u) = (a,a',b)\cos(r\Omega t+s\phi+A),$$
$$v = c\sin(r\Omega t+s\phi+A),$$

for any given r,s,A, where U' is the *total* potential. Substitution in LTE then gives b and c explicitly in terms of a'; linear relations of form

$$a = \Lambda(a')$$

give a single differential equation with respect to the variable $\sin\theta$ to be solved for a', and hence a and the general solution (ζ, u, v).

First species

In considering the first species, (the *long-period* tides), Laplace allowed for a variation of ocean depth with latitude:

$$D(\theta) = D(1 + q\sin^2\theta),$$

formally similar to the earth's ellipticity. For this case an exact solution for $a(\theta)$ was possible as a series of zonal harmonics of co-latitude θ. However, the value of q was numerically very small, so the ocean depth was nearly constant. He then employed arguments involving friction proportional to velocity, to postulate that only an equilibrium response was relevant for this species of tide, that is, one whose surface elevation ζ is proportional to

$$\frac{f_1(\theta, \Theta)}{g(1 - 3/5\rho)},$$

where f_1 is the spatial form of the first species defined in Appendix C. This result, suitably modified to allow for the presence of continents and for the earth's elasticity, accords with modern measurements of the broad-scale components of the long-period tides. However, we now know that the component with largest amplitude, whose period is only about 13.8 days, does exhibit a frictional time-lag relative to the static form, as well as some irregular features of short wavelength.

Second species

Laplace's second species of tide is responsible for the so-called *diurnal inequality* in predominantly semidiurnal tides. He was curious to find an explanation why this inequality was generally such a small fraction of the main tidal amplitude along Atlantic coasts, much smaller than an *equilibrium* tide would suggest. The appropriate part of the forcing potential U is here

$$(K/2)\sin 2\theta \sin 2\Theta \cos(\Omega t + \phi - \Phi).$$

Laplace discovered that for diurnal speeds (close to Ω) there exists a simple solution valid for any depth parameter q:

$$\zeta = \frac{qK\sin 2\theta \sin 2\Theta}{2qg(1 - (3/5\rho)) - \Omega^2}\cos(\Omega t + \phi - \Phi).$$

If the depth is strictly uniform, $q=0$, so the tidal elevation ζ is zero over the whole sphere. However, the amplitudes of the current components b, c, are not zero, so the tide would consist of currents without changes in elevation. (Heuristically, this solution may be visualised as a spherical ocean of uniform depth rotating with speed Ω about an axis oblique to the earth's axis.)

Laplace thought that such a solution for a suitably small depth variation q might account for the small observed diurnal inequality. We now know the true cause to be more complicated. The Pacific, Indian and Southern Oceans all have regions of large diurnal inequality, but this was hardly known at the time of the *Mécanique Céleste*.

Third species

The third species being semidiurnal, with forcing potential U equal to

$$(K/2)\sin^2\theta\sin^2\Theta\cos(2\Omega t+2\phi-2\Phi),$$

Laplace computed the appropriate solution in considerable detail, in the hope of a direct comparison with observations. This time, he simplified the calculation by setting the self-attraction potential δU and the variable depth parameter q to zero, so the depth D is considered constant. For such conditions the differential equation for the vertical amplitude parameter a reduces to

$$x^2(1-x^2)\frac{\partial^2 a}{\partial x^2}-x\frac{\partial a}{\partial x}-a(8-2x^2-\beta x^4)+4Cx^2=0$$

where $x=\sin\theta$, $\beta=4\Omega^2a^2/Dg$, $C=(K\sin^2\Theta)/g$. (Laplace used $\mu=\beta/2$.) Solutions to the above equation were expanded in the form

$$a=A_1x^2+A_2x^4+A_3x^6+\cdots$$

with the coefficients A_n for $\beta=40, 20, 10$, and 5. ($\beta=20$ corresponds to a depth of 4.4 km, which Laplace called '1 lieue' – in English 'a league'.) Coefficients up to A_{13} were found to be necessary for satisfactory convergence of the above series. Spring tide ranges were simulated by adding the lunar to the solar tide with Bernoulli's ratio 5/2 (Chapter 5).

Results were strongly dependent on ocean depth, a fact not previously suspected under the equilibrium hypothesis. For $\beta=40$ (depth 2.2 km), the spring range at the equator ($x=1$) came out to be 19.85 *pieds* (6.4 m) with inverted phase (i.e. Low Water at Transit). The range decreased to zero at 16° latitude; for higher latitudes the phase was direct. Enough was known of tidal ranges at the equator for Laplace to deem this result inadmissible, so the ocean depth had to be greater than 2.2 km, in his estimation.

For $\beta=20$, (i.e. depth 4.4 km), the equatorial range was 4.86 *pieds* (1.6 m) and again inverted, in this case up to 37° latitude. This again was ruled inadmissible. (Ironically, 4.4 km is now known to be very close to the mean depth of the oceans, and the known semidiurnal phase at the equator is closer to 180° than to 0, but the range does not decrease systematically with latitude.)

For $\beta = 10$, (i.e. depth 8.8 km), the equatorial range of the solution was 30.09 *pieds* (9.7 m) with direct phase; this range was too large. Finally, for $\beta = 5$ (17.7 km), the equatorial range was 5.14 *pieds* (1.7 m) with direct phase. This last case was deemed by Laplace to be acceptable, with reservations.

Taken at their face value, the above results are hardly reassuring as to the value of the heavy algebra involved in their calculation. But it was really the over-simplified assumptions of a global ocean with constant depth, not the theory, which was at fault. What Laplace was showing, without perhaps fully realising it, was that the wave structure of the tides in the real ocean is far more complex than anyone had hitherto suspected.

Semi-empirical analysis

Laplace devoted many pages of Books IV and XIII of the *Mécanique Céleste* to the analysis of tidal observations, particularly those from Brest. His objects were, partly to find empirical constants which relate the observed tidal heights and times to the potential U, and partly to derive a definitive value for the moon:earth mass ratio. Preliminary to this, he made more explicit calculations of the frequency-structure within his three species in terms of the mean motion $m = \delta\Phi/\delta t$ of the luminary and the inclination of its orbit. For the second species, spherical geometry gave for the diurnal potential

$$U_1 = K \sin\theta \cos\theta \sin\varepsilon \, [\cos\varepsilon \sin(\Omega t + \phi)$$
$$- \cos^2(\varepsilon/2) \sin\{(\Omega - 2m)t + \phi\}$$
$$+ \sin^2(\varepsilon/2) \sin\{(\Omega + 2m)t + \phi\}]$$

where $K = (3\gamma S\, \varpi^3/2a)$ as before, and ε is the obliquity of the orbit of the luminary relative to the equator, in the region of 23°. The terms of frequency Ω and $(\Omega - 2m)$ are the diurnal harmonics later named by Thomson (Lord Kelvin) K_1 for both sun and moon, P for the sun, and O for the moon respectively, (Chapter 8). The term of frequency $(\Omega + 2m)$ is negligible in this context because of its small multiplier $\sin^2(\varepsilon/2)$.

For the third species, he obtained similarly

$$(K/8) \sin^2\theta [4 \sin^2\varepsilon \cos(2\Omega t + 2\phi)$$
$$+ (1 + \cos\varepsilon)^2 \cos\{(2\Omega - 2m)t + 2\phi\}$$
$$+ (1 - \cos\varepsilon)^2 \cos\{(2\Omega + 2m)t + 2\phi\}].$$

Here, the first two terms are Thomson's semidiurnal harmonics K_2 and S (sun), or M (moon) respectively, and the third term again has negligible magnitude.

Laplace did not expand the parallax factor ϖ^3 in K explicitly, (which would have introduced Thomson's harmonics R,T and L,N), because he stated that, by observation, the tidal response to changes in parallax (of the moon) is merely proportional to the instantaneous value of ϖ^3. Later investigations have shown that the response to the monthly change in parallax is more complicated, but in any case Laplace designed his data analysis to eliminate the effects of changes of parallax which are due to orbital ellipticity.

Omitting, for convenience, the geographical factors $\cos^2\theta$ and 2ϕ, and writing K', ε', m' for the astronomical factors pertaining to the moon, Laplace's model for the semidiurnal elevation at a given place was

$$\begin{aligned}\zeta = {} & 4(Kr^3)\sin^3\varepsilon\ A_0\cos(2\Omega t - P_0) \\ & + 4(K'r'^3)\sin^2\varepsilon'\ A_0\cos(2\Omega t - P_0) \\ & + (Kr^3)\ (1+\cos\varepsilon)^2\ A_S\cos\{(2\Omega - 2m)t - P_S\} \\ & + (K'r'^3)\ (1+\cos\varepsilon')^2\ A_M\cos\{(2\Omega - 2m')t - P_M\}.\end{aligned}$$

Here, r and r' are the ratios of the parallaxes of sun and moon to their respective mean values, and therefore vary slightly about mean values of unity. (A_0, P_0), (A_S, P_S), (A_M, P_M) are arbitrary amplitudes and phase lags of an assumed *admittance* (in modern terminology) at the frequencies 2Ω, $2\Omega - 2m$, $2\Omega - 2m'$, to be determined empirically from the tidal observations. Laplace reduced these six arbitrary parameters to four by further assuming, by a rough argument, that the admittance should be a linear function of (frequency minus 2Ω).

He evaluated all the above parameters from carefully selected observations at the four principal phases of the moon at both solstice and equinox, configurations which sample all the principal phase differences between the three harmonic frequencies involved. He used at least 64 samples for each of the four combinations of pairs, from the later, more extensive, measurements which he instigated at Brest (Chapter 6). Each estimate was suitably adjusted with respect to consecutive data to eliminate diurnal and parallactic inequalities. In reducing these data, Laplace acknowledged the valuable assistance of Alexis Bouvard (1767–1843), an astronomer attached to the Bureau des Longitudes, later to become Directeur of the Observatoire de Paris. (Thomas Young, who tended to disparage the practicality of Laplace's work, referred to his assistant as 'the indefatigable Bouvard' – see Chapter 8.) The employment of a knowledgeable and industrious assistant to reduce large quantities of data became an invariable practice of tidal analysts from that time onwards. Laplace's own expertise in the theory of probability was also usefully applied in the treatment of random errors and tests of significance.

The final results[4] for the semidiurnal tides gave definitive values for the six arbitrary parameters (A, P in the last equations) which could be used for predictions. They also gave a new estimate for the ratio of lunar to solar tidal forces, namely 2.353 compared with Bernoulli's estimate of 2.5 derived from much simpler calculations and relatively little data. Known figures for the mean parallaxes then gave an estimate for the moon:earth mass ratio of 1/75. The latter figure compares plausibly well with the modern value, 1/81.3, and was a considerable improvement on Laplace's earlier estimate, 1/59, derived from the shorter 18th century series of observations.[2]

A similar analysis for the diurnal tides at Brest gave admittance amplitudes two orders of magnitude lower. Laplace took this result as a vindication of his theoretical analysis:

> Ainsi, par l'effet de la rotation de la terre et des circonstances accessoires, le flux diurne est réduit à peu près au tiers, tandis que le flux semidiurne devient seize fois plus grand. Au reste, cette grande difference ne dois point surprendre, si l'on considère que, par Livre IV, la rotation de la terre détruit dans une mer partout également profonde, le flux diurne; et que, si la profondeur de la mer est 1/720 du rayon terrestre, ou d'environ 9000 metres, la hauteur de la marée semidiurne dans les syzygies est de 11 meters.[4]

> (Thus, by the effect of the earth's rotation and other circumstances, the diurnal tide is reduced to about one-third, while the semidiurnal tide becomes sixteen times greater. In fact this large difference is not at all surprising if one considers, as in Book IV, that the rotation nullifies the diurnal tide in a sea of constant depth; and that, with an ocean depth 1/720 of the earth's radius or about 9000 meters, the height of the semidiurnal tide at the syzygies is about 11 metres.)

As a final effort in analysing his Brest data, in Chapter 6 of Book XIII, Laplace differentiated between syzygies at Full and New Moons and between quadratures at first and third Quarters, to extract the leading term of the fourth (ter-diurnal) species. Its amplitude turned out to be 0.12m, with reservations as to accuracy. (The modern figure is 0.015m, an order of magnitude lower; Laplace's result was probably biassed by random noise, of whose danger he was well aware.)

It is a measure of the thoroughness and reliability of Laplace's empirical analysis that the cited formalism for the semidiurnal tides at Brest, together with a similar formalism for the diurnal species and an equilibrium form for the long-period species, were used for the construction of tide-tables by the French hydrographic authorities until about the middle of the 20th century. The amplitude of each tide at Brest is still used in France to determine a certain 'coefficient des marées' as a rough guide to the corresponding amplitude at minor ports on the Atlantic and Channel ports.

Two branches of 19th century research stem directly from Laplace's work on the tides: the hydrodynamic theory of long waves, mostly in seas of less than global dimensions; and the empirical analysis and prediction of tidal observations in terms of a series of harmonic oscillations in time. Only the first category will be considered in the rest of this Chapter; data analysis and prediction will be discussed in Chapter 8.

G.B. Airy – tides in canals

Early in his long career as seventh Astronomer Royal, George Biddell Airy (1801–1892) wrote a treatise called *Tides and Waves* for the *Encyclopedia Metropolitana*.[8] While admiring Laplace's magnum opus, he considered Laplace's global solutions to be too remote from conditions in the real oceans to provide insight of their spatial dynamic behavior. For greater simplicity, Airy studied long waves of tide-like character in narrow canals of variable depth, thereby removing the complication of the transverse accelerations due to rotation.

The classic result, that a wave much longer than the depth travels freely along a canal of uniform depth with speed $\sqrt{(g \times \text{depth})}$, independently of wavelength, had been found by Lagrange,[9] but Airy generalised it to where the amplitude is a considerable fraction of the mean depth, and to nonuniform cross-sections. For large amplitude, he showed that the wave-crest travels faster than the trough, accounting for the observed result that in shallow water the tide rises faster than it falls. He also showed how this situation leads to the presence of harmonic components with twice the argument or half the period.

In treating the action of the tidal forces on such waves in a canal of finite length, Airy showed that the amplitude and phase depend critically upon the depth and the period of the disturbance. As a result, two adjacent canals of different depth could contain quite different configurations of tide, while in either canal the lunar and solar tides would differ. In natural depths of the sea, he showed, in general, that comparison of the lunar and solar tides would tend to overestimate the mass of the moon. Such is the case for Laplace's estimate, 1/75 instead of 1/81.

From a three-dimensional analysis of waves in a canal of finite breadth which is deeper in the middle than at the shores, Airy showed that the tidal crests are convex upstream, and that the currents to the left (right) of mid-section rotate clockwise (anticlockwise), without stagnation; flow is towards the coasts on the rising tide, away from the coasts on the falling tide. Such motion is observed, for example in the English Channel. (J. Proudman later showed that straight crests are also possible in such a channel.[10])

In general, Airy's researches on tides presented a fresh approach. Though limited to two dimensions, his simplified geometry provided heuristic insight into the basic wavelike nature of tides, not easily derived from Laplace's global analysis.

Kelvin waves and Poincaré waves

The famous physicist Sir William Thomson (1824–1907), otherwise known as Lord Kelvin, introduced several important ideas for the understanding of tides and for physical properties of the earth which could be derived from them. His extension of Laplace's harmonic representation of tidal data will be treated in

Chapter 8, and his geophysical ideas in Chapter 10. Here we shall consider Thomson's introduction of a simple solution of LTE which had hitherto escaped attention, using two horizontal dimensions instead of Airy's one.

Avoiding the complications of spherical coordinates, Thomson[11] applied LTE to a flat sea with rectangular axes x,y slowly rotating with anti-cyclonic speed ω ($\Omega\cos\theta$ in our notation). The undisturbed depth of the sea is D, possibly a function of (x,y), and its disturbed elevation above an equipotential surface is $\zeta(x,y,t)$. Thomson also assumed that the magnitudes of $g\partial\zeta/\partial x$ and $g\partial\zeta/\partial y$ are much greater than the external horizontal stresses, as is usually the case in nature, so that waves may progress freely, without much influence from the tidal forces. Replacing Laplace's south and east angular displacements by velocities (u, v) in the (arbitrary) directions (x,y), and setting U to zero for free waves, LTE reduce to the simplified forms:

$$(Du)_x + (Dv)_y = -\zeta_t,$$

$$u_t - 2\omega v = -g\zeta_x,$$

$$v_t + 2\omega u = -g\zeta_y,$$

where I have written ζ_t in place of Thomson's $\partial\zeta/\partial t$, etc.

After noting some basic properties of the *vorticity*, (rotary property), of the motion in polar coordinates, Thomson restricted the argument to constant depth D, for which

$$\nabla^2\zeta = \zeta_{xx} + \zeta_{yy} = (\zeta_{tt} + 4\omega^2\zeta)/gD.$$

For wavelike oscillations of angular speed σ and wavenumbers $\mathrm{i}\alpha$, $\mathrm{i}\beta$:

$$\zeta(x,y,t) = \zeta_0 \exp(\alpha x + \beta y - \mathrm{i}\sigma t),$$

the previous equation gives:

$$\alpha^2 + \beta^2 = (4\omega^2 - \sigma^2)/gD.$$

Thomson sought a solution periodic in x and bounded by the x-axis, implying $v = 0$ along $x = 0$. In fact, for $\alpha = \mathrm{i}m$ (m real) he found that the simple condition

$$\beta = -2\omega m/g$$

has $v = 0$ everywhere, and corresponds to a wave

$$\zeta = h_0 \exp(-2\omega my/\sigma)\cos(mx - \mathrm{i}\sigma t)$$

travelling in the positive x-direction with phase velocity

$$\sigma/m = \sqrt{(gD)},$$

independently of the rotation speed ω.

Such a waveform, with coastline to the right and crest amplitude decreasing exponentially to the left of the direction of propagation (*vice versa* in the southern hemisphere, where $\omega < 0$), was seen to satisfy many of the observed features of tidal waves progressing up the English and Irish Channels and along the west coast of the North Sea. It is now known to be the most characteristic waveform of tides in the deep ocean, with a few exceptions. The wave was later called the *Kelvin wave* after Thomson became better known as Lord Kelvin. In modern oceanography, the term 'Kelvin wave' (or better, 'double Kelvin wave') is also used to describe a type of *internal* wave unrelated to tidal motion, which travels eastward along the equator with amplitude decreasing to both north and south.

A class of waves related to the Kelvin wave, whose crests are sinusoidal instead of exponential (i.e. β imaginary in the above example), was later identified by the French mathematician Henri Poincaré (1854–1912).[12] Its existence requires σ to be greater than 2ω. Now known as *Poincaré waves*, they support rotating currents, nodal points, and other known characteristics of oceanic tides. Geoffrey Taylor (1886–1975) later showed how a combination of oppositely traveling Kelvin waves and an infinite series of Poincaré Waves could validly describe the *reflexion* of a Kelvin Wave at the closed end of a rectangular gulf, roughly simulating the behavior of tides in the southern North Sea.[13]

Thomson's original paper concluded with a solution for tide-like waves in a rotating circular basin of uniform depth. The profile of the wave-crests in this configuration is defined by *Bessel functions*.

Waves of first and second class – Lamb, Margules and Hough

Sir Horace Lamb (1849–1934) was the *doyen* of theoretical hydrodynamics during the first third of the 20th century, but the first edition of his classic treatise was published in 1879, about the same time as the paper by Thomson cited above. It included a comprehensive account of the theory of tidal waves, updated in subsequent editions. The second edition[14] of Lamb's *Hydrodynamics*, published in 1895, extended Thomson's treatment of the flat circular sea (radius b) with rotation ω to the case of parabolic depth:

$$D(r,\theta) = D_0(1 - r^2/b^2).$$

Omitting mathematical details, wave solutions with angular frequency σ, of form

$$\zeta(r,\theta,t) = h_s(r/b)^s\, F_{r,s}(r^2/b^2)\exp i(\sigma t + s\theta) \quad (s = 0, 1, 2, \ldots)$$

were found, where $F_{s,k}$ is a series of hypergeometric functions with integral parameters s and k. The quantity $\sigma' = \sigma/2$ must satisfy a cubic equation:

$$\frac{4\omega^2 b^2}{gD_0}(\sigma'^2 - 1) - \frac{2s}{\sigma'} = k(k-2) - s^2.$$

When $k > s + 2$, this equation always has three real roots σ', of which two have large magnitude and opposite sign, (approximated by ignoring the term $2s/\sigma'$), and the third root has small magnitude and negative sign, (approximated by ignoring σ'^2).

The two waves associated with large frequency $|\sigma'/2|$ rotate in opposite senses and are controlled by gravity in the usual way. These became known as waves of the *first class.* As ω tends to zero, the low frequency solution gives a wave of vanishingly small surface elevation but finite currents, controlled by conservation of vorticity – the rotary part of a wave motion. These became known as waves of *second class.* In Lamb's words and italics, 'when the angular velocity ω is infinitely small, [a second class wave] becomes a *steady* rotational motion, without elevation or depression of the surface.'

A similar distinction in types of wave solution in an atmosphere on a sphere had already been made (1893) by the Austrian meteorologist Max Margules (1856–1920) in the context of air tides, but in 1895 Lamb was unaware of Margules' work. Unlike Lamb, Margules' two classes of solution depended, not on changes in depth, but on the variation of the rotation ω with latitude. Margules[15] called his solutions *Wellen erster und zweiter Art* (waves of first and second kind) and he was the first to identify them. *Art* has a slightly different meaning from *Klasse*, but *class* became the invariable English translation after the work of Hough (below). However, Professor G.W. Platzman of Chicago pointed out to me[16] that Lamb's second-class wave, described above, is unique in being the first example to arise from variations in *depth*, as distinct from variations in *latitude.* Lamb's waves are nowadays called *topographic-Rossby* or *vorticity* waves;[16] they are known to play an important role in the dynamics of tides of long period, and of diurnal tides at latitudes higher than 30° ($\sigma<2\omega$). For further discussion in relation to 20th century observations of tidal currents, see Chapter 13.

The last important solutions of Laplace's tidal equations in the 19th century were those by Sydney Samuel Hough (1870–1923), a Cambridge mathematician, later appointed Astronomer at the Cape. In the first of two long, intensively mathematical papers on the subject[17] Hough re-derived LTE from fundamental principles, an exercise to which others have returned in the 20th century. (The equations as usually written are not quite exact, and one may question whether certain approximations are always valid.) Having verified the usual form of the dynamic equations, Hough proceeded to improve on Laplace's solutions for the three species of tide in an ocean of nearly uniform depth covering the globe. He realised the limitations of relating such a model to the natural ocean, but thought that fresh insight could be brought to natural tides by applying more powerful mathematical tools than were available to Laplace.

Hough devoted much attention to the solutions of Laplace's first species (of long period). His reasons for this were twofold. Firstly, Laplace's assumption, that friction would inhibit all but an equilibrium solution, had been negated by Hough's senior colleague and mentor, G.H. Darwin,[18] of whom we shall hear more in Chapters 8 and 10. Secondly, he thought that the limiting solutions for infinite period, which are nontrivial, might apply to the steady ocean circulation, as the quotation from Lamb might suggest. In fact, he computed the first six *natural periods* of even and odd orders for each of Laplace's depth-parameters (in Lamb's notation) $\beta = 5, 10, 20, 40$, and very accurate solutions for the elevation of the tide of 13.8 days period (Mf), including rigorous allowance for the gravitational self-attraction of the tide itself. In the limit of infinite period, Hough obtained some suggestive results for steady equatorial currents, but he had to admit failure to account for the evidently dominant effects of variable sea-water density and continental coasts, compared with which the tidal effect is quite negligible.

In the second of his pair of papers, Hough extended his rigorous analysis to the tides of species 2 (diurnal) and 3 (semidiurnal) in a global ocean.[17] The analysis of Part I had been confined to axi-symmetric (zonal) waves; in Part II he had also to include variations with longitude. He employed expansions in tesseral (or spherical) harmonics, which proved to converge much more rapidly than Laplace's series in powers of $\sin^2\theta$, contrary to expectations which had been expressed by Airy and Kelvin. (His solutions, still employed today in the theory of air tides, were later called *Hough functions.*[19]) Resonant free wave periods again divided naturally into those of gravity waves (roughly 12 hours), and vorticity waves (several days) whose elevation tends to zero with the rotation ω. Hough called these 'waves of first and second class', apparently unaware of the similar nomenclature introduced by Margules[15] five years previously. (Margules' work was slow to be recognised outside his native Austria.) The Margules/Hough waves of second kind/class differ from Lamb's example, cited above, in that they depend for their existence on variation in the Coriolis parameter rather than variation in depth. The two types of variation are however mathematically analogous, as was realised by later oceanographers.

The free periods of second-class waves computed by Hough for constant depth included one of exactly a sidereal day, entirely without surface deformation. He recognised this as the example discovered by Laplace, who had suggested that it accounts for the small amplitude of the diurnal tides. In fact, the existence of this resonant period precisely at the harmonic constituent K_1 of the generating potential (Figure 8.3) prevented Hough from calculating the tidal response at K_1. But he did compute the tidal response for the lunar diurnal constituent O_1, and also for the semidiurnal constituent M_2. He similarly identified a free period of half a sidereal day (K_2), but only for specific values of the depth parameter β close to 10 (8.9 km) and 40 (2.25 km). The figures varied slightly when a small latitudinal variation of depth was allowed for. However, we now know that such fine tuning to resonance in a realistic ocean is rendered insignificant by the presence of friction, and the resonant periods themselves are different anyway.

I should perhaps apologise to the more mathematically inclined reader for the above superficial summary of Hough's erudite two-part treatise. The physical results of Hough's investigations were not very impressive; the historical significance of his work lies mainly in his mathematical analysis of the two classes of wave structure and his general solutions for waves on a spherical sea of uniform depth. A full account of these theoretical developments would require more space than this book can afford. The above account provides an adequate culmination to research of the 19th century inspired by Laplace's treatise.

Notes and references

1. Laplace, P.S. Recherches sur plusieurs points du Système du Monde. *Mém. Acad. roy. des Sciences*, **88**, 75–182, 1775 and **89**, 177–264, 1776. Also in *Oeuvres*, **9**, 69–183 (1775) and 187–310 (1776), Paris, 1893.
2. Laplace, P.S. Mémoire sur le flux et reflux de la mer. *Mém. Acad. des Sciences*, 45–181, 1790. Also in *Oeuvres*, **12**, 3–126, Paris, 1893.
3. Laplace, P.S. *Traité de Mécanique Céleste*, **2**, Livre 4, 1799. Also in *Oeuvres*, **2**, Paris, 1878.
4. Laplace, P.S. *Traité de Mécanique Céleste*, **5**, Livre 13, 1825. Also in *Oeuvres*, **2**, Paris, 1878.
5. Bowditch, N. *Mécanique Céleste* by the Marquis de Laplace, translated with a Commentary, Volumes 1–4, Boston, 1829–1839. (Reprinted, with Vol. 5 in the original French, Chelsea Pub., Boston, 1966.)
6. D'Alembert, J.R. Reflexions sur la cause générale des vents – (pièce qui a remporté le prix imposé par *l'Académie Royale de Berlin* pour l'an 1746), Paris, 1747.
7. Coriolis, G. Mémoire sur les équations du mouvement relatif des systèmes de corps. *J. Ecole Roy. Polytechnique*, **15**, (Cahier 24), 142–154, 1835.
8. Airy, G.B. Tides and Waves, *Encyclopedia Metropolitana*, **5**, 241–396, London, 1845.
9. Lagrange, J.L. Mémoire sur la théorie du mouvement des fluides. *Nouv. Mém. Acad. Sciences et Belles Lettres*, **4**, 695, (*Oeuvres*, **1**, 747), 1781.
10. Proudman, J. Tides in a channel. *Phil. Mag.*, **49**, 465–475, 1925.
11. Thomson, W. On gravitational oscillations of rotating water. *Proc. R. Soc. Edinburgh*, **10**, 92–100, 1879. (*Math. & Phys. Papers*, **4**, 141–148, 1910.)
12. Poincaré, H. *Leçons de Mécanique Céleste*, Tome III – *Théorie des marées*, Gauthier-Villars, Paris, 469pp., 1910.
13. Taylor, G.I. Tidal oscillations in gulfs and rectangular basins. *Proc. London Math. Soc.*, **20**, 148–181, 1921.
14. Lamb, H. *Hydrodynamics*, 2nd Edition, Cambridge Univ. Press, 604pp., 1895. Also in 6th (last) Edition, *ibid.*, 738pp., 1932.
15. Margules, M. Luftbewegungen in einer rotierenden Sphäroidschale (II Theil), *Sitzungsberichte der Kaiserlichen Akad. Wiss. Wien*, **102**, 11–56, 1893. (Engl. transl. B. Haurwitz, Air motions in a rotating spheroidal shell, Boulder, Colorado, Nat. Cent. Atmos. Res., *Tech. Note* 156, 174pp., 1980.)
16. Platzman, G.W. *The Rossby wave,* (Symons Memorial Lecture). *Q. J. R. Met. Soc.*, **94**, 225–248, 1968. The early history of waves of second class, including the initiatives of Margules, Lamb and Hough, is summarised in the Appendix to this paper, pp. 245–246.
17. Hough, S.S. On the application of harmonic analysis to the dynamical theory of the tides, I – On Laplace's 'oscillations of the first species', and on the dynamics of ocean currents, II – On the general integration of Laplace's dynamical equations. *Phil. Trans. R. Soc. London*, **A**, Part I – **189**, 201–257, 1897; Part II – **191**, 139–185, 1898. (The phrase 'harmonic analysis' in this title refers to spherical-, not time-harmonics as used in Chapter 8.)
18. Darwin, G.H. On the dynamical theory of the tides of long period. *Proc. R. Soc. London*, **41**, 337–342, 1886. Also in *Sci. Papers*, **1**, 366–371, 1907.
19. Siebert, M. Atmospheric Tides. *Advances in Geophysics*, **7**, 105–182, 1961. (The phrase 'Hough functions' is introduced on p.150.)

8

Local analysis and prediction in the 19th century

The reader of the previous chapter will have noticed a marked increase in the subtlety of mathematical argument applied to tidal dynamics during the 19th century. Indeed, the quest for exact solutions to Laplace's equations had by 1900 become an esoteric discipline in its own right, of great interest to mathematical specialists but quite remote from practical experience of the natural tides. At the same time, great advances were also being made, on the one hand in the art of collecting and analysing tidal data from specific locations, and on the other, in the accuracy with which predictions could be computed. Such activity may be less interesting to the mathematician but it is more useful to practising navigators and dockmasters and to the publishers of tide-tables. This chapter will review the progress in that field.

In a history of tide prediction technique, improvements to which are relatively infrequent, it is hard to avoid giving the impression that each improvement is the last word on the subject. I have hitherto described advances in technique by Philips, Flamsteed, Bernoulli (Chapter 5), Cassini (Chapter 6) and Laplace (Chapter 7). Each of these rectified some deficiency in formulation and probably seemed to satisfy current needs, but none was perfect. For example, Laplace's formulation was the first to include the diurnal inequality, vital for predicting tides at most Pacific ports, but he incorrectly assumed an instantaneous response to changes in parallax. Quasi-random errors due to weather impose a threshold to the accuracy of any prediction, but a method such as Laplace's will admit some errors in the tides themselves which should in principle be possible to avoid. At the same time, practical considerations such as increasing drafts of ships and the cost of dredging channels made new demands on precision.

Renewed British interest

After Laplace, most of the initiative for analysis and prediction technique came from Great Britain and the United States of America. A short review of the status of tidal research in each of those two countries about that time is therefore appropriate here. In Britain, little interest was shown during the 18th century, except for Maclaurin's essay for the French prize of 1740, (Chapter 6). The articles entitled *TIDE* in the first two editions (1769–1783) of *Encyclopedia Britannica* were anonymous. The publishers of the third edition (1797) invited John Robison, the Secretary of the Royal Society of Edinburgh, to write the *TIDE* article, among several other articles. John Robison (1739–1805) was a worthy mechanical philosopher with some sea-going experience, but he is not now known for any tangible contribution to tidal theory or practice. His article, mainly based on Newton's and Euler's theories, was reprinted in the 4th (1810), 5th (1815) and 6th (1822) editions, but a supplement to the 6th edition, dated 1824, was written instead by Thomas Young.[1,2]

Thomas Young

Young (1773–1829), a versatile physicist, best known for his discovery of optical interference, is important to our subject as a lone pioneer in the revival of interest in tidal theory in Britain after a century of neglect. His lectures on mechanics to the Royal Institution of London from 1802 to 1807 included the basics of tides, and he developed a wave theory of tides which in some ways foreshadowed Airy's monograph.[3] His theory was weakened by his having ignored some essential aspects of Laplace's theory (known to Young), such as the conditions for mass conservation and for motion on a rotating earth; it was more directly related to Euler's theory, by then outdated. Young's principal innovation was a serious discussion of the effects of a resistance to tidal flow proportional to the square of the speed, which he rightly considered to be more physically realistic than the mathematically convenient linear resistance law which Laplace had assumed.[4]

The editor of Young's Collected Works, George Peacock, added a footnote to one of Young's principal papers on tides,[4] quoting some correspondence from Sir George Airy which is revealing of both men's work:

> You ask my opinion on Dr. Young's researches on tides; as far as they go they are capital: when I was writing my article I totally forgot Dr. Young, although I well know that in writing on *any* subject it is but ordinary prudence to look at him first. When I came to look at him, I was surprised to find that he has clearly enough shown the difference of positive and negative waves, and also the difference of free oscillations and forced oscillations: and that he has hinted at the cause of the rapid rise of river tides as distinguished from their slower fall. All these were great points with me, quite original to myself. There is one of mine, however, which he has not got, namely the effect of friction in producing an apparent retardation of the day of spring tides ...

It will be recalled that Airy's monograph was published by the *Encyclopedia Metropolitana* of London.[3] Young's *TIDE* article survived reprinting through the 7th and 8th editions of the *Britannica*, but was finally supplanted in the 9th edition (1888) by one by G.H. Darwin, a widely acknowledged authority on the subject.[2]

In his article for *Nicholson's Journal*,[4] Young made the following pregnant remarks relevant to the main subject of this chapter:

> There is indeed little doubt, that if we were provided with a sufficiently correct series of minutely accurate observations of the tides, made not merely with a view to the times of low and high water only, but rather to the heights at intermediate times, we might form by degrees, with the assistance of the theory contained in this article, almost as perfect a set of tables for the motions of the ocean as we have already for those of the celestial bodies, which are the more immediate objects of the attention of the practical astronomer. There is some reason to hope that a system of such observations will speedily be set on foot by a public authority ...

This seems to be the first suggestion that something better could be done than the long accepted tradition of recording extrema. It almost certainly led to the invention of automatic continuous tide recorders in the 1830s, and to the practical application of harmonic analysis in the 1870s. Around 1830, Young's revival of interest in tides was taken up with enthusiasm by John Lubbock and William Whewell of Cambridge.

Growth of organisation in the USA

Before Independence, the coastal States of North America relied on Greenwich for lunar ephemerides, but local theoreticians could use the ephemerides to compute simple HWT predictions for the major ports. These tide-tables were published, among much other information, in pocket-sized almanacs. The British Library holds a copy of *Russell's American Almanack* for 1780 (the fourth year of Independence), containing ephemerides calculated by 'that learned and ingenious Master-Piece in Astronomy, Benjamin West' including once-daily figures for 'Full Sea' (i.e. HWT) at Boston Harbor. Another *American Almanac*, containing figures for HWT at Philadelphia, New York and Boston for each 'Day of the Moon's Age', was produced in 1783 by one 'Father Jacobus Bumbo'. Several almanacs of a similar nature were published in the USA in the period 1775–1825.

A more impressive series entitled *The American Almanac and Repository of Useful Knowledge* was published by Gray and Bowen of Boston for several years from 1830.[5] Its annual volumes contained elaborate astronomical ephemerides modeled on the British *Nautical Almanac* and they usually included a once-daily tidal HWT for Boston, New York and Charleston and a list of time differences to be applied to other east coast ports. A curious addition was a table of spring tide altitudes based on Laplace's formula for Brest (France), with proportional factors to apply to ports of the USA – analogous to the French concept of *Coefficients de Marée* – Chapter 7. The astronomical part of the *American Almanac & RUK* was later incorporated in the *American Ephemeris and Nautical Almanac* from the 1840s, without tidal information. Tide-tables were not substantially improved until the following decade, when the *United States Coast Survey* had become well organised.

With its huge coastline, soon to include the Pacific coast, the United States government saw the need for a nationally funded *Coast Survey* to measure and advise on the hydrography of all its ports and coastal waters. The Survey was authorised by Thomas Jefferson in 1807, but did not get properly funded until 1832, with F.R. Hassler as its first head. Both Hassler and his successor, the polymath scientist A.D. Bache, were assiduous in setting up recording stations for tidal elevations and currents at Atlantic, Pacific and Gulf ports, providing the basis for better predictions in due course. Alexander D. Bache, (1806–1867), a great-grandson of Benjamin Franklin, headed the Coast Survey for 24 years from 1843, and used his considerable influence in American scientific circles to build up the Survey into a major intellectual force, with commitments in geodesy and other earth sciences. The *Tide Division* of the Coast Survey was formed in 1854; in 1867 it took responsibility for the official production of all the national tide-tables for the USA.[(6)]

This organisational structure under federal funding was quite new in tidal science, hitherto led by the initiative of a few solitary academics. By contrast, before about 1870 most British tide tables were produced by commercial entrepreneurs, details of whose methods were often shrouded in secrecy.

J.W. Lubbock's synthetic analytical method

To return to our main subject, J.W. Lubbock perfected the *synthetic* method of analysis and prediction during the 1830s. John William Lubbock (1803–1865) was a practically minded mathematician, who also published important researches on the tide-related subjects of gravitational attraction of ellipsoids and on the theory of the moon's orbit. His *synthetic* method (a name bestowed later by G.H. Darwin to distinguish it from partially analytical methods such as Laplace's) was an elaboration of a device first introduced by Jacques Cassini (Figure 6.3), to express parameters of tidal extrema as arbitrary multi-variable functions of lunar age, parallax, declination, and the time of year.

Cassini had had only a few years of data at his disposal, and so he had restricted the lunar age to the times of syzygy and quadrature and virtually averaged over the time of year. Lubbock, on the other hand, treated lunar age as a continuous variable and hence drew on a much larger portion of the available data, even more than Laplace. He also took the time of year as a separate variable, in order to distinguish purely solar effects from that of lunar declination, which is partially dependent on the season. Basically, it is the latter distinction which demands the use of data covering a full 18.6 year cycle of the moon's nodes. Lubbock had access to such data from London Docks and Liverpool, as mentioned at the end of Chapter 6. As an additional refinement, Lubbock distinguished between upper and lower lunar transits, in order to take account of the diurnal inequality, especially for Liverpool where it is stronger than at London. His analyses were applied to the times and heights of all available extrema, as separate parameters.

Lubbock's approach was to compile all his 13,000 data pairs against each of the chosen astronomical variables in narrow intervals. For example, all HW heights would be compiled against each half-hour of lunar transit time, then against each arc-minute of parallax, then against each degree of lunar declination. Within each of the compiled arrays, the data varies with other factors, but their mean value over 19 years may be taken as representative of the variable in hand in its chosen interval. The functional dependence on all variables could thus be estimated, and combined as a product for prediction, using computed lunar and solar ephemerides.

The above procedure required a great deal of numerical book-keeping and use of tables. Much of the work was done under contract, with costs defrayed by a grant from the British Association for the Advancement of Science (BAAS), which for nearly 90 years from 1832 patronised research towards better tide predictions. The data reduction itself was organised by one Joseph Dessiou (1797–1842), a tidal specialist from the Admiralty Hydrographic Office. It was Dessiou who wrote the report on the tides at Liverpool communicated to the Royal Society by Lubbock in 1835.(7) The Admiralty started to publish tide-tables in 1833.

Lubbock's 'synthetic' method proved to give better predictions than any previous method for London and Liverpool. It quickly superseded the secret methods which had been devised for commercial tables without much scientific scrutiny, and became the standard basis of tide-tables for British ports for the rest of the century. Even when the more flexible *harmonic method* had become standard for the overseas ports of the British Empire, Lubbock's method, once set up, was still preferred by the Admiralty for home use, well into the 20th century.

Analyses by Samuel Haughton

The Reverend Samuel Haughton (1821–1897) of the *Royal Irish Academy* deserves a brief mention here because he contributed several papers demonstrating the analysis of tidal observations from 12 ports round the coast of Ireland and from 9 places in or near the Canadian Arctic. The Arctic observations had been taken by naval ships wintering in the ice during the 'Northwest Passage' expeditions of 1848–1878. Haughton's scientific work ranged widely from theoretical mechanics to geology and medicine. His papers on tides make no references to Lubbock's work or to any other author, except for occasional reference to a few results of Laplace and Airy. He had the generosity to tabulate all the HW/LW data at his disposal, after which his papers present analyses of Laplacian type, leading to estimates of diurnal and semidiurnal Establishment and Age at each site and of the moon's mass and orbital eccentricity. In noting the remarkably large Age of the diurnal tide (4–6 days) all round Ireland, he used a formula of Airy to conjecture that the ocean should be 11–12 miles deep, or at least that a channel of that depth extend from the Antarctic to the British Isles. Another approach gave more realistically 3.5 miles, but Haughton was unable to account for the differences.[8] Despite his evidently great interest in tidal data, Haughton's results had no lasting influence and his analytical methods were soon outdated.

The automatic tide recorder

The invention of the recording tide-gage was a conceptual advance as important as any new theoretical idea. First stimulated by the writings of Thomas Young,[1] it was strongly urged by John Lubbock in search of more 19-year records from major ports, and by his colleague William Whewell who required simultaneous records of less duration from all oceanic coasts. (Whewell's work will be discussed in Chapter 9.)

Records from automatic gages were active at all hours of day and night, and they provided freedom from the subjective elements of visual observations. Hence, they changed the very concept of tidal variation from a daily sequence of extrema to a continuous process in time. In due course, this new look led to the application of Fourier's Theorem, which is the basis of the *harmonic method* of tidal analysis.

Invention of the first tide-gage is usually attributed to an engineer named Henry Palmer, who in 1831 described a device to the Royal Society of London.[9] Airy[3,Plate 2] shows a short graphical record of the tide at London Docks for 4–5 October 1828, thought to have been produced by Palmer's gage, but there is no record of the device having been permanently installed or put to regular use. In any case it was rapidly succeeded by other engineering designs based on similar principles.

Figure 8.1 shows the first known continuous sea level record to cover a full spring-neap cycle of 15 days, 6–21 September 1831, at the Navy Dock at Sheerness (Thames Estuary). The recording instrument was described as a 'rebuilt box-gauge' by one J.Mitchell, a civil engineer employed to maintain the Dock.[10] A box-gage was a float-cum-stillingwell device with a vertical rod mounted on the float for visual reading against a fixed vertical scale, derivative from the type of instrument advocated long ago by Sir Robert Moray (Chapter 6). Mitchell had mechanically coupled the motion of the vertical rod to that of a pen, recording at a scale of 1/30 on a cylindrical drum, rotated by clockwork at one revolution per day. When removed from the drum after 15 days, the continuous trace of the pen appeared as in Figure 8.1, a sequence of twice-daily oscillations retarded each day by about 45–50 minutes, with progressive changes of amplitude. Four successive charts, covering 6 September to 4 November 1831, are held in the Archives of the Royal Society. An elaborate sketch of the gage and its housing may be found in the first volume of the *Nautical Magazine*.[10]

Figure 8.2 shows some instrumental details of another gage, constructed by Thomas G. Bunt in 1837, as described in a paper by Whewell.[11] Its principle of operation is similar to those of Palmer and Mitchell, and modernised versions

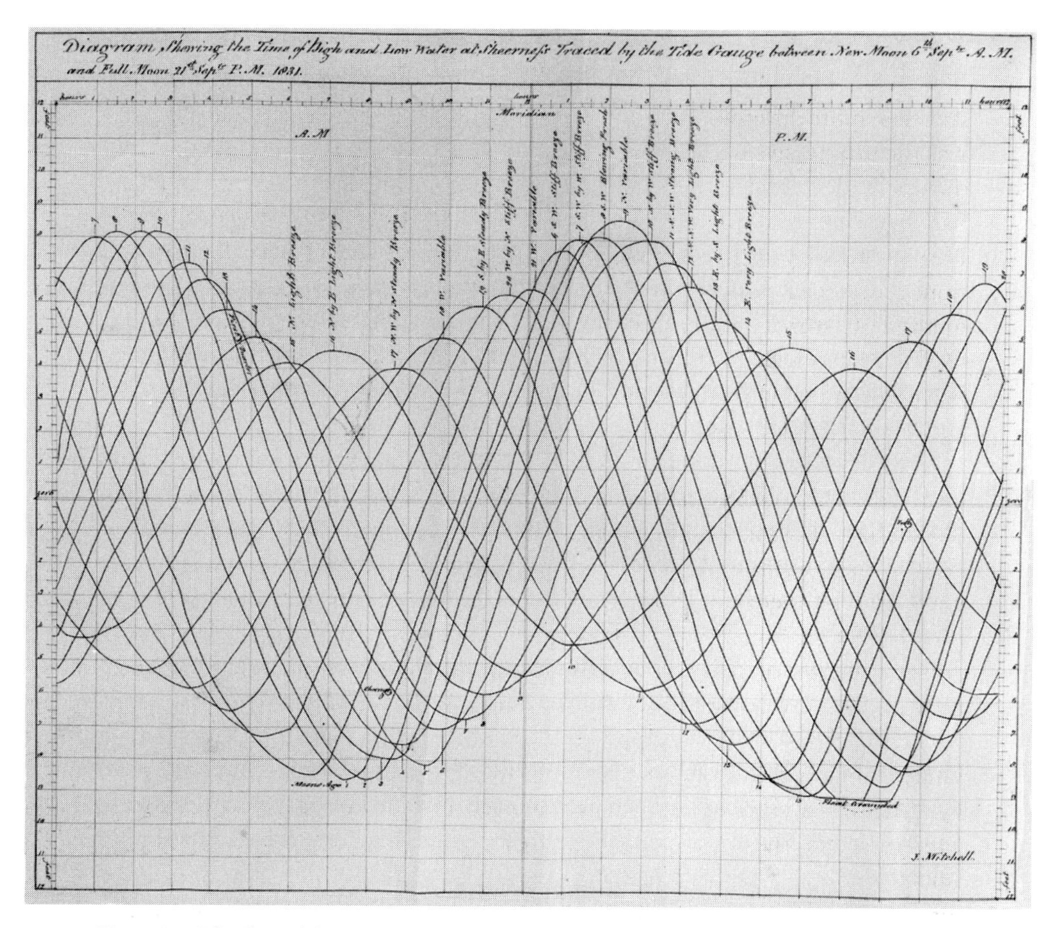

Figure 8.1. The first of four consecutive 15-day tide traces from Sheerness, recorded by Mitchell's gage (see signature, bottom right), held in the Archives of the Roya l Society at the instigation of J.W. Lubbock. At the time of recording, the chart was wrapped round a cylindrical drum rotated by clockwork every 24 hours. The time scale extends from midnight to midnight with noon ('Meridian') at mid-point. The vertical scale is from 12 feet above to 12 feet below a datum close to mean water level, but there is evidence on 7th and 8th Sept p.m. that the float grounded on the base of the stilling well at −9 feet on extreme Low Water springs. The trace starts about 8.30 a.m. on 6th Sept and ends about 9.40 p.m. on 21 Sept, 1831. Irregularities in the trace are caused by variable weather conditions. This is the first continuous record of a spring–neap cycle of the tide in existence. (Photograph courtesy of the Royal Society.)

are still in use today. Bunt's gage was erected on the River Avon below Bristol, where the tides are among the greatest in the world. From the appearance of the recording chart in Figure 8.2, it evidently recorded the tide only for a few hours before and after HWT. It is said to have operated continuously from 1837 to 1872. The idea quickly spread among maritime nations, and by 1850 automatic tide-gages were in regular use at many ports of Britain, France, India and the USA.[12]

Advances in lunar theory

Another, quite different, contribution to the full harmonic expansion of the tide was the progressive refinement of the mathematical theory of the moon's orbit during the 18th and 19th centuries. The history of lunar theory, as it is called, has points of similarity with the history of tidal theory. Both originated from Newton's theory of gravitation and both were motivated by the needs of navigation; the precise position of the moon relative to the stars provided a viable method of estimating real time at a standard meridian and hence the longitude of the observer. Both theories developed slowly over two centuries, and have been accelerated in the last quarter of the 20th century by the needs of modern geodesy and geophysics.

The pioneers in the development of lunar theory between the time of Newton and about 1850 were, in chronological order of publication: Clairaut, Euler, d'Alembert, Laplace, Damoiseau, Lubbock, Hansen, and de Pontécoulant. Four of these authors (Newton excluded) have appeared in these pages as pioneers of tidal theory. The principal difference between the two developments was that lunar theorists were striving for an accuracy of about 10^{-7} radians in angular position and parallax, whereas it would be useless to predict tides to an accuracy better than $10^{-3} \times$ amplitude or period, respectively.

Clearly, there is no place in this book for an account of the intricacies of lunar theory. A thorough account of 18th and 19th century theories was given by E.W. Brown,[13] a leading lunar authority at the start of the 20th century. More recently, the geophysicist Sir Alan Cook has made a more up-to-date survey.[14] However, it is useful to sketch an outline of the most commonly used formulation, in order to show its application to the harmonic development of the tide.

Most of the complications arise from perturbations of the moon's orbit due to the sun, and from the moon's inclined, rotating plane. Smaller perturbations due to the major planets and the earth's nutation are necessary for astronomical precision, but are usually neglected in tidal theory. The four principal angular arguments of harmonic expansion are:

A, A' = moon's, sun's mean anomaly (distance from perigee),

F = moon's mean angular distance from its orbital node,

D = moon's mean longitude minus sun's mean longitude.

(The word 'mean' refers to the steady secular progression, free of perturbations; 'longitudes' are in the ecliptic plane, measured from the equinoctial point.) A, A', F, and D are all well defined by numerical expressions linear in time, with slight variation of their constants on a time scale of a century. Similar expressions exist for the mean longitudes of perigee and perihelion, enabling easy conversion of A, A' to mean longitudes relative to the equinoctial point.

The moon's *true* longitude λ, normalised parallax π, and sine latitude σ, are expressible in the forms (after de Pontécoulant[13]):

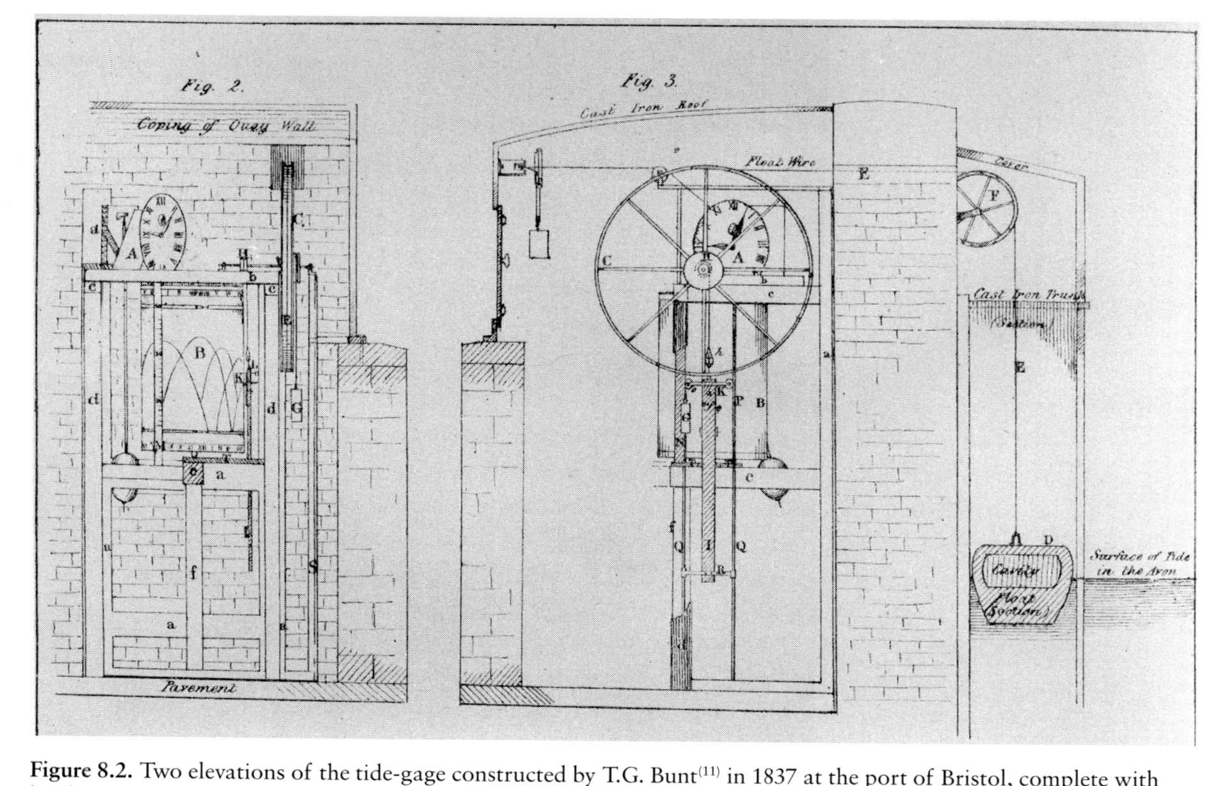

Figure 8.2. Two elevations of the tide-gage constructed by T.G. Bunt[11] in 1837 at the port of Bristol, complete with brickwork and roofing. A taut wire (E) attached to a float (D, extreme right) causes rotation of the large pulley wheel (C, centre), which is converted to a reduced-scale vertical linear motion of a marker (K). A chart mounted on a 24-hour rotating drum (B) records the trace of sea level against time, indicated by a pendulum clock (A). G is a counterweight to prevent backlash in the mechanism. As the trace sketched on the chart suggests, only the upper half of the tidal variation was recorded at this site. The gage is said to have been in continual operation for at least 35 years. (Photograph courtesy of the Royal Society.)

$$\lambda = L + \sum_{n=1} f_n(e) \sin nA + \sum_{n=1} c_n \sin P_n,$$

$$\pi = 1 + \sum_{n=1} g_n(e) \cos nA + \sum_{n=1} d_n \cos P_n,$$

$$\sigma = \gamma \left[\sin F + \sum_{n=1} e_n \sin Q_n \right],$$

where L is the moon's mean longitude, increasing uniformly in time, and P_n and Q_n are all specified in the form

$$iA + jA' + kF + mD,$$

i, j, k, m being small positive and negative integers including zero. In the first two expressions, the first summations on the right hand side are the classically known expansions of undisturbed elliptical motion with coefficients f_n, g_n simple functions of eccentricity e; the second summations are solar perturbations. In the expression for σ,

$$\gamma = \tan(\text{inclination}) = 0.0895$$

and the summation includes both elliptical and solar terms.

The main objective of 19th century lunar theorists was to obtain definitive values for the coefficients c_n, d_n, e_n. De Pontécoulant, for example, expressed these coefficients as power series in m = ratio of mean motions (sun:moon); e,e' = lunar, solar eccentricity, a/a' = ratio of mean orbital distances. His expressions are adequate for ordinary tidal purposes, but their convergence was slow; definitive numerical values for astronomical precision were eventually calculated about the turn of the century by E.W. Brown, from the formulation of G.W. Hill.(13)

The position of the sun, ignoring planetary perturbations and earth nutation, could be easily calculated to better than 10^{-3} accuracy from the formulae for elliptical orbits with A, e replaced by A', e', and with latitude $\sigma = 0$.

The various combinations of A, A', F, D occurring above embody a rich assortment of harmonic frequencies. These are carried over into the right ascension and declination (Appendix A) from standard trigonometric transformations, involving the earth's obliquity, also well defined. Finally, reference to points in fixed earth coordinates introduces the sidereal day as the high frequency harmonic element, in the manner of Laplace's tidal development, (Chapter 7).

The harmonic analysis of tides – William Thomson (Lord Kelvin)

First to realise the practical potential of applying a full harmonic analysis to tide records was Sir William Thomson, whose later study of tidal wave propagation was discussed in Chapter 7. Thomson was of course aware of the excellent synthetic analysis of High and Low Waters by Lubbock, but the stage was now (1867) set for analysis of the whole profile of variation as a continuous function of time. There was every chance that such an analysis would yield good predictions after only a year, or possibly a month of data, without waiting for the long multi-year series required by Lubbock's method. This would greatly accelerate the acquisition of reliable constants from new ports.

To underline the importance of the necessary work, Thomson in 1867 appointed a formal *Committee for the purpose of promoting the extension, improvement and harmonic analysis of tidal observations*, which he chaired on behalf of the BAAS.[15] This committee at first comprised 28 members, including the President and 16 Fellows of the Royal Society, the Astronomer Royal, an Admiral, and various interested naval officers and civil engineers. (Numbers were later reduced.) There is little record of what tangible contributions this distinguished assembly made, other than to give weighty approval. The Annual Reports of the BAAS from 1868 to 1876 contain detailed reports of progress by Thomson himself with the aid of an able computer seconded to the project by the Nautical Almanac Office, named Edward Roberts (1845–1933). Roberts had a good background of astronomical knowledge and mechanical skills, making him an ideal choice for the work in hand, which later involved the construction of mechanical analog computers. A small team of assistants was assigned to him. Expenses were paid by the BAAS.

Thomson's first report to the Committee started:[15]

> The chief, it may almost be said the only, practical conclusion deducible from, or at least hitherto deduced from the dynamical theory is, that the height of the water at any place may be expressed as the sum of a certain number of simple harmonic functions of the time, of which the periods are known, being the periods of certain components of the sun's and moon's motions. Any such harmonic term will be called a tidal constituent, or sometimes, for brevity, a tide.

It is worthwhile to outline Thomson's proposed procedure because, although somewhat old-fashioned in the light of modern practice, it embodies all the essential features still in use today. In particular, the Kelvin/Darwin symbols for the principal harmonic constituents, J, K, L, M, N, O, P, Q, R, S, T are now deeply embedded in the tidal literature (regrettably, some would say).

Thomson borrowed a concept, suggested by Laplace, of *astres fictifs* (fictional celestial bodies) moving in the equatorial plane, each with a constant angular speed, heuristic equivalents of terms in Laplace's expansion of the tide potential. Referring to Laplace's concept as 'a beautiful synthesis of the complex dynamical action to which the tides are due', he translated the term literally as *fictional stars*, but we may as well here call them *satellites*. Each satellite induces one or more harmonic components relative to earth coordinates. One satellite, fixed with respect to the equinoxes, produces the so-called *sidereal tides*. Its Greenwich Hour Angle is $-\Omega t$ in the notation of Chapter 7, and it generates constituents of speed Ω and 2Ω. (They are not strictly 'sidereal', since the equinoxes precess, but the point is understood.) Thomson gave the symbol K to this satellite, and the symbols M, S to satellites at the mean moon and mean sun, speeds $(\Omega t - s)$, $(\Omega t - s')$, respectively, where s and s' are the mean orbital speeds of sun and moon, (Thomson's σ, σ').

With the additional symbols p, p' for the mean longitudes of lunar perigee and earth perihelion, respectively (Thomson's ϖ, ϖ'), the Table below lists all satellite symbols with their angular speeds specified or implied by Thomson.[15] The right hand column refers to purely solar terms; bracketed symbols were added in 1870:

K	Ωt		
L	$\Omega t - (s+p)/2$	(R)	$\Omega t - (s'+p')/2$
M	$\Omega t - s$	S	$\Omega t - s'$
N	$\Omega t - (3s-p)/2$	(T)	$\Omega t - (3s'-p')/2$
O	$\Omega t - 2s$	(P)	$\Omega t - 2s'$
(Q)	$\Omega t - 3s + p$		

The speeds shown above are of course diurnal; in addition all satellites except O,P,Q produce semidiurnal constituents of twice the angular speeds shown. Satellites O,P, and Q (assumed to have some positive declination) produce only diurnal constituents at the given speeds. K produces both diurnal and semi-diurnal constituents. Satellite M produces small tides of diurnal and terdiurnal speeds if the 3rd degree terms of the potential are taken into account. Otherwise, parallax introduces two other constituents close to the speed of M, at $(\Omega t - s \pm p)$, with rather small amplitude.

The frequencies involving p or p' arise from the elliptic inequalities in the orbits of moon and earth. These had not been recognised as distinct harmonic terms by Laplace, who assumed them to conform to a static response to changes in parallax ϖ. Thomson more correctly treated the elliptic terms L and N (and Q) as independent harmonics. However, he admitted that modulations due to the slow regression of the moon's nodal axis could justifiably be treated by equilibrium theory (that is, assumed to occur in the ocean just as they are computed in the tide potential).

Generally, if ω_k denotes the angular speed (i.e. frequency) of the kth listed satellite, the given record of sea level was expressed as:

$$z(t) = A_0 + \sum_k \left[\sum_{m=1}^{8} (A_m^{(k)} \cos \omega_k t + B_m^{(k)} \sin \omega_k t) \right].$$

Here, A_0 is an arbitrary datum level and, if sufficiently large in amplitude, the terms $A_1^{(k)}$, $B_1^{(k)}$ define a diurnal constituent, $A_2^{(k)}$, $B_2^{(k)}$ define a semidiurnal constituent, and terms of higher order, chiefly 3,4,6,8, are *compound tides* or *overtides* due to the nonlinear effects of shallow water. Overtides for the satellites M and S appeared prominently in Thomson's first analysis of sea level at Ramsgate recorded in the year 1864. He also recognised a cross-term MS at the sum of the speeds of M and S, which he called the *Helmholtz luni-solar quarter-diurnal* tide,[15] after an analogous effect in sound waves identified by H. von Helmholtz (1821–1894). Many similar cross-terms were later found necessary for tides in coastal seas and estuaries; the reference to Helmholtz was dropped by later authors.

The method of data analysis proposed by Thomson[15] and tested by Roberts for several years was, in brief, to isolate the influence of each satellite by an elementary form of digital filter. This filter, of little interest today, was of a type known in *periodogram analysis*, whereby the z-value at each of the 24 hours of the satellite-day (solar hours for S, lunar hours for M, and so on) was averaged over a great many days. The resulting 24 average values were nearly independent of the effects of other satellites, and could then be expressed in a series of eight pairs $A_m^{(k)}$, $B_m^{(k)}$, by classical Fourier analysis. Small corrections were finally applied to allow for imperfect filtering. Phase-lags $\arctan(B_m/A_m)$ were easily adjusted to refer to the times of Greenwich transit instead of local time, if preferred.

The initial analysis for Ramsgate (eventually extended over ten years) and Roberts's analyses of Liverpool and Bombay[15] were confined to the satellites K, L, M, N, O and S. Later analyses by Roberts[16] covered the full set K–T including P, Q. Additionally, a satellite for the small diurnal elliptic term with argument $(\Omega t + s - p)$ was labelled J, exhausting any further logical extension of alphabetic characters. Greek letters were added to accommodate the leading solar perturbations of the lunar orbit. The *variation* produced a small constituent μ at argument $(\Omega t - 2s + s')$, and the *evection* produced ν at $\Omega t + s' - (3s + p)/2$ and λ at $\Omega t - s' - (s - p)/2$. (See Appendix A.) The last three satellites all produce semidiurnal harmonics at twice the named frequencies. Perturbation harmonics of diurnal frequency were later identified, and were similarly given Greek letters. Comparisons of the resulting harmonic syntheses with the data and from one year to another were encouragingly close.

Harmonic analysis under G.H. Darwin

Having fairly launched the principles and practice of harmonic tidal analysis, leaving further reduction of data in the able hands of Edward Roberts, Sir William Thomson turned his attention to the construction of a mechanical tide predictor (the subject of a later section) and then moved on to other subjects of research. Towards the end of the 1870's, Thomson's place as head of the BAAS Committee for tide prediction was taken by Professor G.H. Darwin of Cambridge. By then, the Committee was greatly reduced in size, and its only remaining founder-member was Professor J.C. Adams, the Cambridge mathematical astronomer, famous as a lunar theorist (see Chapter 10).

George Howard Darwin (1845–1912), the second son of the famous biologist, was drawn into tidal theory by the inferences which might be drawn from them concerning the physics of the earth and the acceleration of the moon's orbit. In particular, both he and Thomson had hoped for a global picture of the long-period tides in order to determine the earth's elasticity (Chapter 10). Roberts's reports to the BAAS had included estimates of the leading long-period constituents, but except for the annual term Sa, which evidently had a thermal origin, the estimates of the lunar monthly and fortnightly tides varied randomly from year to year on account of low amplitude and high noise level. Darwin thought that a systematic revision of the analysis procedure might produce a clearer picture. Accordingly, he recast Thomson's framework with special attention to the effects of the moon's orbital inclination.

Darwin's 1883 report to the BAAS[17] gave such a systematic account of the harmonic constituents and their modulations that it was accepted as the definitive model for several decades to come. It remains in many respects the basis for modern tide-table construction. He discarded the concept of *astres fictifs*, derived from Laplace, as having outlived its heuristic purpose, but he retained Thomson's symbols for the constituents themselves, adding numbers in suffix to denote their species where ambiguous (Figure 8.3). He did this partly in deference to Thomson's initiative, and partly because the notation had already become standard in the printed computation forms organised by Roberts for the *Survey of India,* the first major beneficiary of the work.

With excrescences such as OO, 2N, μ or 2MS, and some acknowledged constituents for which no symbol was assigned, Darwin's partly inherited notation is far from ideal. Eventually, seven more Greek letters were added. In the Preface to his first report[17] he admitted strong objections raised by Adams, who would have preferred complete replacement of the old notation. There is no record of what alternative system Adams had in mind.

Darwin's most important contributions to the subject were his formulae for the dependence of each harmonic constituent on the inclination I of the moon's orbit to the equator. I varies between about 28°.6 and 18°.3 in the nodal cycle of period 18.6 years, and it can be expressed in terms of the longitude of the node. For example, Darwin's expression for the coefficient of the fortnightly tide Mf is

$$0.5\,(1-2.5e^2)\sin^2 I$$

where $e=0.0549$ is the eccentricity. By taking into account the value of such an expression at the central time of the data being analysed, the derived amplitude and phase may be related to the mean value of the coefficient, to be taken as standard. The same expression is then applied to predict for another epoch.

The constant factors of Darwin's coefficients were derived by him from greatly simplified formulae for the moon's orbit. Advanced lunar theory gives slightly different numerical values, as Adams pointed out in a few cases. A more exact harmonic expansion of the potential was made in 1921 by A.T. Doodson (Chapter 11), but Darwin's expressions were more than adequate for the numerical technique of his day.

Like later practitioners of the harmonic method, Darwin was not pleased

[A.] *Schedule of Notation.*

Initials	Speed	Name of Tide
M_1 M_2 M_3 &c.	$\gamma-\sigma-\varpi$, and $\gamma-\sigma+\varpi$ $2(\gamma-\sigma)$ $3(\gamma-\sigma)$ &c.	Principal lunar series
K_2	2γ	Luni-solar semi-diurnal
N	$2\gamma-3\sigma+\varpi$	Larger lunar elliptic
L	$2\gamma-\sigma-\varpi$ and $2\gamma-\sigma+\varpi$	Smaller lunar elliptic
	$2\gamma+\sigma-\varpi$	
2N	$2\gamma-4\sigma+2\varpi$	Lunar elliptic, second order
ν	$2\gamma-3\sigma-\varpi+2\eta$	Larger lunar evectional
λ	$2\gamma-\sigma+\varpi-2\eta$	Smaller lunar evectional
O	$\gamma-2\sigma$	Lunar diurnal
OO	$\gamma+2\sigma$	
K_1	γ	Luni-solar diurnal
Q	$\gamma-3\sigma+\varpi$	Larger lunar elliptic diurnal
	$\gamma-\sigma-\varpi$ included in M_1	Smaller lunar elliptic diurnal
J	$\gamma+\sigma-\varpi$	
	$\gamma-4\sigma+2\varpi$	Lunar elliptic diurnal, second order
	$\gamma-3\sigma-\varpi+2\eta$	Larger lunar evectional diurnal
S_1 S_2 S_3 &c.	$\gamma-\eta$ $2(\gamma-\eta)$ $3(\gamma-\eta)$ &c.	Principal solar series
T	$2\gamma-3\eta$	Larger solar elliptic
R	$2\gamma-\eta$	Smaller solar elliptic
P	$\gamma-2\eta$	Solar diurnal
Mm	$\sigma-\varpi$	Lunar monthly
Mf	2σ	Lunar fortnightly
Sa	η	Solar annual
Ssa	2η	Solar semi-annual
MSf	$2(\sigma-\eta)$	Luni-solar synodic fortnightly
MS	$4\gamma-2\sigma-2\eta$	Compound tides
μ or 2MS	$2\gamma-4\sigma+2\eta$	Compound tides
2SM	$2\gamma+2\sigma-4\eta$	Compound tides
MK	$3\gamma-2\sigma$	Compound tides
2MK	$3\gamma-4\sigma$	Compound tides
MN	$4\gamma-5\sigma+\varpi$	Compound tides

Figure 8.3. G.H. Darwin's basic set of harmonic tide constituents with initials based on Thomson's original notation and a few additions. The 'speeds' (frequencies) are defined in terms of $\gamma = 1$/sidereal day, $\sigma = 1$/sidereal month, $\eta = 1$/year, $\varpi = 1$/(perigee period, 8.85 years), – Ω, s, s', p in text notation. Initials ν, λ, μ are linear harmonics of the potential arising from the solar perturbations known as 'evection' and 'variation'. The last seven constituents, with two upper-case letters in their initials, have compound frequencies arising from nonlinear hydrodynamic processes.

with the numerical treatment necessary to deal with the awkward constituents known as M_1, nominally at the mean of the speeds of O and K_1, and L_2, at the mean of the speeds of M and K_2. Both constituents include important harmonic lines separated from the nominal speed by the speed of perigee, that is by ± 1 cycle in 8.85 years. These lines are too close to be resolvable in the standard period of analysis (1 year), but too far apart for their modulatory effect to be assumed constant. Unless one has 9 years of data at one's disposal a compromise formula has to be accepted; fortunately the amplitudes of both terms are fairly small. To complicate the situation, there exist theoretically small lines at exactly the nominal speeds of M_1 and L_2, arising from the 3rd degree expansion of the potential. Darwin added: 'If further examination ... should show that the tide M_1 is in reality regular, it should be introduced.' This remark was prophetic. Nearly a century later, electronic computers revealed that the spectral line at exactly one cycle per lunar day (i.e. at the nominal speed), is in fact the dominant contributor to M_1 in the seas round northwest Europe.

Harmonic analysis under W.E. Ferrel

Benjamin Peirce (1809–1880), who succeeded A.D. Bache as superintendent of the US Coast Survey in 1867, was a distinguished Harvard Professor of mathematics and astronomy. He had previously served as consultant to the US Navy for the *American Ephemeris and Nautical Almanac*, and from 1852 as director of longitude observation for the Coast Survey. Thomson and Roberts mention Peirce as having supplied trustworthy tide records from the Pacific coast.[16]

It was under Peirce's direction from 1867 to 1874 that the Coast Survey published its first national tide-tables, and a research group was set up to provide an informed scientific approach to problems concerning the tides along the coastline of North America. Foremost among early members of this tidal research group was the mathematician William Ferrel (1817–1891), who had served for 10 years at the Office of the *American Nautical Almanac* – see the brief historical survey by Hicks.[6]

Ferrel's researches on tides were contemporary with, but independent of Thomson's and Darwin's work, being largely extensions of the works of Laplace, Young and Airy. His main work is compiled in a voluminous Report of 1874.[18] The Report is difficult reading, on account of Ferrel's propensity to introduce many new symbols without recapitulating their meanings at any stage, probably to economise on space. After a historical summary of theoretical work on tides since Newton, Ferrel developed his own harmonic expansion of the tide potential, using a combination of Laplace's species separation with current lunar theory, for which he had expert advice from Peirce. His expansion included 8 long period terms, 10 diurnals and 16 semidiurnals, without involving compound terms. Separate terms with arguments involving the longitude of the node were included; these are only implicit in Darwin's expansion. Not being hampered with Thomson's alphabetic symbols, Ferrel used a simple numerical suffix to distinguish each harmonic term.

Adding frictional terms (linear and cubic) to Laplace's tidal equations, Ferrel solved the equations in simple canal-like configurations to produce idealised forms for the variation of amplitude ratio and phase lag (i.e. *admittance*) with frequency, dependent on frictional parameters. The results have little meaning in modern terms, since we know that a variety of other causes affect admittance, but they made a novel approach to the interpretation of harmonic constants. Near the sidereal harmonics K_1, K_2, which involve both the moon and the sun, interpretation also depends on the assumed mass ratio; Ferrel included this ratio as one of the unknown parameters – an extension to Laplace's empirical method for obtaining the mass ratio.

Ferrel then devised a method for extracting leading harmonic constants from a long series of observed High and Low Waters, and applied it to 19 full years of the data from Brest (1812–1830) four years longer than Laplace had used. His best result for the mass ratio, taking full account of his frictional theory, was 1/78, which is closer to the modern value, 1/81.3 than Laplace's result, 1/75. Omitting friction gave a less good estimate. He obtained an even better estimate, 1/81.7, from a 19-year series of data from Boston Harbor. We now know that any estimate of the moon:earth ratio derived from tidal observations is limited by the distortion of the solar tide in the ocean by the thermal tide in the atmosphere. In any case, the method of using tides to estimate the mass ratio was soon to be superseded by astronomical methods involving perturbations to the earth's position at a close approach of the minor planet Eros.

The mechanical tide predictor

When it came to preparing tide-tables by the harmonic method, the summation of long series of terms of the form $h \cos (jt + k)$ using a table of cosines and hand computation proved to be intolerably tedious and prone to human error. The additional task of computing the extrema from an evenly spaced time series was also considerable. Sir William Thomson soon hit upon the idea of an analog machine to perform the computations automatically, related to a device used in telegraphy by Sir Charles Wheatstone (1802–1875), the experimental physicist. The machine's principle was quite simple, but skilled precision engineering was needed to construct one of reliable accuracy.

A vertical simple harmonic motion of appropriate speed, amplitude and phase was imparted to each axle of a number of coplanar pulley wheels, whose diameters and spacing were such that a flexible wire or metal tape could be threaded vertically from one pulley to the next, (Figure 8.4). One end of the wire being held fixed, and the other end counterweighted to keep it taut, the vertical motion of the free end was (twice) the sum of the motions of the pulleys. The summed motion was usually transferred to a pen which plotted a timed trace at reduced scale on a moving chart.

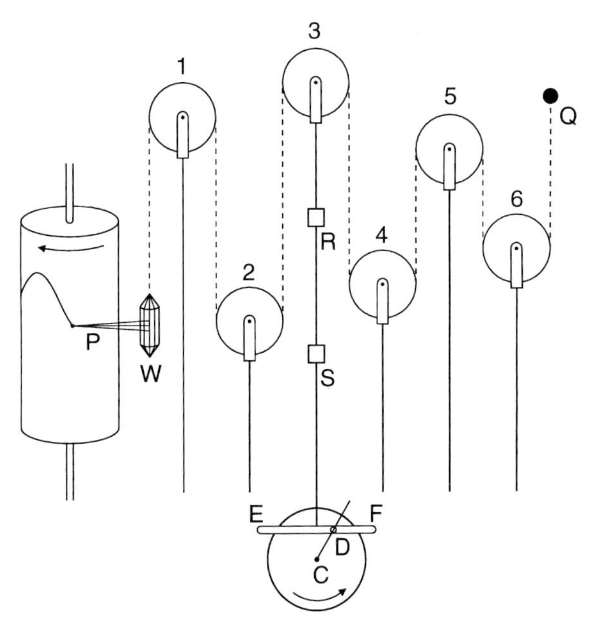

Figure 8.4. Sketch to show principle of the Tide Predicting Machine, originally invented by William Thomson (Kelvin). The centres of pulley-wheels 1–6 oscillate vertically with preset speeds by the mechanism shown in detail below pulley no. 3. A bearing D attached to a steadily rotating wheel centre C is free to slide in a horizontal slot EF at the base of a rod attached to pulley 3, maintained in the vertical by guides R and S. The amplitude of the movement is set by adjusting the length of the crank arm CD; its phase relative to the wheel C is also adjustable. A wire with end fixed at Q, threaded over pulleys 1–6 as indicated by broken lines, is kept taut by a weight W, whose vertical movement is thus twice the sum of the pulley movements. A marker P attached to (or near) W makes a continuous trace on a chart on a rotating cylinder on the extreme left. In practice, many more than six oscillating pulleys were used – see for example Figure 8.5.

The first operational Tide Predicting Machine (TPM) was constructed in 1872–73 to Thomson's design by the *A. Légé Engineering Company* of London. Expenses were covered by the BAAS, so this pilot instrument became known as the *British Association Machine.*[(19)] It summed ten harmonic components, including two compound tides, M_4 and MS_4, which were necessary to delineate the asymmetry of tides in shallow locations. The basic design was Thomson's, but the gear ratios which determine the speeds of the constituents were worked out by Roberts. Unlike Figure 8.4, the motions of the pulley axles were at first circular, thus having horizontal as well as vertical components. This introduced slight errors; all subsequent TPM's had strictly vertical movements.

At this time, and until about 1920, tide-tables for all British ports were computed by the synthetic method of J.W. Lubbock, which is well enough suited to tidal regimes with only weak diurnal inequality. The harmonic method made a better account of the diurnal tides, so the first countries to benefit from harmonic prediction were the more distant parts of the British Empire such as India and Australia, whose tides have strong diurnal inequality. Within six years of the completion of the British Association Machine, the *Survey of India* paid for a second, more elaborate, TPM to be constructed by Légé & Co for dedicated application to Indian ports.

The *India Office* TPM was designed by Edward Roberts himself, on contract to the India Office; it was originally known as the *Roberts Tide Predicting Machine*,[19] but that name was later applied to his *Universal Tide Predictor* of 1906. For the India Office machine, Roberts corrected some design defects of the BA Machine and extended the number of harmonic constituents to 24. Figure 8.5 shows line drawings of the front and side elevations of this elaborate product of Victorian technology. Roberts used his machine to compute tide-tables for Indian ports and later for other countries of the British Empire from 1879 to 1903. He then, in 1906, designed a new TPM for 40 constituents, which he called a *Universal Tide Predictor*, for his own commercial use. The latter TPM won a 'Grand Prix' at the Franco-British Exhibition of 1908. For its subsequent history, see Chapter 11.

The US Coast Survey (from 1878 called the *Coast and Geodetic Survey*) followed with interest the development of the TPM in Britain. In 1882, the Washington engineers Fauth & Co. completed for the Survey a novel machine to a design by William Ferrel. The Ferrel TPM was the first to compute the *derivative* of the predicted tide (dz/dt) simultaneously with $z(t)$ itself. It did this by duplicating the pulley movement $h\cos(jt+k)$ at each speed j with an amplitude jh and phase $(k+90°)$. The zeros of dz/dt gave the required times of extrema, which were read directly from a dial. Nineteen constituents were included. The Ferrel TPM was intensively used for all the Survey's tide-tables until 1912, when it gave place to a grander machine with 37 components.

The second USCGS machine had been designed in principle about 1894 by R.A. Harris, of whom we shall hear more in Chapter 9. After long delays, it was constructed in the Survey's own workshops under the direction of their chief mechanical engineer, E.G. Fischer, and finally commissioned in 1912. Apparently, there was a long and bitter in-house dispute between Harris and Fischer as to who should take the greater share of the credit.[20] (In the printed discussion following Thomson's paper of 1881 to the ICE[21] there is evidence that there had been a mild dispute of similar nature between Thomson and Roberts.) The machine in question, officially known as *US Coast & Geodetic Survey Tide Predicting Machine no.2*, is fully described in ref. (19). It was very accurate and versatile, and it remained in constant use until finally superseded in 1966 by the electronic computer.[6]

Four more constructions bring the inventory of these specialised machines up to ten by about the end of World War I.[19] A 17-component TPM had been constructed in the 1880's by Kelvin & White, Sir William Thomson's own company for the manufacture of marine scientific equipment. Eventually, in 1901 it passed into the hands of the French Service Hydrographique for general application to French colonial ports. Another Kelvin TPM was sold to the

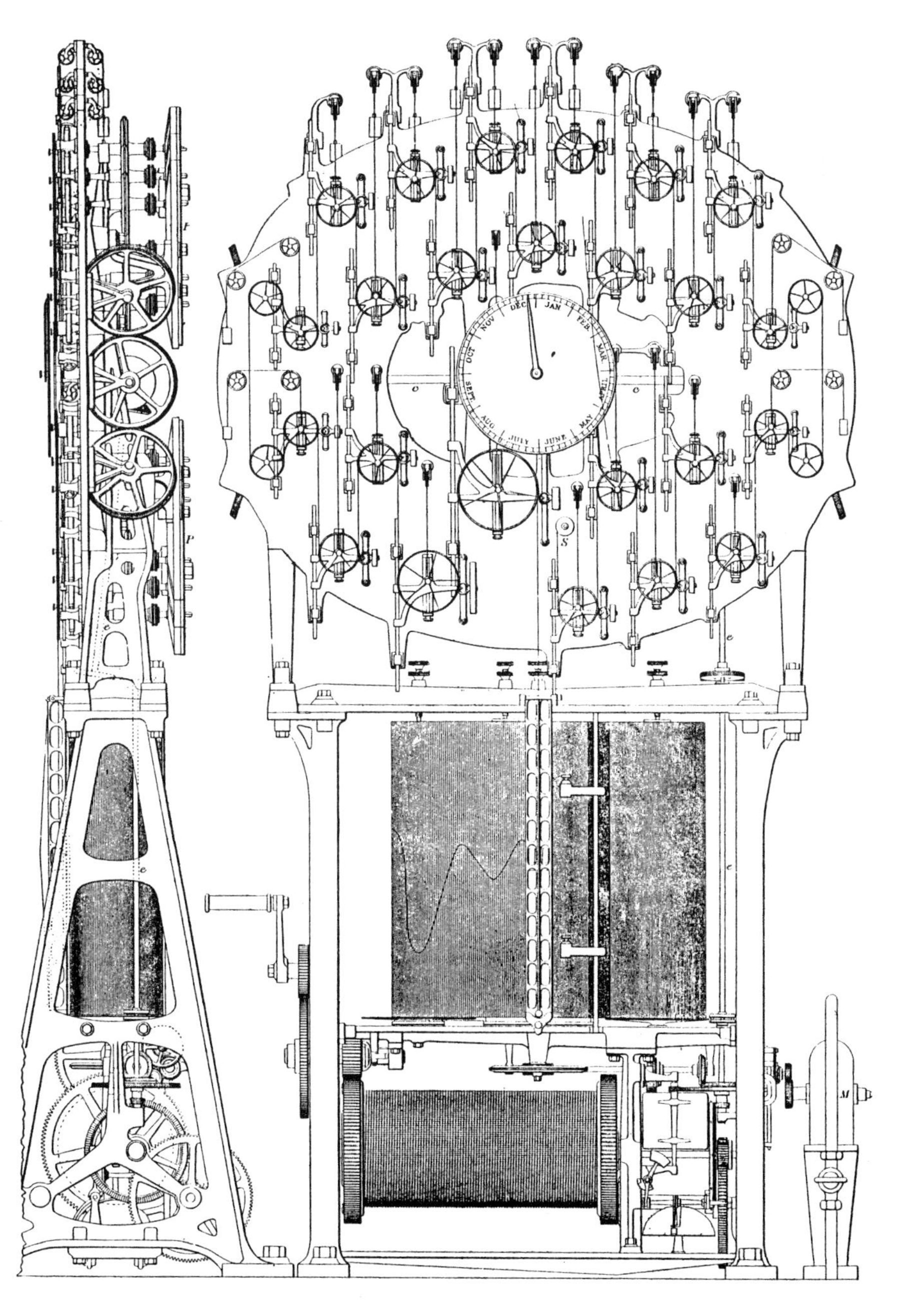

Figure 8.5. Two elevations of the 'India Office Machine', constructed by A. Légé & Co., London, in 1877–79 to a design by Edward Roberts. As shown, this machine summed 20 harmonic constituents; the number was later increased to 24. The right hand elevation shows only the 'front' of the machine, containing 20 vertically sliding pulleys, wire and recording drum – cf. Figure 8.4. Each moving pulley has its own tension wire and weight to reduce backlash. The central dial shows the date of the prediction. The cranks which move the pulleys up and down with their complex gearing are on the 'back' of the machine, viewed side-on to the left. The whole gear train, including the recording drum, was set in motion by the crank handle shown below the front elevation.

Japanese Government in 1914. A 16-component TPM based on the Kelvin design was constructed in Argentina in 1918.

A 20-component TPM, on similar principles to USCGS no.2, was built by the German instrument firm of Otto Töpfer & Sohn in 1915–16 for the naval *Kaiserliche Observatorium* at Wilhelmshaven, and was later transferred to the Deutsche Seewarte at Hamburg. An original feature of the German TPM was the use of four axles made to turn at the fundamental mean speeds Ω, s, s', p of earth, moon, sun and lunar perigee, respectively. All constituent speeds were generated by simple geared combinations of these four, as in the combinations of γ, σ, η, ϖ listed in Figure 8.3, thus avoiding the multitude of awkward and approximate gear ratios used in other machines. The same principle of operation was later used in the gigantic German TPM of 1938 – Figure 11.2.

The history of the second half of the era of the TPM, lasting roughly from 1920 to 1965, will be described in Chapter 11 in the context of the Liverpool Tidal Institute.

Notes and references

1. Young, T. *Tides* – article in *Encyclopedia Britannica,* 6th edition (supplement), **6**, 658–675, 1824. (reprinted in 7th – 1842 and 8th – 1860 editions; also in *Works* (Ed. G. Peacock), **2**, 291–335, 1855. (The references to Laplace and Bouvard cited in the text are from another article by Young in *Brande's Q. J. Sci.,* **17**, 295–315.) *Works,* **2**, 336–358, 1824.)
2. Information on the history of *TIDE* articles in *Encycl. Britannica* was provided to the author by Professor G.W. Platzman from the complete set of editions at EB headquarters in Chicago.
3. Airy, G.B. Tides and Waves, *Encyclopedia Metropolitana,* **5**, 241–396, London, 1845.
4. Young, T. A theory of the tides, including the consideration of resistance. *Nicholson's Journal,* **35**, 145–159 and 217–227, 1813. Also in *Works* (Ed. G. Peacock), **2**, 262–290, 1855.
5. *The American Almanac and Repository of Useful Knowledge,* vol.1, Gray & Bowen, Boston, 1829. (As well as the ephemerides for 1830 mentioned in the text, this initial volume includes a history of English Almanacs, an account of a series entitled 'Poor Richard's Almanac', previously published by Benjamin Franklin in Philadelphia, and geographic/economic statistics pertaining to USA and other countries.)
6. Hicks, Steacy D. The Tide Prediction Centenary of the U.S. Coast & Geodetic Survey. *Int. Hydrog. Rev.,* **44** (2), 121–131, 1967.
7. Lubbock, J.W. Discussion of tide observations made at Liverpool. *Phil. Trans. R. Soc. London,* **125**, 275–299, 1835, and **126**, 57–73, 1836. Also: On the tides of the port of London, *ibid.,* **126**, 217–266, 1836.
8. S. Haughton's papers analysing the tidal observations made under the direction of the Royal Irish Academy in 1850–51 appear in: *Trans. R. Irish Acad.,* **23**, 35–139, 1854; in *ibid.* **24**, 195–211, 1864, and **24**, 253–350, 1866. Haughton's papers on Arctic tidal data appear in: *Phil. Trans. R. Soc. London,* **153**, 1863; **156**, 1866; **165**, 1875; **169**, 1878; and in: *Trans. R. Irish Acad.,* **30**, 1893–95.
9. Palmer, H. Description of a graphical register of tides and winds. *Phil. Trans. R. Soc. London,* **121**, 209–213, 1831.
10. Anonymous. Voyages and Maritime Papers, 1 – The Tide Gauge at Sheerness. *Nautical Mag.,* **1**, 401–404, 1832. The article is centered on a detailed sketch of the apparatus and its housing executed by 'Mr.Mitchell, civil engineer', with a footnote reference to the influence of a paper by 'Mr.Lubbock'. The original conception and choice of site at Sheerness were due to a land surveyor named J.A. Lloyd, FRS, who wished to determine

the mean level of the sea in relation to up-river locations. Lloyd's paper of 1831 describes the whole surveying exercise (Chapter 9[12]) and gives a short description and location map of the installation.

11. Whewell, W. Description of a new tide-gauge, constructed by Mr. T.G. Bunt, and erected on the eastern bank of the River Avon in front of Hotwell House, Bristol. *Phil. Trans. R. Soc. London*, **128**, 249–251, 1838.
12. Matthäus, W. On the history of recording tide gauges. *Proc. R. Soc. Edinburgh*, Section B, **73**, 3, 25–34, 1972.
13. Brown, E.W. *An Introductory Treatise on the Lunar Theory*. Cambridge Univ. Press, xvi + 292pp., 1896. (Reprinted Dover Pubs., New York, 1960.)
14. Cook, A. *The Motion of the Moon*. Adam Hilger, Bristol, x + 222pp., 1988.
15. Thomson, W. Report of the Committee for the purpose of promoting the extension, improvement and harmonic analysis of Tidal Observations. In: Report of the 38th Meeting of the BAAS, 1868, pp. 489–510, John Murray, London, 1869.
16. *Ibid.* Reports of the Meetings of BAAS for 1870 (pp. 120–151), 1871 (pp. 201–207), 1876 (pp. 275–303). (Though nominally authorised by Thomson, much of this reported material was specifically drawn up by Edward Roberts, who carried out most of the analysis described.)
17. Darwin, G.H. The harmonic analysis of tidal observations. (Report of a Committee consisting of Professors G.H. Darwin and J.C. Adams for the harmonic analysis of tidal observations. Brit. Assoc. Report for 1883, pp. 49–118) Reprinted in *Scientific Papers*, **1**, 1–70, Cambridge Univ. Press, 463pp., 1907.
18. Ferrel, W.E. *Tidal Researches*. Appendix to Coast Survey Report for 1873, Washington, D.C., 268pp., 1874.
19. Anonymous. *Tide Predicting Machines*. Internat. Hydrogr. Bureau, *Special Pub.* no. 13, 110pp. + 34 plates, 1926.
20. Copies of numerous letters by R.A. Harris and E.G. Fischer, many of them irate, some protesting to the Superintendent, USCGS, others printed in *Science* and *Scientific American* between 1911 and 1914, were passed to the author by Steacy Hicks (US National Ocean Survey) in 1982. Harris wanted the machine to bear his name alone, as with the 'Ferrel TPM'. Although sometimes referred to as the 'Harris-Fischer TPM', USCGS with Fischer's acquiescence put an end to the dispute by removing both personal names.
21. Thomson, Sir William. The tide-gauge, tidal harmonic analyser and tide predicter. (With lengthy discussion on pp. 28–74.) *Proc. Inst. Civil Engineers*, London, **65**, 4–74, 1881.

9

Towards a map of cotidal lines

This chapter describes another line of enquiry which began to take shape in the first half of the 19th century. Its principal early protagonist was William Whewell, who was mentioned in association with J.W. Lubbock in Chapter 8. The basic question was: can we begin to map the progression of High Water times across seas and oceans, using only empirical data from coastal tide-gages? If so, one could begin to understand the wavelike properties of tides, their directions of propagation and standing oscillations. Such a question had actually been posed two centuries previously by Sir Francis Bacon (Chapter 4), but it was barely mentioned in writings of the 17th and 18th centuries, being overshadowed by the generalised analytical approach of Newton and his successors.

We have seen that right up to the end of the 19th century, theoretical analysis was restricted to an ocean covering the globe or to hypothetical narrow canals and flat seas. The growing body of coastal data, even by 1800, showed that the natural tides are strongly affected by the presence of the continental masses and have complex structures which could hardly be understood from such simplified models. A more heuristic approach was seen to be desirable.

Interaction between Whewell and Lubbock

The Reverend William Whewell (1794–1866) is best described as a scientific philosopher, having successively occupied Chairs of mineralogy and of moral philosophy at Cambridge. He was later Master of Trinity College. His most respected book is called *A History of the Inductive Sciences*,[1] and he was instrumental in introducing both the Natural Sciences Tripos and the Moral Sciences Tripos into the university curriculum. Whewell's interest in the tides was stimulated by Lubbock's papers around 1831–32,[2] whose views are often reflected in Whewell's own writings on the subject. At the 1832 meeting of the newly formed *British Association for the Advancement of Science* (BAAS),

Whewell reported on mineralogy and Lubbock reported on tides. As a strong supporter of the BAAS, Whewell helped Lubbock to obtain substantial grants from that body towards Lubbock's reduction of data from London and Liverpool, the first of many grants awarded by the BAAS for tidal research.

Lubbock is sometimes credited with the invention of the term *cotidal lines* (as would appear from its first appearance in his paper of 1831). However, in his 1832 report to the BAAS,[(3)] he writes: '... at a series of points which form the crest of the tide-wave, and which I have called, *at the suggestion of Mr. Whewell*, cotidal lines.' (my italics). So the two men must have conferred privately on the subject about that time. The title of Lubbock's 1831 paper to the Royal Society: *On the tides of the port of London*,[(2)] seems inappropriate for a discussion of the tides in the seas surrounding Britain including the Atlantic Ocean, but the paper indeed contains maps showing the known values of HWT at Full Moon and Change *(establishments)* on all the coastlines, together with tentative straight lines joining points of equal times. On Lubbock's own admission, these maps are little more than slight elaborations of sketches used by Thomas Young in lecture-notes for the Royal Institution in 1807.[(4)] They also have some affinity with Edmund Halley's much earlier tidal chart of the English Channel.[(5)]

Lubbock, a fluent reader of French, was responsible for the translation of the French word *Etablissement* into English tidal literature. *Etablissement* was used by Lalande in the tide sections of his *Astronomie* (1781 – see Chapter 6), and the word appears in much earlier *Traités de Navigation*,[(6)] but Lubbock's introduction of *establishment* in his paper of 1831[(2)] was the first use of the English word. Though used freely by Whewell, Airy and Bache, it was not liked by British seafarers. In a *Sailors' Word Book* of 1867,[(7)] (written around 1858), Admiral W.H. Smyth describes *establishment* as 'An awkward phrase lately lugged in to denote the tide-hour of a port'. We shall see shortly that Whewell made the concept even more awkward by defining two possible variants of its meaning. However, it was superseded before the end of the 19th century by the more precise concept of harmonic phase-lag, and then soon became obsolete.

Whewell's enterprise – the world oceans

Whewell took over the subject of mapping cotidal lines from Lubbock with enthusiasm. Between 1833 and 1850 he presented a series of 16 long and rather verbose papers on cotidal mapping to the Royal Society[(8)] and several summary reports of progress to the BAAS. He exercised the pioneer's privilege of coining new words and phrases appropriate to his subject. Many failed to stick; some phrases of Whewell's origin still occasionally used are: *age of the tide, luni-tidal interval, semi-menstrual inequality*.

On the terminology of *establishment*, Whewell objected to the conventional idea of HWT at Full and Change of the moon as represented in all extant compilations, on the grounds that the High Water referred to is not generally

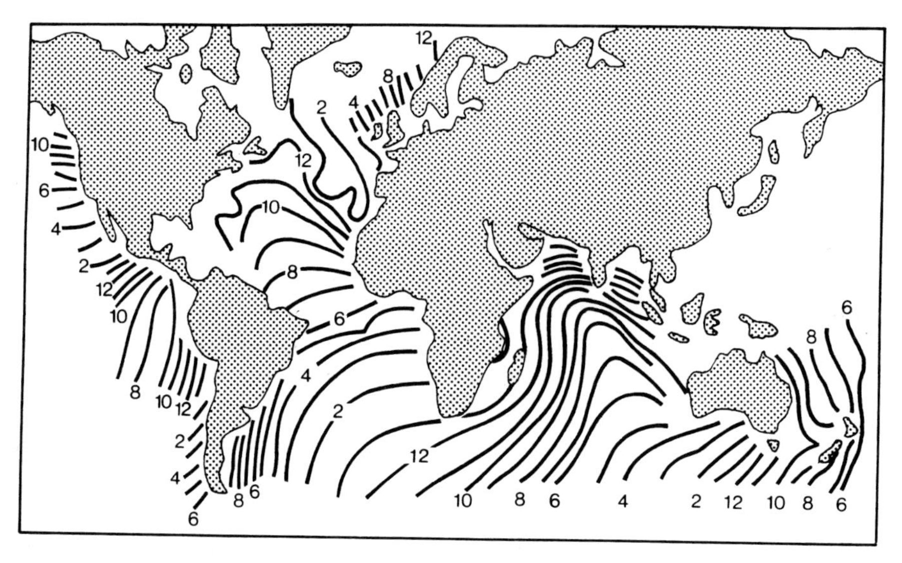

Figure 9.1. William Whewell's cotidal map for the parts of the world ocean where data was available in the 1830's, redrawn from Airy's copy[11] of Whewell's final draft of 1836.[8] The numbers shown are hours of *Vulgar Establshment* referred to the Greenwich meridian. Thus, High Water is supposed to occur at the same Greenwich time along any one of the contours. A steady wave-like progression in a generally northwesterly direction from New Zealand to the Norwegian Sea is suggested.

the *spring tide* associated with the syzygy. He named such an interval the *vulgar or common establishment*, and expressed a preference for a *corrected establishment* defined as the *average* daily luni-tidal interval over a semi-menstrual cycle. The latter quantity, usually about a half-hour less than the former, corresponds more closely with the harmonic phase lag G (for M_2) in later terminology.

Whewell started by sifting through the considerable quantity of 'vulgar' establishments listed by Lalande, Young, and in various navigational guides. Priorities of reliability had to be assigned, and he quoted some examples of absurd differences reported from neighboring sites. Some of these were due to varying notions of which event should be timed, High Water, Slack Water, or even Low Water. The remote island of Tristan da Cunha in 37° S was reported to have a tidal range of 8–9 feet (2.6 meters), compared with 2–3 feet (0.8 meters) at St. Helena; 20th century measurements showed that the mean range at Tristan is only about 0.5 meters. The higher value must have contributed to Whewell's erroneous belief in a belt of large tides in the Southern Ocean from which the tides in other oceans were thought to emanate.

Whewell's interpretation of the more reliable data was directly or implicitly derived from the assumption of purely progressive waves. This was more or less justified in the wider parts of the English Channel and the German Ocean (as the North Sea was then called). In the Atlantic it was fortuitously suggested by the northward progression of tide times along the western coasts of Africa and Europe. Data from the western Atlantic coasts was at first sparse, but reliable enough from Florida northward to rule out a simple picture of progression. Whewell explained the behavior along the North American coast by postulating that the tidal wave should progress faster in deep than in shallow water, to produce cotidal lines convex in the direction of propagation. This permitted a much more acute angle between cotidal lines and the coast. Taking account of bays and promontories, some cotidal lines could be nearly parallel with the coast, as observed off the USA. (Figure 9.1)

A progressive-wave interpretation for the Indian Ocean is now known to be quite inadequate, but Whewell managed to bend cotidal lines to accord with the available data, giving plausible consistency with his Atlantic picture. Data from the Pacific were so fragmentary that his interpretation was confined to a series of short lines nearly normal to the coasts of America and New Zealand. Whewell's initial cotidal map for the world ocean was presented in his first paper of 1833.[8] By his own admission, it was entirely preliminary and tentative, what nowadays might be called a 'strawman', to stimulate discussion. He later (1836) suggested small modifications, especially near the coast of North America; these were incorporated in an 'improved' world map by G.B. Airy in his celebrated treatise on *Tides and Waves* (Chapter 7). Figure 9.1 is transcribed from Airy's version of Whewell's world map.

Tides of the German Ocean (North Sea)

The much greater density of observations around Britain and adjacent coasts allowed surer interpretation on a smaller scale, but Whewell was puzzled by the feature at the southern end of the German Ocean, where the tide progresses to the south on the coast of East Anglia but to the north on the opposite coast of Holland, only 150km away. The phenomenon was evidently related to the meeting of two branches of the Atlantic tide, one from Orkney, the other up the English Channel, but it was quite unlike any wave motion known at the time. For a dynamic explanation it required Thomson's 'Kelvin wave' (Chapter 7) of 45 years later.

Whewell's 1833 cotidal map of the European seas[8] was also tentative. He was doubtful about comparing data from different epochs, and he expressed a plea for a special period of simultaneous observation at many ports. From these, the spatial pattern of each successive tide could be observed on the same day, preferably in calm weather conditions.

For such an undertaking, Whewell had an invaluable ally in Captain (later Admiral Sir Francis) Beaufort, recently appointed as Hydrographer of the Navy. Francis Beaufort (1774–1857) was the most dynamic of all British Hydrographers, determined to expand the Admiralty charts to cover the world and to establish his Service as a reliable seaman's authority in matters of weather and tides.[9] (The 'Beaufort Scale' of wind speed at sea remains standard to the present day.) As well as helping Lubbock by providing one of his officers, named J.F. Dessiou, to assist him to reduce his long series of data from

Liverpool and London, (Chapter 8) Beaufort understood the importance of his friend Whewell's endeavors, and used the wide influence of his Office to arrange for simultaneous observations round the coast to be carried out with remarkable promptness.

At Beaufort's command, tides were observed simultaneously at over 100 British coast-guard stations for two weeks in June 1834. The following year, with an amazing display of international scientific cooperation, for which Whewell also had diplomatic assistance from the Duke of Wellington, another observational exercise was achieved along the entire coast of northwest Europe and eastern America for three weeks in June 1835. The latter exercise covered 101 ports of seven European countries, 28 in America from the mouth of the Mississippi to Nova Scotia, and 537 ports of the British Isles including Ireland. Captain Beaufort allocated three technical clerks from the Hydrographic Office to assist Whewell in reducing the great quantity of data which resulted.

The success of these enterprises forced Whewell to produce his 'definitive' cotidal map of the north European seas in his sixth Royal Society tide paper of 1836.[8] The left hand map of Figure 9.2 is copied from the eastern part which includes the German Ocean. The original map shows contours for both 'corrected' and 'vulgar' establishments, with a note that the latter may be taken as roughly a half-hour later than the 'corrected' times. Figure 9.2 (left) shows only the 'vulgar' times, in order to afford easy comparison with Airy's modified cotidal map of the same area (right). The contentious feature of Whewell's map was his representation of the tidal regime in the Southern Bight as a complete counter-clockwise rotation through all phases.

Points of no-tide, and Airy's objection

Although he did not attempt to plot contours of amplitude, Whewell recognised that his rotary system implied a *point of no-tide* (his phrase) in the center of the Bight. He sought observational evidence, and was again successful in persuading Captain Beaufort to mount a remarkable exercise to record the variations in soundings at 30 minute intervals on a shallow bank in 52°27.5′N, 3°14.5′E during two daylight periods in August 1840. The exercise was performed under the expert command of Captain William Hewett, RN, who had a longboat doubly tethered to act as a stationary mark, while sailors in another boat maintained a fixed position relative to it by rowing against the tidal stream and repeatedly carried out careful soundings by logline. Between 5.30 a.m. and 5.30 p.m. on 25 August, in exceptionally calm conditions, the soundings were found to vary by no more than a foot (0.3m) in about $18\frac{1}{2}$ fathoms, (34m).[10] Such a variation was near the limit of observational accuracy, and effectively corroborated the existence of Whewell's *no-tide point* somewhere in the vicinity. To Captain Beaufort's chagrin, Captain Hewett, his ship (HMS *Fairy*) and its crew were lost in a violent storm later the same year. Hewett's letter to Beaufort

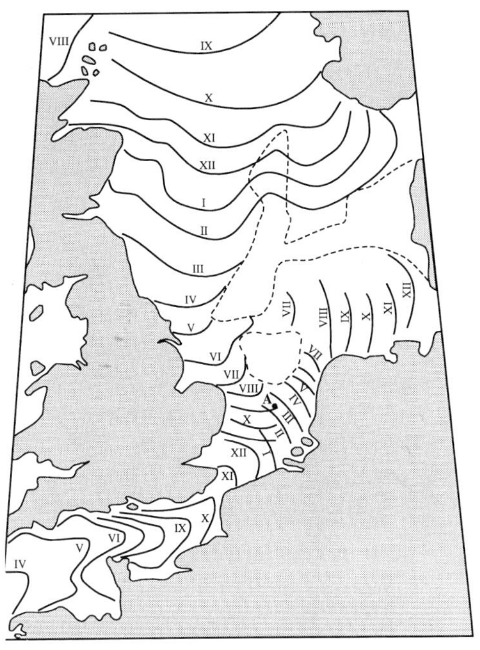

Figure 9.2. Two early interpretations of the Whewell/Beaufort tidal data round the coasts of the English Channel and German Ocean (North Sea). Left: Cotidal map of *Vulgar Establishments* (hours – Roman numerals) by Whewell (1836),[8] showing a prominent center of rotation ('no-tide point') between Holland and East Anglia, and a less clearly defined rotary area west of Jutland. Right: Airy's version of (1842)[11] in a slightly different projection, in which the cotidal lines are interpreted as being strongly influenced by shallow banks (dotted lines), and the southern rotary system is replaced by a pattern of interfering waves. The point marked 'A' in Airy's diagram is the position of Captain Hewett's measurements of very low amplitude.[10] Comparison with the tidal map properly computed by Proudman and Doodson in 1924 (Figure 11.4) shows Whewell's interpretation to be more nearly correct then Airy's.

recounting the details of the experiment was published posthumously as a report to the BAAS in 1841.[10]

The Astronomer Royal, G.B. Airy, would have none of this. Airy was obviously a far more accomplished mathematician than Whewell, and he observed with irony, and use of the royal 'we':[11] 'Although our mathematical acquaintance with the motion of extended waters is small, we have little hesitation in pronouncing this to be impossible.' Airy's own version of the regime in question (Figure 9.2, right) has a strange interference pattern between two wave trains, as is sometimes seen in wind waves in the lee of an obstruction. In fact, the analogy of interfering wave trains is irrelevant when the waves have exactly the same period and length, because they merge into a single wave of more complex shape. Airy thought that extensive shallows such as Wells and Dogger Banks (dotted outlines shown), were somehow responsible in separating the tidal waves. He ridiculed the idea of no tide at Hewett's point, ('A' in Figure 9.2), because tidal currents were experienced there. (In fact, currents at nodal points are the norm in later hydrodynamic theory.)

Whewell later retorted in his 13th tide paper (the *Bakerian Lecture* for 1847)[8] that his concept of a no-tide point was never intended to apply to other than the rise and fall, and was quite consistent with the existence of tidal currents, as is indeed the case. Comparison with a properly computed tide map of the region (Figure 11.4) shows that Whewell was right in every respect, including the existence of another rotary system west of Jutland which he also suggested. Rotary tides, later named *amphidromic systems* by the American hydrographer R.A. Harris (see below), were eventually found to be prevalent in all seas and oceans. Nevertheless, Airy's version of the tide map for the 'German or North Sea' was copied in official atlases for at least 80 years, until Whewell's version was finally vindicated by the calculations of J. Proudman and A. Doodson (Chapter 11).

Diurnal tides and mean tide level

The simultaneous data sets from 1835 afforded the opportunity to study the progression of the diurnal inequality in HW times and heights. The diurnal inequality had hitherto been thought of as a more or less similar perturbation over all the oceans, rather like the monthly (*semi-menstrual*) inequality. Whewell soon found that the diurnal effects obey no simple law but are extremely variable in their relationship to the principal tides. In particular, he noticed that the diurnal *age*, that is, the time delay of maximum inequality after maximum lunar declination, varied from practically zero on the Atlantic coast of North America to 3–6 days on the European and North African coasts, quite unlike the age of the twice-daily tide. (This phenomenon still lacks an easy explanation. In mathematical terms, it is thought to be accounted for by a different mix of weakly excited modes of oscillation at the higher and lower diurnal frequencies; tidal *age* is a measure of the increase of phase lag with frequency.)

Whewell discussed his findings in several papers from 1836 onwards. He quickly realised that a diurnal tide exists as a complex wave in its own right, and he sought records from places such as the Indian and Pacific Oceans where it has large amplitude. However, owing largely to the restrictive and irregular sampling at times of twice-daily extrema, he was unable to present a sensible mapping of the diurnal inequality, either across oceans or around Britain. Such mapping had to await the introduction of diurnal *harmonic constants*.

As a final example of Whewell's perceptive discoveries from the tide records at his disposal, he made the important observation in 1837,[8] that the *mean* height of HW and LW over a few days at any place was relatively constant, independently of variations in range. He suggested that *mean tide level* (similar but not identical to *mean sea level*) could be used as a datum in measuring vertical elevation across land, much more suitably than extreme Low Water, which was then in common use for map-making.

The idea was not new; mean tide level at Sheerness recorded by Mitchell's gauge had been used in 1831 as a datum for a levelling exercise between London Bridge and the sea.[12] A scientifically controlled experiment in the use of *mean sea level* was conducted in 1842 by Colonel Thomas Colby, RE, Director of the Trigonometrical Survey of Ireland, with advice from G.B. Airy. That exercise employed carefully timed visual tide records from 22 stations round the coast of Ireland. The tides were analysed by Airy himself in a long paper,[13] which also gives a detailed account of the levelling results. Thereafter, reference to mean sea level became standard practice in geodetic levelling, until the recent introduction of geocentric vertical fixing by satellites.

Captain Fitzroy on ocean tides

As part of Captain Beaufort's promotions, copies of Whewell's 1833 paper were sent to the Navy's overseas surveyors, including Captain Fitzroy of HMS *Beagle*. Robert Fitzroy (1805–1865) is most widely remembered as the leader of the surveying cruise which took Charles Darwin round the world as naturalist in 1831–36. Besides being an expert surveyor, Fitzroy also earned a reputation as a 'meteorological statist',[14] the inventor of a popular barometer, and as an amateur natural philosopher. His three-volume *Narrative* (Volume 3 by Darwin) of his ten years of voyaging[15] includes essays on several diverse subjects, for example a theory of the biblical Deluge. The Appendix to Volume 2 – a separate book – contains a lengthy chapter on ocean tides, prompted by Whewell's paper.

With due deference to Whewell's scholarship, Fitzroy confessed to being unconvinced by the hypothesis of tidal waves progressing northwards from the Southern Ocean, chiefly because of the 'trifling rise of tide' in most of the South Atlantic, and the slow progression of phase along the coast of southwest Africa. (Fitzroy also claimed that the tide progressed towards the southwest from Pernambuco to the Rio Plata, but by modern knowledge that is manifestly incorrect.) Seeing that the moon appears to traverse the oceans from east to west, he proposed an alternative hypothesis, that the tide performs stationary oscillations between the east and west sides of the ocean basins.

To support his hypothesis, Fitzroy quoted a paragraph from an essay published in 1837 by the mathematician Charles Babbage (1792–1871) in an appendix to a philosophical discourse on Science and Religion.[16] In the context of a theory of the cause of raised beaches, Babbage's essay had suggested that, if the zonal tidal wave had a wavelength equal to twice the width of the basin, it could resonate locally to produce a stationary oscillation of large amplitude, with opposite phases at opposite coasts and a nodal point in the middle. Such a mechanism had been postulated even earlier by Thomas Young, not to mention Galileo's 'water-barge' analogy, (Chapter 4). Fitzroy felt that it provided a better description of tidal behavior in the Atlantic than Whewell's hypothesis.

Fitzroy went on to cite many examples in the Atlantic and the Pacific where, in the same zonal strip, the time of HW is later on the western side than on the eastern by the equivalent in time of the difference in longitude. For a width of 90° this means that HW in the west is simultaneous with LW in the east, as required by Babbage's proposition. Fitzroy showed that this was indeed the case in some localities, for example in 40° S between Valdivia, Chile and North Island, New Zealand. He also claimed that Whewell had used seriously erroneous data for the Atlantic islands of Cape Verde, Madeira and St. Helena. However, in the latter two cases, Fitzroy's own figures can now be seen to be inaccurate.

Despite some inconsistencies in Fitzroy's arguments, Whewell's faith in the progressive theory was shaken. In the 1847 Bakerian Lecture,[8] after reference to Fitzroy's comments, he admitted: 'I do not think it likely that the course of the tide can be rightly represented as a wave travelling S-N between Africa and America. We may much better represent it as a stationary undulation, of which the middle space is between Brazil and Guinea in which the tides are very small, as at St. Helena and Ascension ...' However, he never published a revised world cotidal map incorporating the latter ideas. The intention of Whewell's 1833 map had been to stimulate controversy, and in that respect it had served its purpose. But it was reproduced as illustration for many years by several later authorities, including Sir George Darwin, for want of anything better.

Cotidal mapping by A.D. Bache

Alexander D. Bache was introduced in the previous chapter as the second Superintendent of the Coast Survey, who set up tide recording stations round the extensive coastlines of the USA He also subjected the records of the various tide regimes to elaborate analysis, leading to plausible predictions by methods of Lubbock's type, and local cotidal maps in the spirit of Whewell. Bache's work in the 1850's showed that the Pacific and Gulf coasts required different approaches from the Atlantic because of their strong diurnal inequality. Like Whewell, he realised that *establishment* has little meaning in such cases. He devised his own methods of separating diurnal and semidiurnal waves, (without going as far as a harmonic analysis), and he produced the first tentative cotidal map of the diurnal tide in the Mexican Gulf at the greatest declinations of the moon.[17]

Bache's semidiurnal map of the Atlantic coastal area[17] showed only slight seaward extension. Except for important variation north of Cape Cod, his cotidal lines had quite small inclination to the coast; his line for XII hours (Greenwich) closely hugs the coast from Florida to Long Island (New York). This property is consistent with Whewell's interpretation (Figure 9.1) and also with more advanced maps such as Harris's (Figure 9.4).

With less data available from the Pacific, Bache mapped its 'approximate cotidal lines' (his phrase) with small coastal inclination, as in the Atlantic. In fact the change in establishment from San Diego (California) to Cape Disappointment (Washington) is a little less than 3 hours. His interpretation was, however, quite different from Whewell's (Figure 9.1), whose cotidal lines are almost all normal to the American coast. Neither author had much data to guide him; modern computed cotidal maps are closer to Whewell's.

Somewhat incidentally to our main theme, Bache was the first to recognise *tsunami* waves in automatic tide-gage records from the Californian coast on 23 December 1854 as having originated from a violent earthquake in Japan about $12\frac{1}{2}$ hours earlier.[18] From the time of travel of these waves across the northern Pacific, he estimated the mean depth of that ocean to be 2100–2500 fathoms (roughly 4.2 kilometers), a plausible value.

The work of Rollin A. Harris

No serious attempts were made to construct new oceanic cotidal maps for about half a century after the pioneering work of Whewell, Airy and Bache. Several hydrographic surveyors continued to determine tidal constants in seas where they were unknown, and the introduction of harmonic analysis in the 1870's (Chapter 8) provided a more rational basis for mapping tidal phases at both semidiurnal and diurnal periods.

First to embark on a radical new approach to cotidal mapping was Rollin A. Harris (1863–1918), who joined the tidal research group of the US Coast & Geodetic Survey in 1890, four years after the retirement of William Ferrel. Harris was mentioned briefly in Chapter 8 as the co-inventor of a tide-predicting machine, and I have referred in Chapter 1 and elsewhere to his comprehensive history of tidal science to the end of the 19th century. The History effectively formed the Introduction to Harris's five-part *Manual of Tides*, published by the C&GS between 1897 and 1907.[19] Parts IV-A and B of the Manual describe (A) the theory, and (B) the execution of a novel scheme for the construction of cotidal maps in all oceans.

Harris's scheme was a brave attempt at a tractable formulation of the idea suggested by Young, Fitzroy and others, and eventually admitted by Whewell, that the tides are generated as stationary waves in resonant basins between the continental boundaries. A full mathematical expression of this idea is very complex, but Harris greatly simplified it by assuming that resonance occurs in strips of ocean (of any orientation) with half a natural wavelength or a multiple thereof between coasts. Wavelength was determined by an assumed free-wave speed $c=\sqrt{(g\times \text{depth})}$; enough was then known of ocean bathymetry to make this possible. With a tide of period T in a strip of uniform depth h, a half-wavelength thus equals $L_{\frac{1}{2}}=cT/2$. More generally, with variable depth $c(x)$ at longitudinal distance x, $L_{\frac{1}{2}}$ would be determined numerically as the solution of

$$t(L_{\frac{1}{2}}) = T/2; \quad t(l)=\int_0^l c(x)\,\mathrm{d}x.$$

Nodal lines, transverse to the strip, occur at $L_{\frac{1}{4}}$, defined similarly. Some degree of depth-averaging transverse to the strip is implied. The Coriolis force due to earth rotation was ignored.

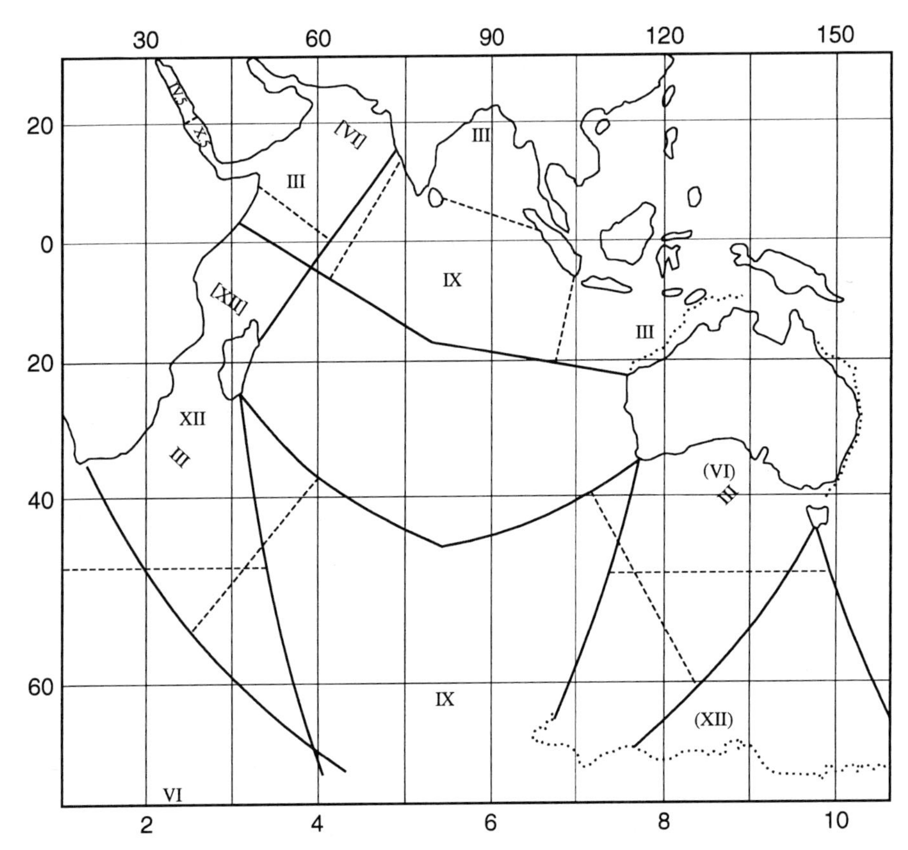

Figure 9.3. R.A. Harris's scheme of resonating basins in the Indian Ocean, bounded by solid lines. Broken lines represent nodal zones in individual basins; Roman numerals denote the phase in lunar hours. Dotted lines are continental shelf-breaks of uncertain position. Comparison with Harris's global representation of the cotidal lines (Figure 9.4) shows that his amphidromic centers (nodes) vaguely approximate to intersections of the nodal zones. However, his four points of convergence: (1) between India and Madagascar; (2) southeast of Madagascar; (3) west of Australia; (4) close to Sri Lanka (former Ceylon), are roughly confirmed by later more sophisticated computations. From *Manual of Tides*, Part IVB, p. 357.[19]

The manner of application is well enough illustrated by Harris's chosen partitioning of the Indian Ocean for semidiurnal tides (Figure 9.3). Solid lines mark the boundaries of resonating strips, some of which overlap; that is, they have resonant waves in two orientations. Nodal lines at $L_{\frac{1}{4}}$ and occasionally $L_{\frac{3}{4}}$ are represented by broken lines. Strips with only one nodal line have half-wavelength resonances, two nodal lines indicate a full-wave resonance. The central zone between Madagascar and West Australia is not a resonant strip, but is supposed to contain tides interpolated between the resonating areas. Where the nodal lines in overlapping strips intersect or come close together, as at about 65°E on the equator, a no-tide point results, round which the cotidal lines rotate through all directions. Harris coined the phrase *amphidromic system* for such a feature – from Greek *amphi* = around, *dromos* = running – (Part IV-B, p. 326). (Abbreviation to *amphidromy* or *amphidrome* came much later.)

It is noteworthy that Harris's amphidromic systems, arising from intersecting stationary waves without rotation, differ in principle from Whewell's no-tide point in the North Sea which is largely due to the effects of earth rotation. Both causes are possible in principle, but total neglect of rotation was a serious weakness in Harris's formulation.

The unbracketed roman numerals in Figure 9.3 are the Greenwich times of High Water of each resonant system in lunar hours; they change by 6 hours across a nodal line. Harris computed these times by resolving the horizontal component of the tide-raising force along the axis of the strip at all hours; the time at which the *virtual work* over the strip is zero was taken as the hour of HW (or LW) – see Part IV-A, Chapter VI. That these times accord broadly with observed values of coastal HW times is a remarkable vindication of Harris's insight.

After extremely laborious calculations along those lines over all oceans excepting the Arctic, Harris was able to construct (with some hand-interpolation) a complete set of cotidal lines for M_2 in the world ocean, as shown in Figure 9.4. His map differs radically from Whewell's map (Figure 9.1) in the appearance of several mid-ocean amphidromic systems, one on the equator of the Indian Ocean, two in the North Atlantic, and three in the Pacific. There is also some localised crowding of cotidal lines which indicates near-amphidromy. With the exception of the no-tide point southwest of New Zealand, all these amphidromic systems with their senses of rotation are broadly confirmed by modern computations based on far more elaborate dynamic principles. Others now known, south of Mexico and near Antarctica, Harris failed to spot owing to lack of data.

Part IV-B of Harris's Manual also contains painstakingly produced cotidal maps for nearly all the marginal seas, extrapolated from his oceanic system as progressive waves in the manner of Whewell, and in some cases admitting no-tide points. For such seas he was greatly aided by the appearance by 1900 of quite a substantial number of harmonic constants at coastal tide recording stations. Diurnal cotidal maps could have also been produced by his method, but apart from a sketched diurnal amphidromic system in the mid Atlantic, Harris did not elaborate that subject, possibly because of adverse criticism of his theoretical principles.

Hard-core mathematicians found Harris's scheme full of unwarranted assumptions. Despite having no alternative solutions of their own to offer, many of his critics refused to accept his results. Foremost among these was Sir George Darwin himself, who gave a polite but condemnatory review[20] of the purely theoretical Part IV-A of the Manual, published three years before IV-B. Darwin's review starts and ends:

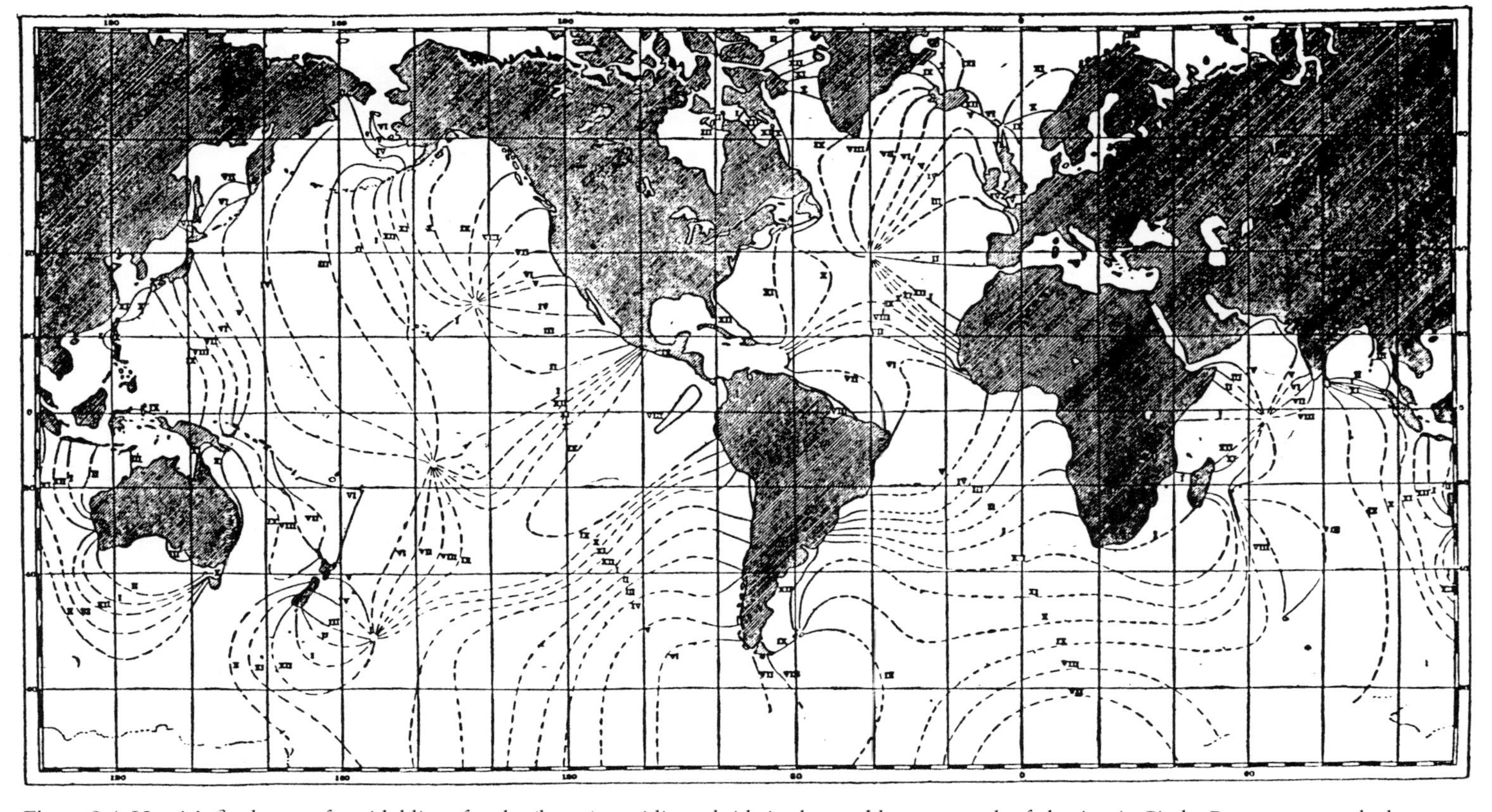

Figure 9.4. Harris's final map of cotidal lines for the (lunar) semidiurnal tide in the world ocean south of the Arctic Circle. Roman numerals denote lunar hours of High Water relative to the Greenwich meridian. Their qualitative accuracy may be assessed by comparison with the empirical map for M_2 drawn some 40 years later by G. Dietrich, using a much greater quantity of harmonically analysed data, (Figure 11.5).

> Mr. Rollin Harris has done so much good work in preparing his Manual of Tides for the United States Coast Survey, that it is an ungrateful task to find oneself constrained to criticise adversely his recently published part IV-A of that treatise. I shall pass over many points of interest which occur in the earlier portions of the book, because the discussion of them is apparently designed to lead up to a new theory of oceanic tides. That theory, to which I shall confine my attention, depends on a proposition that it is possible to dissect our oceans into a number of basins in which the oscillations are virtually independent of one another and are almost unaffected by the diurnal rotation of the earth. . . . I can, in conclusion, only express a hope that I am not doing an injustice to Mr. Harris in dissenting so absolutely from his views. No one would have welcomed more warmly than I a new clue to our treatment of this difficult problem. I venture to express my admiration at the courage of the attempt, and although, as I think, it is a failure, yet it may inspire others to more successful attacks.

Between the quoted paragraphs, Darwin exposed a long list of weaknesses and *non sequiturs* in Harris's theoretical arguments, acknowledging support of his criticism from the distinguished mathematical physicist, Professor A.E.H. Love.

After the appearance of the maps themselves in Part IV-B, however, commentators took a more lenient view. The French mathematician Henri Poincaré, who devoted the third tome of his *Leçons de Mécanique Celeste*[21] to tidal theory, concluded it with a long Chapter XVI on Harris's cotidal maps, in which he examined all their salient features and showed that their dynamic properties were at least qualitatively in keeping with elementary theory. As a general observation, Poincaré remarked (p. 403):

> Sa manière de voir s'écarte beaucoup de celle de Whewell et ne se heurte pas, dans ses principes généraux, aux même objections essentielles. Il est vraisemblable que la théorie définitive devra emprunter à celle de Harris une part notable de ses grandes lignes.
>
> (His way of seeing things differs greatly from that of Whewell and his general principles do not run into the same essential objections. It is very likely that the definitive theory [of the ocean tides] will take a large part of its outline from the theory of Harris.)

Poincaré's generous review must have done much to restore Harris's self-confidence to continue with his studies of Arctic tides.

Much later, in reviewing the state of the science of tide-mapping at the dawn of electronic computers, the redoubtable English tidalist A.T. Doodson went even further in praise of Harris's achievement:[22]

> The wave theory [of Whewell and Airy] implied that in a wide gulf the wave must either circulate within the gulf or else have all its energy dissipated at the closed end of the gulf. It is to Harris's lasting credit that he utilised the concept of standing oscillations about nodal lines, and the influence of depth upon tidal oscillations. . . . His theory, as a theory, was not very well received, but the charts he produced, as far as the oceans are concerned, were much in advance

of those previously existing, and his work was a great stimulus to others. For the first time the predominance of amphidromic systems in the oceans was made evident.

Early mapping of the Arctic Ocean

Towards the end of his career, Harris took a great interest in the Arctic Ocean. The bathymetry was not known well enough to apply his stationary wave calculations with confidence, and the data from the coasts and islands were too sparse and unreliable to serve entirely for empirical guidance. After careful examination, he considered that the tidal data, together with reported drift-currents, indicated the presence of an uncharted polar landmass. This may, he suggested, correspond to land supposedly sighted from a height by the explorer Robert Peary in 1906, and provisionally named by him *Crocker Land*.(23)

After 'Crocker Land' had been deemed nonexistent, on examining the soundings taken during Peary's polar journey of 1909, Harris admitted that the obstruction in question could be an underwater ridge. One of the major bathymetric features of the Arctic Ocean, now known as *Lomonosov Ridge*, was for a time unofficially named *Harris Ridge* – see review by Zetler.(24) It is not now thought to have a great influence on the tide. Harris's final solution for the Arctic cotidal map was a wave of translation entering from the Atlantic via the Greenland Sea, but he left some large areas of his map blank.(24)

Arctic bathymetry was considerably improved during the early decades of the 20th century as a result of the Norwegian scientific voyages of Nansen in the *Fram* and Amundsen in the *Maud* (results published some years after the expeditions themselves). The oceanographer Harald Sverdrup took extensive measurements of tidal currents on the Siberian shelf from the *Maud*, discussed in Chapter 11, and was able by subtle dynamic analysis to map the tidal behavior of most of this large area, showing cotidal lines nearly parallel to the coast.(25) His map was extended to the whole Arctic Ocean with similar characteristics by Sverdrup's colleague F.E. Fjeldstad.(26)

During the 1920's, Arctic tide researches were taken up with new techniques of computation by Albert Defant (1884–1974) and by Robert von Sterneck (1871–1928), both of Austria. The essence of their computing scheme, to be described more fully in Chapter 11, was to treat seas of elongated shape as canals of varying depth and cross-section, along which the one-dimensional wave equation could be solved numerically. The transverse slope across each section was then derived by simple application of the Coriolis stress to the longitudinal currents. The latter procedure was able to produce amphidromic systems at the head of gulfs such as the Adriatic Sea. Defant treated the combined Atlantic and Arctic Ocean in this manner, at the risk of gross oversimplification. His cotidal map of the area(27) showed several of the features of Harris's map of the Atlantic, but suggested at least two amphidromic systems in the Arctic which were not present in the maps of Harris or Fjeldstad.

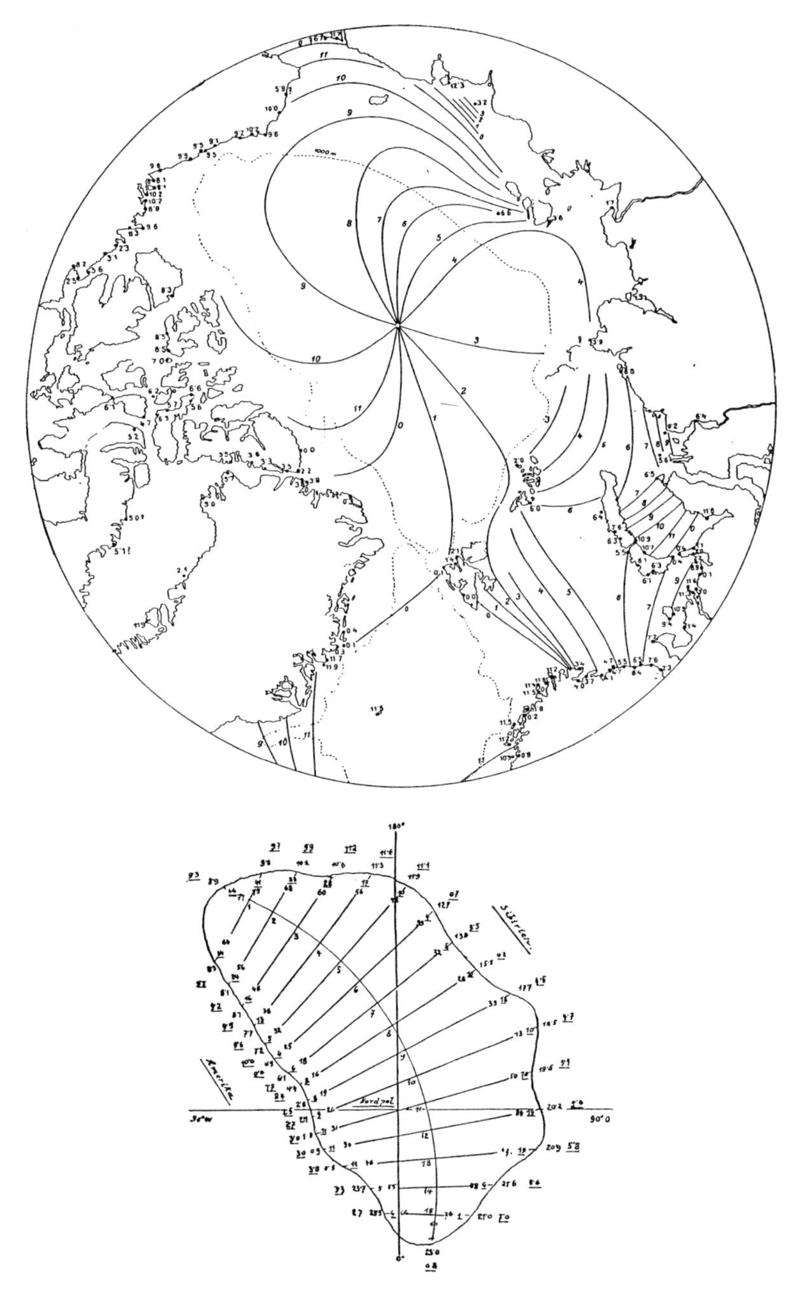

Figure 9.5. Above: Cotidal map by Robert von Sterneck for the M_2 tide in the Arctic Ocean, computed in 1928 by an approximate dynamic model with data from surrounding coasts and islands. Integer numerals attached to the lines are Greenwich phase-lag in lunar hours (that is G°/30). Numbers associated with black circles are observed phases to one decimal. Dotted lines are approximate 1000 m depth contours. Below: Sterneck's computational grid for the above map, showing 15 stepped sections between the end points, used for integration of the one-dimensional dynamic equation. The closed contour here approximates to the 200 m shelf-break. Numbers round the contour are the assumed tidal times (outside) and amplitudes (inside), with underlined figures for M_2, the others for K_1.

Sterneck performed calculations similar to Defant's for the deep central basin of the Arctic Ocean by itself, as a tidal system co-oscillating with the known Atlantic tide between Greenland, Spitzbergen and Norway. He obtained just one prominent amphidromic system between the Pole and Alaska,[28] which he linked by hand to the coastal regime of Sverdrup (Figure 9.5). Sterneck's amphidromy corresponds with one of Defant's, and explains some features of Harris's suspected obstruction in the same area. More recent computations confirm Sterneck's main system, but add a number of minor amphidromic systems round the coast.

Enough has been said of work in the Arctic Ocean up to 1928, to indicate the emergence of new schools of tidal study in various European centers during the early 20th century. The overall pattern of research into tidal mapping up to about 1950, including the rigorous mathematical studies of J. Proudman and A. Doodson of the Liverpool Tidal Institute, will be described in Chapter 11.

Notes and references

1. Whewell, H. *History of the Inductive Sciences*, 3 vols, London 1837. (Remarks on the history of tidal science, with reference to the work of Whewell and Lubbock, are included in Vol. 2, pp.111–113 and 246–253.)
2. Lubbock, J.W. On the tides on the coast of Great Britain. *Phil. Mag.*, **9**, 333–335, 1831. Related paper, containing the first English use of 'establishment': On the tides of the port of London, *Phil. Trans. R. Soc. London*, **121**, 379–416, 1831.
3. Lubbock, J.W. Report on the tides, British Assoc. Report, (Oxford Meeting), Vol. 2, 189–195, London, 1832.
4. Young, T. *A Course of Lectures on Natural Philosophy and the Mechanical Arts*, Vol. 1, London, 796pp., 1807. (Young's sketch of cotidal lines round the British Isles is reproduced in Lubbock's *Phil. Mag.* paper of 1831[2].)
5. Halley, E. Tidal Chart of the English Channel – see Chapter 6, Ref. 14.
6. Bouguer, Jean. Traité complet de la Navigation, Paris, 1706. (Ch. VI, Leçon 32, pp.131–137, has: 'Nouvelles Remarques sur les Marées, et Table de leur *établissements*'.)
7. Smyth, W.H. *Sailor's Word Book – an alphabetical Digest*. London, 1867. (Written about 1858.)
8. Whewell, W. Essay towards a first approximation to a map of cotidal lines. *Phil. Trans. R. Soc. London,* **123**, 147–236, 1833. *Ibid.* Papers with related titles, including coastal observations, in volumes for 1834, 1835, 1836 (3), 1837 (3), 1838, 1839 (2), 1840 (2), [*Bakerian Lecture*, 1847, pub. 1848], 1850.
9. Ritchie, G.S. *The Admiralty Chart*, Ch. 13: *High Noon; Beaufort as Hydrographer*. Hollis & Carter, London, 388pp., 1967. (See also: Friendly, A. *Beaufort of the Admiralty*, 219–295, Hutchinson, London, 1977.)
10. Hewett, W. 'Letter from the late Capt. Hewett to Capt. Beaufort, R.N. (referred to in a communication by Professor Whewell)'. British Assoc. Rep. for 1841 Meeting, Pt. 2, 32–35, 1842.
11. Airy, G.B. Tides and Waves, *Encycloped. Metropolitana*, **5**, Art. 525–528, London, 1845.
12. Lloyd, J.A. An account of operations to determine the difference of level between London Bridge and the sea. *Phil. Trans. R. Soc. London*, **121**, 167–197, 1831. This exercise motivated the construction at Sheerness of the first operational tide-gage. See Ch. 8, Ref. 10, and Figure 8.1.
13. Airy, G.B. On the laws of the tides on the coasts of Ireland, as inferred from an extensive series of observations made in connection with the Ordnance Survey of Ireland. *Phil. Trans. R. Soc. London*, **135**, 1–124, 1845.

14. Charnock, H. Fitzroy – Meteorological Statist. *Proc. R. Soc. Edinburgh*, Section B, **72**, (1), 115–122, 1972.
15. Fitzroy, R. Narrative of the surveying voyages of HMS *Adventure* and *Beagle*, 1826–1836. 3 vols, London, 1839. (Appendix to vol. 2, 277–297.)
16. Babbage, C. *The Ninth Bridgewater Treatise – a Fragment*. John Murray, London, 2nd Edn., 1838. (The only references to tides are in Appendix K: 'On the elevation of beaches by tides'. pp. 248–251, whose second paragraph is quoted *in toto* by Fitzroy(15).)
17. Bache, A.D. Reports to the American Association for the Advancement of Science (AAAS): Physics of the Globe: 8th Meeting, 1854 – Atlantic coast; 9th Meeting, 1855 – Pacific coast; 10th Meeting, 1856 – Gulf of Mexico northern coast.
18. Bache, A.D. Note on earthquake waves on the west coast of USA, 23 and 25 December, 1854. *Proc. AAAS*, 9th Meeting, pp. 153–160 (with diagrams), 1855.
19. Harris, R.A. *Manual of Tides*; Part IVA – Outlines of tidal theory; Part IVB – Cotidal lines for the world, US Coast & Geodetic Survey, Washington, D.C., 1901(A) and 1904(B).
20. Darwin, G.H. A new theory of the tides of terrestrial oceans. (Review of Harris(19) – Part IVA.) *Nature*, **66**, 444–445, 1902.
21. Poincaré, H. *Leçons de Mécanique Céleste,* Tome 3, *Théorie des Marées*, Gauthier-Villars, Paris, 469pp., 1910.
22. Doodson, A.T. *Oceanic Tides*. Advances in Geophysics, **5**, 117–152, Academic Press, New York, 1958.
23. Harris, R.A. Appendix I, pp. 299–307 in: Peary, R.E., *The North Pole*, Stokes, New York, 315pp., 1910.
24. Harris, R.A. *Arctic Tides*, USCGS Report, 105pp., US Govt. Printing Office, Washington D.C., 1911. (Harris's work is generously reviewed, with copies of his cotidal maps, in: Zetler, B.D. '*Arctic Tides* by Rollin A. Harris (1911) revisited', *EOS*, **67**, (7), pp. 73 and 76, American Geophys. Union, 1986.)
25. Sverdrup, H.U. Dynamics of tides on the north Siberian Shelf. *Geofysiske Pub.*, **4**, (5), 75pp., Oslo, 1926.
26. Fjelstad, J.E. Literature on the tides of the Arctic Ocean, *Naturen*, **47**, 161–175, Bergen, 1923.
27. Defant, A. Die Gezeiten Atlantischen Ozeans und des Arktischen Meeren. *Annalen Hydrogr*. **52**, 153–166, 177–184, 1924. (A summarised English account of all of Defant's works on tides is in his *Physical Oceanography*, Vol. 2, Pergamon Press, Oxford, 598pp., 1961.)
28. Sterneck, R. von, The tides of the Arctic Ocean, (translated from the original German in *Annalen Hydrogr*., 1928) *Int. Hydrogr, Rev*., **5**, (2), 207–213, 1928.

Contributors to the science of tides in the 19th to early 20th centuries. (From left to right), *top:* William Whewell (1794–1866); Rollin A. Harris (1863–1918); *bottom:* William Thomson, Lord Kelvin (1824–1907); George Howard Darwin (1845–1912). (Photograph of R.A. Harris by courtesy of the United States National Ocean Service (NOAA); others by permission of the President and Council of the Royal Society, London.)

10

Tides of the geosphere – the birth of geophysics

We now return for the last time to the late 18th to 19th centuries to consider the early development of research, not into the oceanic tides themselves, but into physical phenomena which are closely related to them. It was a period when scientists first became aware that there exist tide-like oscillations in other levels of the geosphere – in the atmosphere and in the upper part of the earth's crust known as the lithosphere, and by implication in the body of the earth. I will call these *air tides* and *earth tides* respectively. Such oscillations are not easily observable in the sense that ocean tides are, but refined instruments with which they could be detected were also being developed. Detection prompted questions about causes and relation to other physical properties of the geosphere. An important group of questions centered on the role of oceanic tidal friction in accounting for the observed secular acceleration of the moon.

The second half of the 19th century saw the emergence of *geophysics* (originally known as *physical geology*) as a recognised research domain, embracing all the earth sciences from seismology to aeronomy, with geodesy, physical oceanography and tropospheric meteorology occupying middle ground. Links to the physics of the solar system, including other planets and their satellites, provided external constraints. Various branches were pioneered by noted scientists in France and the German speaking countries, but the acknowledged 'father of geophysics' – a phrase applied by Harold Jeffreys in the dedication of the first edition (1924) of his famous book *The Earth* – was Sir George Howard Darwin of Cambridge University, whom we have encountered in previous chapters.

We must also remember the many seminal ideas introduced by Sir William Thomson (Lord Kelvin) and by Thomson's teacher, William Hopkins.(1) Indeed, since practically every problem on which Darwin worked had previously been enunciated by Thomson, the latter amply merits the title *grandfather of geophysics*. In his Preface to the first volume of his collected scientific papers(25) Darwin wrote:

> The whole of my work on oceanic tides ... sprang from ideas initiated by Lord Kelvin, and I should wish to regard this present volume as being, in a special sense, a tribute to him. Early in my scientific career, it was my good fortune to be brought into close personal relationship with Lord Kelvin. Many visits to Glasgow and to Largs have taught me to look up to him as my master, and I cannot find words to express how much I owe to his friendship and to his inspiration.

George Darwin's name is especially associated with oceanic tides, but only the first of the five volumes of his scientific papers is devoted entirely to tides *per se*. His delightful semi-popular book, *Tides and Kindred Phenomena in the Solar System*[(2)] gives a broad picture of his range of interests. It is clear from this book that he regarded the tides of the geosphere as given oscillations which could be used to determine other properties of the earth. This new approach to tides is the one largely followed in the present chapter, although Darwin did not contribute actively to our first two subjects.

I shall treat the various tidal subjects related to geophysics individually in the following sections of this chapter. Each subject will be considered from its earliest origins in the 18th–19th centuries, and followed as far as a convenient pause, usually in the earliest decades of the 20th century. Their later developments will be taken up again in the following chapters when suitable contexts have been reached. I will take air tides and magnetic tides further into the 20th century, in order to reach a significant stage of development without having to return to them in a later chapter. However, air tides will be briefly revisited in Chapter 15 in the context of air/ocean interaction.

Tides in the atmosphere

Tidal oscillations in a medium other than the ocean were first observed in surface air pressure. The mercury barometer was invented by Torricelli in 1643. Travelers soon noticed that the barometric variation in the Tropics is much quieter than in European latitudes, but since the barometer was habitually read only once or twice each day, nobody noticed a small regular variation of 12 hours period. If someone did, it was attributed to instrumental errors due to solar heating.

English authorities on air tides who have written about the history of the subject – Wilkes, Chapman – have confessed to being at a loss to name the first observer, or have left the matter in obscurity. From my researches, it turns out that the first person to observe the 12-hourly oscillation in surface air pressure was a young French 'naturalist' (in the wider sense of the 18th century) named Robert de Paul, chevalier de Lamanon (1752–1787), who accompanied the explorer J-F. Galaup de La Pérouse in his ill-fated voyage of 1785–88. The entire expedition perished in the South Pacific in 1788, and in the previous year Lamanon had been among a party of men massacred by the natives of the island of Tutuila (Samoa) – La Pérouse's 'Maouna'. However, Lamanon made his historic barometric observations in September 1785, and was able to send an account to the Secretary of the *Académie Royale* shortly afterwards. Because of the lack of recognition of Lamanon's careful work in the literature, (except by Humboldt,[3] who was also interested in Lamanon's magnetic observations), I reproduce in Figure 10.1 the first three pages of his Report, as recorded in the posthumous accounts of the voyage.[4]

Figure 10.1 contains several points of interest, but the text is too long for a verbatim English translation. The opening paragraph reflects a preconception concerning the moon's influence on the atmosphere, which we may ignore. The second paragraph quotes instructions advised by the *Académie* before the voyage, in which one may clearly detect the influence of Laplace. Because of the extremely small amplitude predicted by the gravitational theory, the *Académie* (Laplace) strongly advised that the barometer should be read hourly *on land* near the equator. In fact, Lamanon found that a tidal variation was clearly detectable at sea, using a good instrument. His care in the prior selection of a good marine barometer is described on the second page shown. On the third page one sees that he recorded for a continuous period of 75 hours while crossing the Atlantic equator, the now familiar rise and fall between 4h and 10h a.m. and p.m. in local solar time, with double amplitude generally greater than 1 'ligne' of mercury (2.8 millibars) to the nearest tenth part of a ligne (12 lignes = 1 inch).

Maxima occurred close to 10 a.m. and 10 p.m. local time, and this was later found to be the case at all longitudes, indicating a double wave of global scale leading the sun by about two hours. The oscillation merited the name 'tide', even though no relation to the moon was evident. Figure 10.2a shows a recent recording of air pressure at St. Helena (16° S); the 12-hourly oscillation is still obvious at this latitude, and a smaller diurnal tide is also evident.

In Chapter 35 of his famous paper of 1776, Laplace had been interested in applying his tidal theory (Chapter 7) to *le flux et le réflux de l'atmosphère*, because the air conforms more closely to the concept of a thin sheet of fluid covering the globe than the oceans do. He recognised an important difference, that the density of the air varies with pressure and with radiant heat, but he simplified the physics by assuming uniform temperature, isothermal changes of pressure, and absence of vertical acceleration. Under these assumptions Laplace showed[5] that the atmosphere is subject to the same dynamic equations as he had derived for the ocean, but with an *equivalent depth*

$$D' = p_0/g\rho_0,$$

where p_0 and ρ_0 are the mean pressure and density at ground level. (D' is also called the *scale height* of the atmosphere; its value is typically about 10 km.)

Laplace's dynamic equations for the atmosphere allowed for a tidally moving ocean surface as lower boundary. This factor became important in later, more precise, calculations of the lunar air tide, (see Chapter 15), but since Laplace's

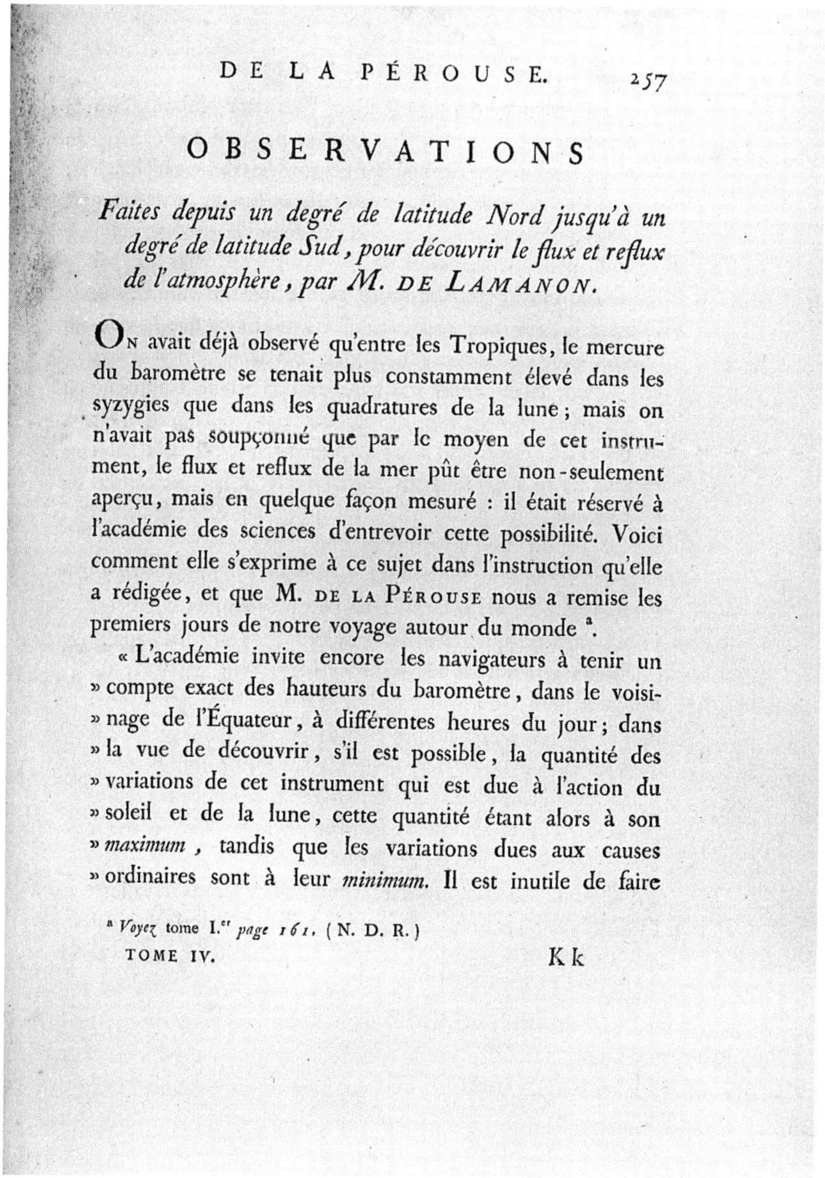

DE LA PÉROUSE. 257

OBSERVATIONS

Faites depuis un degré de latitude Nord jusqu'à un degré de latitude Sud, pour découvrir le flux et reflux de l'atmosphère, par M. DE LAMANON.

ON avait déjà observé qu'entre les Tropiques, le mercure du baromètre se tenait plus constamment élevé dans les syzygies que dans les quadratures de la lune ; mais on n'avait pas soupçonné que par le moyen de cet instrument, le flux et reflux de la mer pût être non-seulement aperçu, mais en quelque façon mesuré : il était réservé à l'académie des sciences d'entrevoir cette possibilité. Voici comment elle s'exprime à ce sujet dans l'instruction qu'elle a rédigée, et que M. DE LA PÉROUSE nous a remise les premiers jours de notre voyage autour du monde [a].

« L'académie invite encore les navigateurs à tenir un » compte exact des hauteurs du baromètre, dans le voisi-» nage de l'Équateur, à différentes heures du jour ; dans » la vue de découvrir, s'il est possible, la quantité des » variations de cet instrument qui est due à l'action du » soleil et de la lune, cette quantité étant alors à son » *maximum*, tandis que les variations dues aux causes » ordinaires sont à leur *minimum*. Il est inutile de faire

[a] *Voyez* tome I.er *page 161*. (N. D. R.)

TOME IV. Kk

Figure 10.1. The first three pages of chevalier Paul de Lamanon's report of his discovery of a tide in barometric pressure, with the advice of L'Académie Royale des Sciences. The observations were made in September 1785 on board *La Boussole*, commanded by La Pérouse himself, between 1°N and 1°S of the Atlantic Ocean. The readings (3rd page) are in units of 1 ligne in the height of mercury, to the nearest one-tenth. A ligne was one-twelfth of a 'pouce' (1″), roughly the equivalent of 2.8 millibars. From Vol. IV of *Voyage*.[4] (Photograph by courtesy of the Royal Society.)

258 VOYAGE

» remarquer que ces observations délicates doivent être » faites à terre, avec les plus grandes précautions ».

Ayant entendu la lecture de cet article dans une séance particulière de l'académie, j'avais fait construire par le sieur FORTIN, un excellent baromètre propre à apercevoir un 50.e de ligne de variation. M. LAVOISIER m'avait indiqué cet artiste intelligent. On a cru que je me servirais de cet instrument, construit exprès, et c'est pour cela que l'académie a dit dans son instruction, que cette observation devait être faite à terre : mais ayant trouvé à Brest un baromètre marin selon la méthode de M. NAIRNE, décrit dans les Voyages du célèbre COOK, j'ai vu qu'il remplissait toutes les conditions nécessaires pour faire en mer des observations exactes. Quelque grand qu'ait été jusqu'à présent le roulis du vaisseau, le mercure est resté immobile; la bonne suspension du baromètre, et le tube capillaire qui est adapté au tube ordinaire, en sont la cause : avec le nonius qui y est joint, on peut apprécier les variations d'un 10.e de ligne.

En observant tous les jours ce baromètre, au lever du soleil, à midi, et à son coucher, j'ai remarqué que depuis le 11.[e] degré 2′ de latitude Nord, jusqu'à 1 degré 17′, il affectait une marche très-régulière : il était toujours à son maximum d'élévation vers le midi; il descendait ensuite jusqu'au soir, et remontait pendant la nuit. Nous étions à la latitude de 1[d] 17′ le 27 septembre.

Le 28, avant le jour, je commençai les observations pour lesquelles je m'étais préparé la veille, et je les continuai

Figure 10.1 (*cont.*)

DE LA PÉROUSE. 259

d'heure en heure, jusqu'au 1.er octobre à six heures du matin, c'est-à-dire, pendant plus de trois jours et trois nuits. (M. Mongès eut la bonté de les faire pour moi pendant les six heures de mon sommeil.) Je crus devoir observer en même temps le thermomètre en plein air, le thermomètre attaché au baromètre, et l'hygromètre à cheveu. Je fis aussi plusieurs colonnes pour la direction du vent, celle du vaisseau, et pour le chemin que nous parcourions, estimé par le loch. Je profitai de cette occasion pour observer la température de l'eau de la mer à toutes les heures, et l'inclinaison de la boussole.

Les résultats de ces observations m'ont paru très-curieux: le baromètre est monté d'heure en heure, pendant six heures, et il est ensuite descendu pendant six heures, pour remonter pendant les six heures suivantes, et ainsi de suite, comme on peut le voir par la table suivante, extraite de mon journal.

Le 28 septembre..	de 4h à 10h matin....	monté de.....	1l	2/10.
	de 10 à 4 soir.....	descendu.....	1	2/10.
	de 4 à 10 soir.....	monté.......		9/10.
Le 29..........	de 10 à 4 matin....	descendu.....	1	3/10.
	de 4 à 10 matin....	monté.......	1	5/10.
	de 10 à 4 soir.....	descendu.....	1	3/10.
	de 4 à 10 soir.....	monté.......	1	
Le 30..........	de 10 à 4 matin....	descendu.....		7/10.
	de 4 à 10 matin....	monté.......	1	4/10.
	de 10 à 4 soir.....	descendu.....	1	4/10.
	de 4 à 10 soir.....	monté.......	1	
Le 1.er octobre....	de 10 à 4 matin....	descendu.....		8/10.

Kk ij

Figure 10.1 (*cont.*)

own solution for the ocean tide was unrealistic his use of it is scarcely relevant. For an unyielding ocean surface, he deduced an oscillation of pressure on the equator due to gravitational forcing, of amplitude 0.4mm of mercury at the syzygies, equivalent to 0.28mm for the lunar air tide, 0.1mm for the sun alone. Laplace's immediate reaction to the amplitude of about 1.5mm at solar semi-diurnal period observed by Lamanon is not recorded.

The theoretical solar tide amplitude being an order of magnitude less than observed, Laplace deduced correctly that the observed oscillation was thermally driven. For the same reason it was also clear that, in order to test his gravitational theory for the atmosphere, a very long series of air pressures must be spectrally analysed to search for a *lunar* air tide. No such series was available from the tropics, but by the time of writing the last chapter of Book 13 of the *Mécanique Céleste*, he and Alexis Bouvard had analysed 8 years of pressure from Paris Observatory. The result was much lower than the theoretical amplitude, and in fact statistically insignificant.[5] Much later, G.B. Airy obtained a similar result from 20 years of observations at Greenwich, but the lunar tide there was at length extracted in the 20th century by Sydney Chapman, using 64 years of data.[6]

Barometric records from the tropics contain a lunar variation which is more easily extracted, on account of the low weather noise. The existence of a lunar air-tide was first demonstrated in 1842 by Captain Lefroy, Director of a British colonial meteorological observatory at St. Helena, using only a 17-month series at 2h intervals, excluding Sundays. The exercise was extended by Lefroy's successor, Captain Smythe, with an additional 3 years of hourly records. Lefroy's and Smythe's results were published by Edward Sabine in 1847.[7] Their mean amplitude was 0.00183 inches (0.05mm) of mercury, somewhat greater at perigee than at apogee. This is about a fifth of the amplitude predicted by Laplace's theory, and a thirtieth of the observed solar tidal amplitude. Maxima of the lunar air tide at St. Helena coincided with (upper or lower) meridian transits of the moon. Later worldwide assessments gave a global average phase lag of 15° or about half an hour.

Sir William Thomson, as usual, put his finger on the essential problem of the solar air tide in 1882.[8] Since thermal generation has day-night asymmetry, air pressure should exhibit periodicities at a long series of harmonics of a solar day, including the observed second harmonic. But why is the first harmonic so small? Thomson suggested that one should seek a physical mechanism for a strong atmospheric resonance at 12 hours period which inhibits energy at longer periods.

Search for a 12-hour resonance dominated air tide studies for the next 70 years. Though ultimately unsuccessful, the search encouraged atmospheric scientists to refine their physical models of the vertical structure of density, heat absorption and energy dissipation. Horace Lamb adapted Laplace's theory of free waves to accommodate adiabatic motion, more appropriate than the isothermal assumption; Lamb's modified scale depth for a resonance near 12 hours was 8 kilometers, not unrealistic.[9] G.I. Taylor[10] showed that a stratified atmosphere allows a whole series of possible resonant depths, and C.L. Pekeris[11] showed that with a realistic temperature structure there exist two modes of free oscillation, at 10.5h and 12.0h respectively. However, it was difficult to account for a resonance so lightly damped that it magnifies S_2 without apparently affecting M_2.

The resonance hypothesis was finally abandoned in the early 1950's after rocket measurements showed the presence of *diurnally varying* winds at high level, comparable with or stronger than the semidiurnal winds. It was succeeded by one in which the diurnal tide is supposed strong at high level but suppressed at ground level. A dynamic explanation for such a regime appeared in the years 1966–1968. At least four authors discussed the explanation independently. In a lecture delivered in 1985 and finally published in 1996, G.W. Platzman has called attention to a common source of inspiration in the teachings of Professor Bernhard Haurwitz (1905–1986) of the US National Center for Atmospheric Research.[12]

The breakthrough in air tide theory by those who followed B. Haurwitz deserves a short explanatory technical digression. Readers who do not require this should skip to the following section. The modern theory may be described as an extension to Laplace's tidal dynamics, greatly complicated by the vertical structure of the atmosphere's density, compressibility and temperature, and by thermodynamic processes. A standard exposition[13,Ch.3] introduces 110 equations and a practically uncountable number of algebraic symbols in its first 16 pages. In an attempt to make the essentials comprehensible to readers with a little knowledge of fluid dynamics, and at the suggestion of Professor Carl Wunsch, I have added in Appendix D a simplified account of the theory of internal tides in a stratified fluid. Appendix D will also be relevant when we come to consider research on internal tides in the sea in Chapter 13.

In essence, a fluid whose vertical density structure is nonuniform can support waves in its interior. These *internal waves* can propagate both horizontally and vertically, but the law governing their propagation in the vertical is affected by the earth's rotation. If the relevant wave frequency is less than the *inertial frequency* $2\Omega\cos\theta$, vertical oscillations cannot exist, and the energy of the motion is trapped at its level of input. Semidiurnal tidal frequencies are greater than $2\Omega\cos\theta$ at nearly all latitudes except quite near the poles; so semidiurnal air tides can propagate their energy downwards from their thermal source at high level. Diurnal tidal frequencies are greater than $2\Omega\cos\theta$ only at latitudes lower than 30°; they cannot propagate downwards at higher latitudes. Thus, while both diurnal and semidiurnal motions are generated by heating at heights greater than about 50km, and there they have comparable amplitudes, the diurnal variation is trapped at source level in most latitudes and therefore appears much smaller than the semidiurnal tide at ground level.

Figure 10.2b illustrates this property by showing the theoretical variation of amplitude of the diurnal tide with height at a number of latitudes. All curves decrease by two or three orders of magnitude from 100km to zero altitude, but the curves for 15° and 30° latitude have larger amplitudes at the surface, in accordance with the described theory. The corresponding curves for semidiurnal tides (not shown) attenuate much less from top to bottom.

Enough has been said to suggest how observations of simple barometric oscillations, of little practical consequence, stimulated a wide range of dynamical studies of the whole atmosphere. A good account of the work up to about

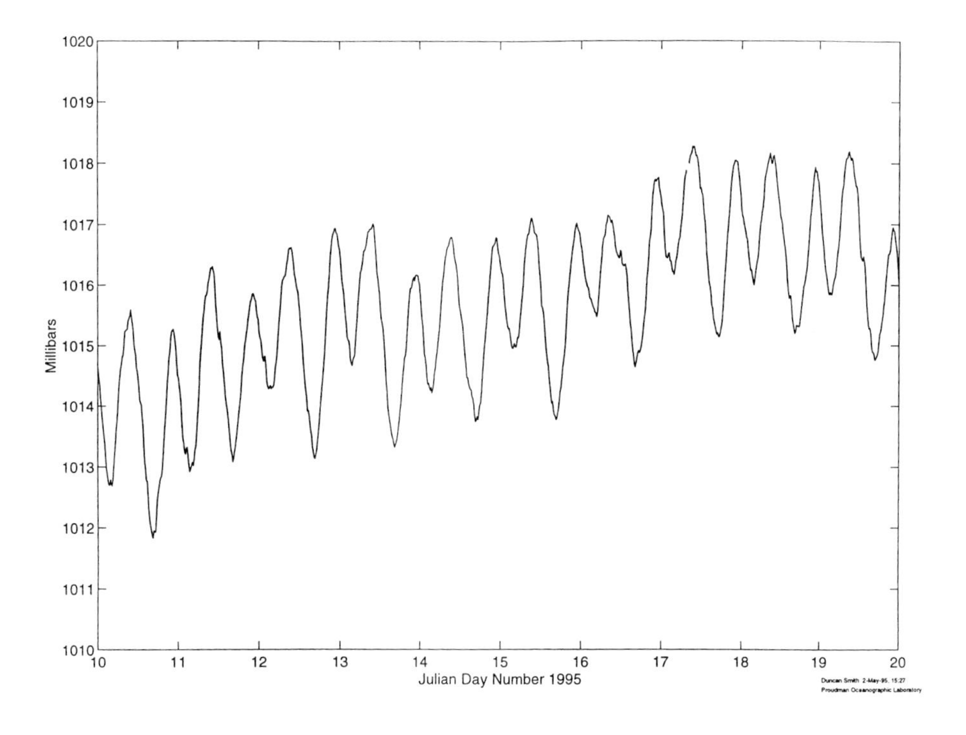

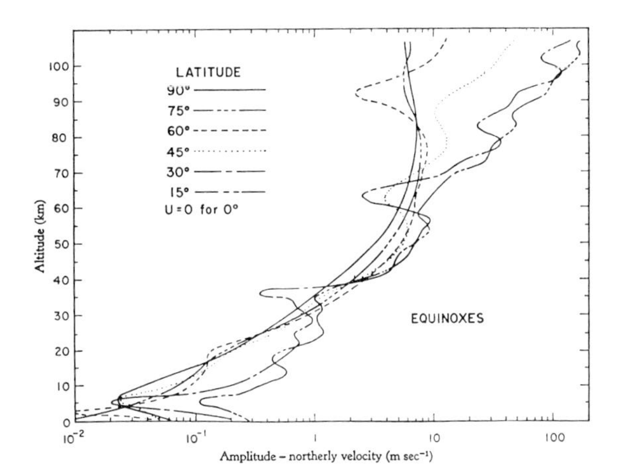

Figure 10.2. (a): A modern digital recording of air pressure near sea level at Jamestown, St. Helena, (16°S, 6°W) on 10–20 January 1995. Universal Time is recorded; local apparent time would be about 23 minutes earlier. The 12-hourly oscillation with mean amplitude 1.1 mb, maxima near 10h and 22h UT, is prominent; a diurnal component with amplitude about 0.6mb and maxima near 02h UT is also clearly discernible. (Record supplied by Robert Spencer, by courtesy of the Proudman Oceanographic Laboratory, Bidston, UK.) (b): Computed amplitude of meridional wind velocity (m/s) of the diurnal air tide at heights 0–100km, for six latitudes from 15° to 90°, North or South, showing the strong attenuation towards ground level. Pressure amplitude would behave similarly. Amplitudes at ground level for latitudes 15° and 30° are greater than for the other latitudes, which are poleward of the 'critical latitudes'. (From ref. 13, by courtesy of Reidel Publishing Co.)

1970 may be found in the now classic book by Chapman and Lindtzen.[13] Since that time, research has focussed on the complication to the theory posed by the presence of steady winds. More recent work has been surveyed by H. Volland.[14]

Magnetic and electrical tidal variations

This subject will be treated more briefly because it entails matters rather remote from our central subject of oceanic tides. Following Gilbert's postulation in 1600 of the earth's magnetic field, attention during the 17th century focussed on the collection of data for the magnetic *variation* (i.e. compass bearing relative to the meridian, known since the 15th century), mainly in the interests of navigation. Secular changes were soon confirmed, and early in the 18th century scientists started to study the temporal changes in the laboratory; they found that there are days of marked high frequency disturbance, interspersed with *quiet days*. During the 19th century, the disturbances were found to be spatially coherent between different observatories and related to auroral activity, later to the appearance of sunspots.

A more or less regular daily variation on 'quiet days' was first noticed by G. Graham in 1722–23.[15] In the following century, daily oscillations were observed in all parameters of the geomagnetic field by the famous German physicists Carl Friedrich Gauss (1777–1855) and Wilhelm Weber (1804–1891).[16] Permanent magnetic observatories were set up in many countries, and their instruments showed that the daily variations were basically caused by fluctuating electrical currents in the conducting layers of the upper atmosphere, now known as the *ionosphere*. An association was also found with electromagnetic oscillations inside the upper layer of the earth's crust, known as the *lithosphere*.[17]

The ionospheric currents were due to the air tides considered in the previous section, and like air tides their spectra contained many harmonics of the solar day. In particular, the first harmonic of 24 hours period was found to be dominant, unlike the tide in surface air pressure. As related above, this is in keeping with the tidal winds of the upper atmosphere. The variations in the lithosphere also led to the use of magnetic tides to learn about the electrical properties of the earth's crust.

Small *lunar* variations in the magnetic field were detected on 'quiet days' by several observers in the early 1850's, including Edward Sabine (1788–1883), Treasurer and later President of the Royal Society.[18] The lunar variations were found to be mainly semidiurnal, but strongly modulated by the solar variation in the ionosphere. They were again largely attributed to the lunar air tide, but induction by oceanic tidal currents also became evident in the 20th century.

Probably the strongest electromagnetic effect of a tidal nature to be directly measured was the difference in electric potential recorded in 1851 by Charlton Wollaston on the first telephone cable across the English Channel.[19] The signal of amplitude about 1 volt was obviously fluctuating in sympathy with the tidal stream in Dover Strait, and was in fact generated by the mass of water moving in

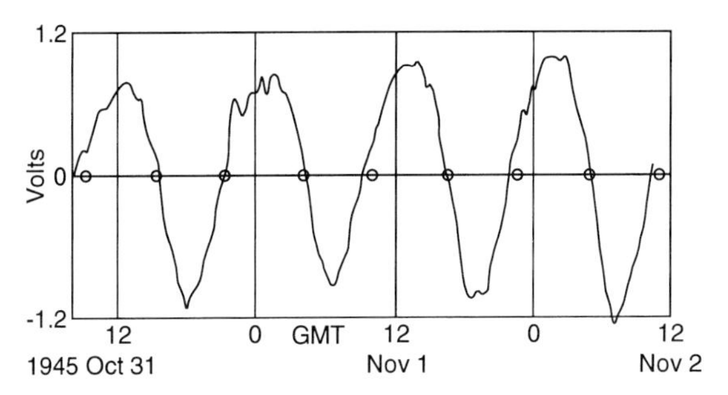

Figure 10.3. Tidal voltage fluctuation recorded on a telephone cable between Cuckmere (Sussex) and Dieppe (Normandy) from 8h, 31 October to 8h, 2 November, 1944. Filled circles are local LWT; half-filled circles, HWT; open circles are approximate times of slack water. (From ref. 20, courtesy of *Nature*.)

the vertical component of the earth's magnetic field. A potential difference of tidal period between England and the Continent had been predicted in 1831 by Michael Faraday as a simple result of his dynamo theory. Faraday had tried unsuccessfully to measure an analogous effect in the River Thames at Waterloo Bridge. When Wollaston showed his records to Faraday, he later (in 1881) recalled:

> Mr. Faraday (I recollect it well as if it were today) said 'Oh, beautiful, beautiful', took down a book from the shelf, and read the following effect – 'Supposing a wire could be suspended from Shakespeare's Cliff [near Dover] to Cape Grisnez [near Calais], the effect of the water running under that wire ought to be apparent', etc.

Tidal fluctuations in potential difference across the English Channel were rediscovered by Post Office engineers D.W. Cherry and A.T. Stovold when restoring telephone cables after World War II.[20] Figure 10.3 shows part of their 1945 recording between Cuckmere (Sussex) and Dieppe (Normandy). During the ensuing decades several uses of such records were made to monitor tide- and wind-driven sea currents, but direct conversion of electrical potential to channel flow was found to be severely limited by the wide horizontal area of integration.[21]

Most recently, electromagnetic signals of tidal nature have been recorded at the bottom of the ocean, using two pairs of electrodes separated by about a meter and a sensitive three-component magnetometer. The recorded signals are partly of ionospheric origin and partly due to direct induction by barotropic tidal flow, (i.e. the depth-mean tidal current). They also afford information about the conducting layers beneath the ocean floor.[22]

Earth tides and rigidity

Unlike the previous subjects of this chapter, interest in the tides of the solid earth arose several decades before instruments were available to measure them directly. This interest stemmed from consideration of the earth's rigidity. Volcanoes and the raised temperature in deep mines had suggested to early 19th century geologists that the earth consisted of a ball of molten rock surrounded by a thin crust. Hopkins, and then Thomson[23] objected to this model because it would imply a pronounced yielding of the earth's surface to tidal forces, which was not apparent. Their principal argument was that the observed precession and nutation of the polar axis against the stars agreed very well with theoretical expectation from a rigid earth.

Precession and nutation are the top-like motions of the spin axis caused by the torque of the sun's and moon's tidal forces on the earth's spheroidal bulge. They were well determined by astronomers and would have quite different values if the earth were effectively fluid. Thomson calculated that, for the surface not to yield unreasonably, the crust would have the stiffness of steel, and therefore be some thousands of kilometers thick. The best way to test this hypothesis, he proposed, was to compare observations of the oceanic tides of long period with their equilibrium values for a perfectly rigid earth. The long-period oceanic tides (Laplace's 'first species') being measured relative to the earth, would be insensible if the earth yielded as a liquid; their actual value looked to be a sensitive measure of earth rigidity.

The principles are best explained in terms of the *Love numbers* h,k introduced by A.E.H. Love in 1911,[24] and now standard. An elastic solid spheroid similar to the earth will yield to an external tide potential U_2 of spherical harmonic degree 2 by a surface tide h_2U_2/g, and the self-attraction of this tide will increase the external potential by k_2U_2. The self-attraction effect had been treated by Laplace (Chapter 7) and others in the 19th century. The magnitudes of the coefficients depend on the rigidity and the distribution of density (assumed spherically symmetric). Similar Love numbers h_n,k_n are defined for higher spherical degree n, but since we shall here consider only the case $n=2$, I shall henceforward omit the suffix.

Using somewhat different notation, Thomson(23) showed that, for the basic case of a uniform sphere of rigidity μ,

$$h=(5/2)(1+\nu)^{-1}, \quad k=(3/5)h, \quad \text{where } \nu=19\mu/2\rho ga,$$

a non-dimensional proxy for rigidity. (ρ,a are the mean density and radius of the earth.) Under static conditions, which we know to hold approximately for the tides of longest period, the tidal displacement of a fluid covering the sphere would be $(1+k)U/g$, and as measured relative to the solid earth, $(1+k-h)U/g$. Thomson took a value of ν equivalent to about 2.05 for the rigidity of steel, for which the above formulae give $1+k-h=0.67$, i.e. two-thirds. He also considered glass, for which $\nu=0.41$, $1+k-h=0.29$. The latter seemed unreasonably small.

Darwin[25]) was the first to examine the amplitudes of the long-period ocean tides from the newly acquired harmonic analyses of the 1870's (Chapter 8). The only long-period constituent of large enough amplitude was the 'fortnightly tide' Mf, although even this required several years of data, especially at mid-latitudes where U/g and hence the tidal amplitude are very small. He also considered the monthly tide Mm. After rejecting several unsuitably sited stations Darwin was left with 33 station-years of data. The best overall estimate of the ratio tide:potential was 0.676 ± 0.076 with a small phase lag indistinguishable from zero, but probably positive. Equating the above ratio to $(1 + k - h)$ clearly substantiated Thomson's rather arbitrary choice of steel as a standard of rigidity.

Others repeated Darwin's exercise later when the fortnightly and monthly harmonics became available from a larger number of stations, with somewhat similar results. Around 1930, S.F. Grace, a student of Joseph Proudman of Liverpool, attempted to evaluate the rigidity factor from the *semidiurnal* tides in the Red Sea, where it was thought that they could be accurately calculated. Grace's result was absurdly low, but he obtained a more realistic result, 0.64, from the very small tides which had been measured by Russian scientists along the shores of Lake Baikal, Siberia.[26] The accepted modern estimates for the Love numbers at semidiurnal, lunar diurnal and long periods, from later knowledge of the density structure of the earth, are $h = 0.609$, $k = 0.302$, giving $1 + k - h = 0.693$.[27] There is a small anomaly near the frequency of the harmonic K_1, due to resonance in the earth's core, to be mentioned again in Chapter 13.

Pendulum measurements of tidal attraction

A more direct experimental approach, far more demanding instrumentally, was to attempt to measure the horizontal component of the tide-raising force itself. Since the force per unit mass is of order 10^{-7}g, changes in the direction of the vertical are required to be measured to an accuracy of 10^{-3} arcseconds. In the 19th century one could only hope to achieve such a measurement by observing the deflection of a damped pendulum of very long period (10 seconds at least), rigidly supported in an extremely quiet environment, with an optical/mechanical device for magnifying minute deflections by a factor of several thousand. Some pioneering measurements of this sort were made by George Darwin with his brother Horace in 1879–80.[2,Ch. VI] After much trial and error they found that the desired tidal signal was completely swamped by seismic noise of various kinds, movement of the supports with changing temperature, and plastic creep of materials.

Successful results were later obtained by other investigators using a *horizontal pendulum*, essentially a long beam, free to swing in a plane normal to a fixed axis which is very slightly inclined to the vertical. Installation in deep boreholes was necessary in order to minimise temperature strain of the local rock.

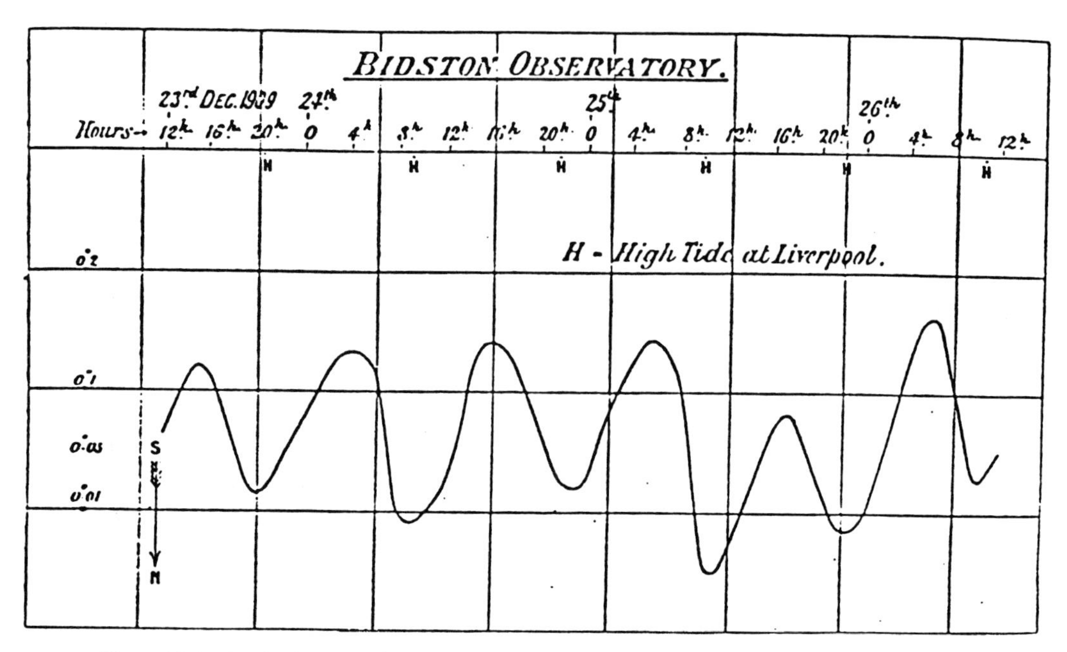

Figure 10.4. North (downward) – South (upward) tilt of the vertical between 23 and 26 December 1909, recorded by J. Milne in a vault below Bidston Observatory near Liverpool Bay. Times of HWT in the local sea are shown by letters H, where the tilt is obviously northward, that is, toward the sea. The vertical scale is graduated in 0.1 arc seconds. (From *Nature*, 82, 427, 1910.)

Notable measurements with a horizontal pendulum were made by Otto Hecker between 1902 and 1909. Darwin comments on them in his 3rd Edition supplements to Chapters VI and VII.[2] The ratio of deflection to the primary tide potential again depends on the factor $(1+k-h)$, for which Hecker obtained estimates in the region of 0.4–0.7. However, further instrumental refinements were still required.

An interesting record from an early horizontal pendulum (1909) by J. Milne is shown in Figure 10.4. It was installed to record N–S deflections of the vertical in a vault at Bidston Observatory, near Liverpool. The obvious tide-related oscillations with amplitude around 0.5″ are considerably greater than those appropriate to the tide-raising force. They are in fact due to the gravitational attraction of the very large sea tide in Liverpool Bay, about two miles to the northwest. After this observation, loading by the oceanic tides was recognised to be a serious contaminant to tidal tiltmeters, even at quite large distances inland, although the loading signal was also of interest *per se*.

Later in the 20th century, instruments were devised to record the minute changes in *gravity* due to the tidal yielding of the earth's crust. The variation is more easily interpretable than tilt of the vertical, and it provides a measure of a different combination of the Love numbers, namely $(1+\mathrm{h}-3k/2)$. Researches with tidal gravimeters will be described in Chapter 13.

Polar motion and the pole tide

Astronomers of the late 19th century were aware of small systematic changes in the apparent latitude of their observatories with respect to the celestial sphere, even when the tidally forced precession and nutation had been allowed for. This *variation of latitude* as it was usually called, of order 0″.1, was attributed to a free *wobble* of the earth's figure about the spatially fixed axis of rotation. It was known from the work of Euler that, if the earth were rigid, the resonant period of such a wobble should be $A/(C-A)$ times the diurnal spin rate, where A and C are the two principal moments of inertia of the earth. This gives a period of oscillation of 305 days. Some astronomers sought evidence for a 305-day periodicity in their records, but without any significant result.

In 1891–2, the American actuary, geodesist and astronomer, Seth Carlo Chandler (1846–1913), made history when he announced that he had analysed the full spectrum of the variation without bias for any pre-conceived period, and found prominent peaks at 365 and 427 days.[28] Chandler discovered this result by the use of a refined theodolite of his own design for determining zenith angles, still known as the *almucantar*. (The almucantar was later superseded by the more refined *Photo-Zenith Tube*, which photographed the images of stars in a free mercury surface.)

The 365-day peak was recognised to be a result of seasonal changes in the geosphere, and the distinguished astronomer Simon Newcomb pointed out with hindsight that the 14-month period was indeed the resonant period, increased from 305 to 427 days by the earth's elasticity.[29] To quote a later formula by Love:[24]

$$\left[\frac{2e.GM}{n^2a^3} - 1\right]\left[1 - \frac{\tau_0}{\tau}\right] = k,$$

where e is the eccentricity of the earth's figure, GM is the product of the universal constant of gravitation and the earth's mass, and τ and τ_0 are the actual (427 d) and 'rigid' (305 d) resonant periods, respectively. So the Chandler period provided another measure of the Love number k, in fact 0.29, a value which accords well with other theoretical estimates.

The importance of these results was recognised in 1899 by the International Geodetic Association in the setting up of six observatories (later reduced to five), all in latitude 39°8′ N and well spaced in longitude, for the express purpose of monitoring the variation of latitude by the same zenith stars. From these and later more sophisticated measurements, the monthly (later daily) positions of the polar axis in geographical coordinates have been recorded ever since as an international service to astronomers and earth scientists.

The polar wobble causes a tide in the ocean known as the *pole tide*, which is due to the shift of coordinates relative to the centrifugal force field. An angular displacement of the pole by m_1 towards Greenwich and m_2 towards 90°E gives rise to a change of primary potential:

$$U_p = -\Omega^2 a^2 \sin\theta \cos\theta \,(m_1 \cos\phi + m_2 \sin\phi),$$

where in our usual notation, (θ, ϕ) are (colatitude, east longitude), Ω is the sidereal frequency of rotation, and a the equatorial radius. (The data are conventionally listed as $x, y = m_1, -m_2$.) U_p has the same spatial form as the diurnal gravitational potential, but at a period of 427 days instead of 1 day. The corresponding equilibrium pole tide is

$$Z_p = (1 + k - h)U_p/g.$$

Its amplitude is of order 5 millimeters. Such a tide had been predicted by Thomson as early as 1876.[30] It was first detected in the sea in 1900 by A.S. Christie.[31] In 1913 H.G. Bakhuyzen determined the pole tide by analysing some 200 years of sea level data from Amsterdam.[32]

The pole tide is too small for practical interest in forecasting sea level, but it has two points of geophysical interest, neither of which is yet fully understood: (1) Z_p is found to have anomalously large amplitude (10–20mm) along the coasts of Holland, Denmark and the western Baltic Sea, but virtually nowhere else; why? (2) The polar motion appears to be damped with a decay time of about 10 periods; is this due to friction in the oceanic pole tide? Both questions have seen many years of neglect punctuated by occasional revival of interest and analysis of new data. Regarding Question (1), recent work (see Chapter 12) has shown a correlation with the pattern of the weather over the North Sea, but why the local atmosphere should behave in such a way is obscure.

Lunar acceleration, earth retardation and tidal friction

These three phenomena are intimately related, but the relationship and its far-reaching implications did not become apparent until Delaunay's paper of 1865[42] and the subsequent analysis by Darwin.[45] The growth of understanding in this field has involved an interesting interplay of observation, hypothesis and conflicting theory – a history in its own right. I shall outline only the developments which have a direct bearing on the tides. The heated debates in mid 19th century between English and French mathematicians over the term due to planetary perturbations are summarised in greater depth in a historical review by Kushner.[33]

Edmond Halley was the first to notice that the times and places of ancient solar eclipses recorded by an Arab astronomer named al-Battani were systematically offset from modern measurements of the moon's longitude by an order of 0.5 degrees or an hour or two of time. At the end of a discursive paper of 1695 on the ruins and antiquities of Palmyra,[34] Halley made a plea to travelers for careful observations of:

> the Phases of the Moon's Eclipses at Bagdat, Aleppo and Alexandria, thereby to determine their Longitudes ... [from which] ... I could then pronounce in

what Proportion the Moon's Motion does *accelerate*, which that it does, I think I can demonstrate and shall (God willing) one day, make it appear to the Publick.

In fact, Halley never published any results of his findings in this area, but his idea was followed up by one Richard Dunthorne (1711–1775) who, in a letter communicated to the Royal Society in 1749, described calculations based on six solar eclipses recorded by Junis, Theon and Ptolemy. From these Dunthorne deduced an apparent acceleration of the moon's longitude at the rate of 10 arcsec century^{-2}.[35] Similar figures were later obtained by Tobias Mayer and by J-J de Lalande.[33]

It is worth anticipating later results here, by pointing out that the apparent lunar acceleration depended on an implicit assumption that the earth's rotation is a reliable basis for time-keeping. Not until towards the end of the 19th century did it begin to seem plausible that the length of the day (l.o.d.) is slowly but steadily increasing in real time. The retardation of the earth more than accounts for the apparent acceleration of the moon. The final outcome was to be that the moon's longitude is really *decelerating* by some $26''c^{-2}$, a quite different result from what was at first supposed.

First to suggest that the l.o.d. may be increasing was the philosopher Immanuel Kant (1724–1804). Kant is of course better known as a metaphysician, but his *Sämmtliche Werke* includes a short article written in 1754, (originally printed in a newspaper, according to Brosche[36]). There, Kant speculated on the slight but continual braking effect of friction in the oceanic tides. The article passed unnoticed at its time of writing, but a certain A.D. Wackerbarth called the attention of the Royal Astronomical Society of London to it in 1867,[37] adding that Kant's idea had anticipated the then celebrated paper by Delaunay,[41] which put the whole matter of lunar acceleration, earth rotation and tidal friction on a proper physical if tentative footing. (The subject had also been been discussed independently in 1848 by Julius R. Mayer.[36,38])

To return to the 18th century, unaware of the tidal issue, astronomers had been puzzling to identify a rational cause for the observed acceleration of $10''c^{-2}$. With his innate belief that every natural phenomenon could be explained in terms of Newtonian mechanics, Laplace discovered a mechanism which seemed to account for the acceleration with amazing precision.[39] In brief, the attractions of the major planets cause the eccentricity of the earth's orbit about the sun to decrease; this in turn reduces the tendency of the sun to lengthen the moon's period. The resulting acceleration of the moon, proportional to m^2, where $m=0.0748$ is the ratio of the mean motions of sun and moon, was, according to Laplace, $10''.6c^{-2}$. The agreement between theory and observation was understandably hailed as a triumph of rational analysis.

In 1820, the Académie Royale des Sciences offered a *Grand Prix de Mathématique* for further advances in lunar theory. The prize was shared between M.C.T. Damoiseau and the Italian astronomer Giovanni Plana,[33] both of whom carried Laplace's perturbation to series of higher powers in m^2. Their results agreed with Laplace's to better than 0″.5.

The Laplace–Plana result, doubly-checked by Damoiseau, seemed perfect until 1853 when John Couch Adams of Cambridge (last encountered in Chapter 8) detected a subtle flaw in Plana's reasoning. The term in m^2, first derived by Laplace, was correct, but the ensuing terms in Plana's power series were seriously in error, owing to a logical fault.[40] The final outcome of Adams' analysis was that the correct numerical value for the acceleration from planetary perturbation was only 5″.7c^{-2}, significantly less than observed.

Adams' criticism caused a furore among French lunar theorists which raged from 1859 to 1866. Adams was no stranger to astronomical rivalry between England and France, having been in 1846 with Le Verrier at the focus of the notorious arguments over priority in the discovery of the planet Neptune. In the present contest, he had a valuable ally in Charles-Eugène Delaunay (1816–1872) of the Observatoire de Paris, who confirmed Adams' result by a different method,[41] but Le Verrier, De Pontécoulant and P.A. Hansen argued in support of Plana and Damoiseau, using the observed figure as a persuasive influence. Hansen eventually came to agree with Adams. Delaunay accused De Pontécoulant of using unsound arguments, rightly as it turned out.[33]

At length the scientific community came to see that the arguments of Adams and Delaunay were unassailable, and the question reverted to how to explain the discrepancy (now only 4 to 5″c^{-2}) between the apparent lunar acceleration and the corrected planetary perturbation. Although anticipated by Kant, J.R. Mayer, and by W.E. Ferrel, (who presented a paper in 1864 which unfortunately was published only in 1866[33]), Delaunay's paper of 1865 was the first widely recognised study of the role of oceanic tidal friction as the major factor in accounting for the whole phenomenon.[42]

The most obvious large-scale effect of tidal friction is to brake the earth's rotation. Friction generally retards the tide, so the bulge of the tide's P_2^2 spherical harmonic appears eastwards of the line of centers of earth and moon. By a well-known diagram, a modern version of which appears in Figure 14.3, the tidal forces on the retarded bulges apply a roughly constant torque of second order of smallness, opposing the rotation. (The torque is supposed transferred to the solid earth through horizontal pressure gradients and currents – the exact physical mechanism is still a little obscure.) The significance of Delaunay's preliminary analysis of this effect lay in his assertion that a negative change in the earth's rotation rate Ω will be interpreted as a positive change

$$-\frac{\Omega}{\omega}\frac{d\Omega}{dt}$$

in the moon's rotation rate ω, if Ω is used as a time standard, as it invariably was then. A 5″c^{-2} increase in ω could thereby be accounted for by a 135″c^{-2} decrease in Ω, that is, by a rate of increase of 0.25 milliseconds per century in the length of day.

In reality, a considerably larger retardation of the earth was necessary, because the moon's longitude is also retarded through conservation of angular momentum and Kepler's third law, but full understanding of the coupled system had to await Darwin's exhaustive treatment. (A simplified account of Darwin's analysis is given in the last section of this Chapter.) As far as Delaunay took the arguments, two weaknesses delayed general acceptance of the tidal friction hypothesis. One weakness was that the theory implied an apparent acceleration of the sun and other celestial objects as well as the moon. This possibility had not been investigated, but acceleration of the sun and of transits of Mercury across the sun's disc relative to the earth's clock was at length demonstrated by P.H. Cowell in a series of papers presented in 1905–1906.[43] (In Cowell's article of November 1905, he ascribed his observed acceleration of the sun to the effect of 'resistance of the ether' on the earth's orbit, but by April 1906 he had become aware of Darwin's work on tidal braking, and correctly deduced a larger increase in l.o.d. than had been thought possible, accompanied by a negative acceleration of the moon.)

It should be added that the earth's rotation is now known to vary from other geophysical causes besides tidal braking.[44] These causes are internal to the geosphere and do not affect the moon's acceleration. The tidal effect is important chiefly as a 'correction' to the observed changes in l.o.d. so that other causes may be clearly quantified.

The other weakness in Delaunay's hypothesis was the inadequacy of knowledge of the oceanic tide itself to quantify the retarding torque. Delaunay's guess at the tidal displacement and phase lag required seemed not too unrealistic, but only in 1921 did Jeffreys and Heiskanen first demonstrate that frictional dissipation in shallow seas was sufficient to account for the observed retardations (see Chapter 11). At any rate, here was the first intimation that the problem of the world cotidal map which engrossed William Whewell and Rollin Harris (Chapter 9) and many later scientists, had implications outside the realm of oceanography.

Darwin's theory of evolution of the lunar orbit

George Darwin was a firm believer in tidal friction as the principal cause of evolution of the orbital mechanics of the earth-moon system on a geological time scale, although, as now seems curious, he considered the energy sink to lie not in the ocean but in the viscous resistance of the solid earth to tidal stresses. (The effect, if true, would be qualitatively the same.) Interior dissipation was suggested by the evident source of heat below the crust and by the (correct) hypothesis that tides in the body of the moon had long ago caused its rotational and orbital periods to equalise. It was only towards the end of his life – he died in 1912 – that radioactivity was discovered to be the source of the earth's interior heat.

1879–80 was a very productive period in Darwin's scientific career, with the

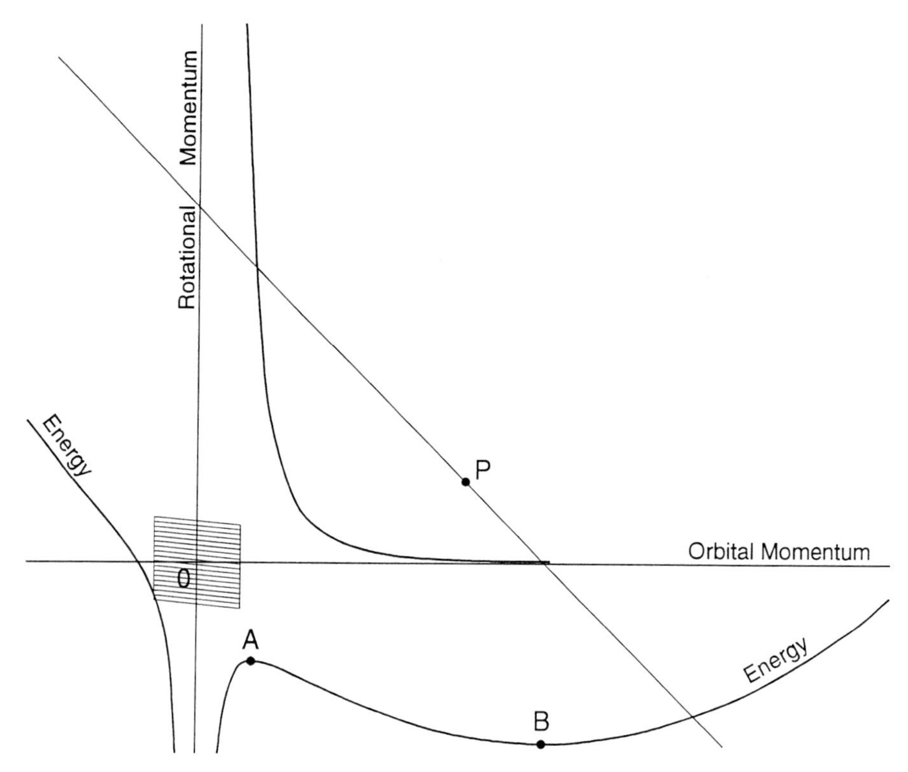

Figure 10.5. From Figure 1 of Darwin (1879).[45] Horizontal axis (x) is scaled orbital angular momentum of a satellite (moon), increasing with the decrease in y – scaled rotational momentum of primary body (earth), due to tidal friction. Since $x+y$ is constant, the downward sloping line is the locus of (x,y) as energy is lost, with P the position at the present epoch. The curve labelled 'Energy', displaced downward for convenience, represents the scaled energy $2E$ corresponding to points vertically above it on the momentum line. A and B are turning points of E, and are where the momentum line intersects the third curve shown, $x^3y=1$. The shaded portion of x, shown only near the origin O, corresponds to the condition of coalescence of satellite and primary. The moon is supposed to have separated from the earth at A and to have been continually progressing towards the point of minimum energy at B, where the day and the month will be equal. (For further details, see text and ref. 45.)

publication of five major papers, among other subjects, on the theory of tidally distorted viscous spheroids and the resulting evolution of their orbits.[45] His work is hard to summarise in a short space in its full generality of eccentric orbits inclined to the ecliptic, but Darwin himself provided an admirable summary of his main thesis in a graphical illustration.[45, paper 5] I will use his diagram to illustrate the essentials of his theory in simplified form.

For a satellite in a circular equatorial orbit of angular velocity ω and distance r from the center of a primary planet, Kepler's third law and the conservation of angular momentum, (both provable from Newton's principles), give respectively:

$$\omega^2 r^3 = G\,(M+m) = K,$$

$$Mm(M+m)^{-1}\,\omega\, r^2 + C\,\Omega = H,$$

where m, M are the masses of satellite and primary and C is the axial moment of inertia of the primary. K and H denote constants easily ascertainable from observation. As a consequence of the first of the above equations,

$$\omega\, r^2 = (K/\omega)^{\frac{1}{3}},$$

so with a suitable choice of units of mass and length, one may write

$$x + y = h,$$

where $x = \omega^{-\frac{1}{3}}$ (also proportional to $r^{\frac{1}{2}}$) and $y = \Omega$ represent the angular momentum of satellite and primary, respectively.

Darwin showed that the total energy, E, of the system is given in the same system of units by

$$Y = 2E = y^2 - 1/x^2 = (H - x)^2 - 1/x^2.$$

Figure 10.5, re-drawn from Figure 1 of his cited paper of 1879,[45] plots the scaled rotational momentum of the planet, y, and twice the total energy of the system, Y, against the scaled orbital momentum x. The curve representing Y is displaced downwards arbitrarily, to avoid confusion with the y-curve. For the earth-moon system, x and y have the same sign (i.e. both have positive rotations), and $h = 4$ approximately. $x < 0$ is possible in other planetary systems. The point P represents the present state of earth and moon.

Since there is frictional dissipation somewhere in the system, Y must decrease with time, and one sees that there are three distinct parts of the energy curve where this is possible, two with decreasing x and one (AB) with x increasing. Each value x has a different ratio $\omega{:}\Omega$, that is, a different ratio of day:month.

Two turning points of energy ($\partial Y/\partial x = 0$) are denoted by the letters A and B. A is a point of instability in that a small perturbation about A may cause x to decrease or to increase, while B is stable. At A and B the day and the month are equal and the satellite rotates with its primary as if the two were rigidly joined. The corresponding points on the momentum line (vertically above A and B) occur where it intersects the curve $x^3 y = 1$, also shown. Darwin named this 'the curve of rigidity' for the reason just given. At the intersection above B, the rotational momentum y is very small but positive. At the intersection above A, the orbital momentum x is so small that the separation distance r between centers (proportional to x^2) is comparable to the sum of the radii (shaded zone); the two bodies then virtually coalesce.

Some time in the remote past, according to Darwin, the moon separated from the earth by centrifugal action. Their positions in Figure 10.5 then corresponded to A. An uncertain physical factor triggered the instability to frictional fall towards B, rather than to a loss of orbital momentum by falling back into the primary. Ever since, the orbital momentum has increased along the line towards P, the moon receding from the earth while the earth's rotational momentum has decreased correspondingly. Both the 'day' and the 'month' have thereby steadily lengthened.

Darwin pointed out that the greatest possible number of days in the month along the momentum line is given by

$$\Omega/\omega = 27(k/4)^4,$$

which in earth-moon units is about 27. We are therefore very close to that maximum now. Days per month have increased from 1 at A to about 27 now, and will henceforward decrease slowly towards 1 at B.

The big question in all this, in fact Darwin's basic source of motivation for his research, was: how long ago was the moon at A? The event at A must have been quite catastrophic, implying a molten earth and some violent rotational instability, perhaps triggered by a celestial impact. Once separated, with rotation speeds beginning to differ, both earth and moon would have experienced enormous tidal stresses with generation of heat by friction. Figure 10.5 suggests deceptively that the time scale from A to P may be calculable, but the dissipation rate in the early millennia is hard to quantify. At the present-day dissipation rate, (now known fairly accurately), AP would have taken some 10^9 years. Darwin estimated the moon's age to be at least 54 million years. He made similar calculations for the observed satellites of other planets. Whatever the perceived outcome in Darwin's day, the subject raised many new questions which are still debated, (see Chapter 15), while setting new constraints to the geological time scale. In this work he was no doubt conscious of the spirit of his father's biological studies; the works of both effectively extended the possible time scale for the origin of life.

Notes and references

1. Kushner, D. Sir George Darwin and a British School of Geophysics, *OSIRIS*, 8, 196–224, 1993.
2. Darwin, G.H. *The Tides and Kindred Phenomena in the Solar System*. 3rd Edn. with Supplements, Murray, London, 437pp., 1911.
3. Humboldt, A. von, and A. Bonpland. Observations faites pour constater la Marche des variations horaires du Baromètre ... In: *Relation du Voyage aux régions équinoxiales du Nouveau Continent fait en 1799–1804*. First Edition: (Paris, 1814–25), Tome III, Ch. XXVI, 270–313; 'New' Edition: (Paris, 1816–1831), Livre IX, Ch. XXVI, 330–478. (This article with tabulated observations is omitted from standard English translations of the *Voyage*. It contains a full history of early observations of the air tides at various heights, including those of the authors. While giving pride of place to Lamanon's observations, it cites earlier observers, some anonymous, who remarked on the regularity of the times of maxima and minima.)
4. Milet-Mureau, A. (ed.) *Voyage de La Pérouse autour du monde*. 5 vols., Paris, 1797. See also: Cartwright, D.E. Robert Paul de Lamanon, an unlucky Naturalist, *Annals of Science*, **54**, (6), 585–596, 1997.
5. Laplace, P.S. Chapter 35 of 1776 paper – see Ch. 7, ref. 1; also *Mécanique Céleste* – see Ch. 7, ref. 3, Livre I, Ch. 8, Section 37 and Livre IV, Ch. 5. (The analysis of barometric records from Paris is described in Livre XIII, Ch. 7.)

6. Airy, G.B. Greenwich Meteorological Reductions, 1854–1873, *Barometer*, London, 1877. Also: Chapman, S.G. The lunar atmospheric tide at Greenwich. *Quart. J. R. Meteorol. Soc.*, **44**, 271–280, 1918.
7. Sabine, E. On the lunar atmospheric tide at St. Helena. *Phil. Trans. R. Soc. London*, **137**, 47–50, 1847.
8. Thomson, W. On the thermodynamic acceleration of the earth's rotation. *Proc. R. Soc. Edinburgh*, **11**, 396–405, 1882.
9. Lamb, H. On atmospheric oscillations. *Proc. R. Soc. London*, **A**, **84**, 551–572, 1910.
10. Taylor, G.I. The oscillations of the atmosphere. *Proc. R. Soc. London*, **A**, **156**, 318–326, 1936.
11. Pekeris, C.L. Atmospheric oscillations. *Proc. R. Soc. London*, **A**, **158**, 650–671, 1937.
12. Platzman, G.W. The S-1 Chronicle: A tribute to Bernhard Haurwitz. *Bull. Amer. Meteorol. Soc.*, **77**, 1569–1577, 1996. (A discussion of the chronology of the discovery, and sometimes later publication, of the concept of *negative equivalent depth* in the atmosphere, with references to the writings of B. Haurwitz (1965), R.S. Lindzen (1966), S. Kato (1966), T.W. Flattery (1967), and M.S. Longuet-Higgins (1968) – publication dates.)
13. Chapman, S. and R. S. Lindzen. *Atmospheric Tides*, D. Reidel Publ., Dordrecht, Holland, 200pp., 1970.
14. Volland, H. *Atmospheric Tidal and Planetary Waves*. Kluwer Academic Pub., Dordrecht, Holland, 348pp., 1988.
15. Graham, G. An account of observations made of the variation of the horizontal needle at London, in the latter part of the year 1722 and beginning of 1723. *Phil. Trans. R. Soc. London*, **32**, 96–107, 1724.
16. Gauss, C.F. and W. Weber. Resultate aus den Beobachtungen des magnetische Vereins in Jahre 1836–41. 6 Hefte, Gottingen and Leipzig, 1841.
17. Schuster, A. (with Appendix by H. Lamb). The diurnal variation of terrestrial magnetism. *Phil. Trans. R. Soc. London*, **180**, 467–518, 1889.
18. Sabine, E. On the influence of the moon on the magnetic declination at Toronto, St. Helena, and Hobarton. *Phil. Trans. R. Soc. London*, **143**, 549–560, 1853.
19. Wollaston, C. Discussion on a paper by A.J.S. Adams in *J. Soc. Telegraphic Engineers and Electricians*, **10**, 50–51, 1881.
20. Cherry, D.W. and A.T. Stovold. Earth currents in short submarine cables. *Nature*, **157**, 766, June 1946.
21. Robinson, I.S. A theoretical analysis of the use of submarine cables as electromagnetic flowmeters. *Phil. Trans. R. Soc. London*, **A**, **280**, 355–396, 1976.
22. Filloux, J.H. Instrumentation and experimental methods for oceanic studies. In: *Geomagnetism*, Ed. J.A. Jacobs, **1**, 143–248, Academic Press, 1987.
23. Thomson, W. On the rigidity of the earth. *Phil. Trans. R. Soc. London*, **153**, 573–582, 1863. (The opening page refers to related papers by Hopkins in 1839–42.)
24. Love, A.E.H. *Some problems in geodynamics*. (Adams Prize Essay, 1911), Cambridge Univ. Press, 180pp., 1911.
25. Darwin, G.H. *Scientific Papers*, 5 vols., Cambridge Univ. Press, 1907–1916. **25a.** Attempted evaluation of the rigidity of the earth from the tides of long period (1883), *Sci. Papers*, **1**, 340–371, 1907.
26. Grace, S.F. The semidiurnal lunar tidal motion of the Red Sea. *Monthly Not. R. astr. Soc., Geophys. Suppl.*, **2**, 273–296, 1930. Also: The semidiurnal lunar tidal motion of Lake Baikal and the derivation of the earth-tides from the water-tides. *Ibid.* **2**, 301–309, 1931.
27. Wahr, J.M. Body tides on an elliptical, rotating, elastic and oceanless earth. *Geophys. J. R. astr. Soc.*, **64**, 677–703, 1981.
28. Chandler, S.C. On the variation of latitude. *Astron. J.*, **277**, 97–101, 1892, and subsequent papers in same Journal. (For a general review of Chandler's work on polar motion, including a picture of the *Almucantar*, see: Carter, W.E., *EOS*, **68**, (25), 593, 603–605, Amer. Geophys. Union, 1987.)
29. Newcomb, S. On the dynamics of the earth's rotation, with respect to the periodic variations of latitude. *Monthly Not. R. astr. Soc.*, **52**, 336–341, 1892.

30. Thomson, W. Presidential Address to the BAAS, 1876 Meeting.
31. Christie, A.S. The latitude variation tide. *Bull. Phil. Soc. Washington*, **13**, 103–122, 1900.
32. Bakhuyzen, H.G.S. Über die Änderung der Meereshöhe und ihre Beziehung zur Polhöhenschwankung. *Vierteljahrschr. Astron. Ges.*, Leipzig, **47**, 218–221, 1913.
33. Kushner, D. The controversy surrounding the secular acceleration of the moon's mean motion. *Arch. Hist. Exact Sciences*, **39**, (4), 291–316, Springer Verlag, 1989.
34. Halley, E. Some account of the ancient State of the city of Palmyra ... *Phil. Trans. R. Soc. London*, **19**, 160–175, 1695.
35. Dunthorne, R. A letter from ... concerning the acceleration of the moon. *Phil. Trans. R. Soc. London*, **46**, 162–172, 1749. (Dunthorne was an oddity among important astronomers in being officially employed as 'Butler' to Pembroke College, Cambridge, while acting as assistant to his friend and mentor, the astronomer Roger Long, D.D.)
36. Brosche, P. Understanding tidal friction: the History of Science *in nuce*. Pp. 442–445 in: Ocean Sciences; their history and relation to Man. (Proc. 4th Int. Congr. History of Oceanography), *Deutsche Hydrogr. Zeitschr.*, Ergängzungsheft, B, **22**, 1990.
37. Wackerbath, A.D. On an astronomical presentiment of Immanuel Kant relative to the constancy of the earth's sidereal period of rotation on its axis. *Monthly Not. R. astr. Soc.*, **27**, 199–200, 1867.
38. Debus, H. 'Discursive paper on Celestial Dynamics' by J.R.Mayer (1848), transl. from German. *Phil. Mag.*, **25**, 387–428, 1863. (The only passages of present interest are in a section entitled: 'The tidal wave', pp. 403–409)
39. Laplace, P.S. Sur l'équation séculaire de la lune. *Mém. Acad. Roy. Sci.* (1786), 235–246, 1788.
40. Adams, J.C. On the secular variation in the moon's mean motion. *Phil. Trans. R. Soc. London,* **143**, 397–406, 1853.
41. Delaunay, C.-E. Sur l'accélération séculaire du moyen mouvement de la lune. *C. R. Séances de l'Acad. Sci.*, **48**, 137–138 and 817–827, 1859.
42. Delaunay, C.-E. Sur l'existence d'une cause nouvelle ayant une influence sensible sur la valeur de l'équation séculaire de la lune. *C. R. Séances de l'Acad. Sci.*, **61**, 1023–1032, 1865. (An English summary of this paper may be found in: Anon. *Monthly Not. R. astr. Soc.*, **26**, 85–91, 1866.)
43. Cowell, P.H. On the secular acceleration of the earth's orbital motion; (and a series of related papers), in *Monthly Not. R. astr. Soc.*, **66**, 3–6, 34–41, 352–355, 1905–1906.
44. Munk, W.H. and G.J.F. MacDonald. *The Rotation of the Earth – a Geophysical Discussion*. Cambridge Univ. Press, 323pp., 1960.
45. Darwin, G.H. *Scientific Papers*, Vol. 2 – Tidal Friction and Cosmogony, (9 papers), Cambridge Univ. Press, 1908. (Paper no. 5, from which Figure 10.5 was taken, originally appeared in *Proc. R. Soc. London*, **29**, 168–181, 1879.)

Contributors to the science of tides in the first half of the 20th century. (From left to right), *top:* Joseph Proudman (1888–1975); Arthur Doodson (1890–1968); *bottom:* Harald Sverdrup (1888–1957); Albert Defant (1884–1974). (Reproductions by courtesy of the following institutions: Proudman, Doodson – Proudman Oceanographic Laboratory, Bidston; Sverdrup – Scripps Institution of Oceanography, La Jolla, (UCSD/SIO Archives); Defant – Institut für Meereskunde an der Universität, Kiel.)

11

Tidal researches between World Wars I and II

I have brought most of the subjects introduced in the last four chapters to a temporary close in the first quarter of the 20th century. Having now left the 19th century behind us, except for an occasional retrospect, it is well to reflect on the changing pace and character of scientific research in general which has become increasingly evident since 1900.

The growth of research in natural sciences

The Industrial Revolution had shown that study of the natural sciences has useful as well as cultural benefits. Associations for the Advancement of Science in Britain and the USA had engendered popular appeal through their annual meetings in the provinces, whereas older scientific establishments such as the Royal and Imperial Societies had by the turn of the century become rather esoteric. Several new scientific societies had been created to foster specific branches such as physics, astronomy, geography, engineering and various life sciences. As a result, more students at universities were opting for courses in science or engineering rather than in the traditional disciplines of classics and divinity. The abler ones stayed on to do research or obtained employment in the research laboratories of industrial enterprises, national institutions and private foundations.

Thus, many younger people were producing serious research papers, hitherto the privilege of a few distinguished professors. At the same time, the professors and other research leaders had able subordinates to work out their ideas for them, and they could influence the direction of group research. Eventually, as all present day scientists are aware, the production of research papers in refereed journals became increasingly competitive and a condition for tenure of employment.

The emergence of physical oceanography

Scientific study of the sea had been stimulated more by the great voyages of exploration than by mathematical theories about tides. Although the earlier voyages from Cook onwards had astronomical and/or colonial objectives, their leaders were also commissioned to record the magnetic field and the weather, and the temperature, depth, currents, birds and fishes of the sea. The new science of *oceanography* grew from their reported observations. The famous voyage of HMS *Challenger* (1872–76)[1] was the first to have purely oceanographic aims, but its emphasis was strongly on the study of marine biology. Nevertheless, the expedition brought large valuable collections of physical facts about winds, currents, depths, sea temperature and salt content. During the early 20th century several European countries sponsored expeditions to observe and study the ocean and its environment.

Through pioneer sea-going physicists such as the American Matthew Fontaine Maury (1806–1873), the German Otto Krümmel (1854–1912), the Swede Vagn Walfrid Ekman (1874–1954), and the Norwegian Harald Ulrik Sverdrup (1888–1957), *physical oceanography* became established as a subject in its own right, of similar stature to its closely related discipline, *meteorology*. The theory of oceanic tides was included as the most established body of relevant knowledge, and enthusiasts such as Sverdrup and Proudman showed that there was still interesting physics to be learned from tidal motions and difficult tidal problems to be solved, as well as the new problems of ocean circulation.

IAPO *Bibliography on Tides*

In 1955, the International Association of Physical Oceanography, (IAPO, later IAPSO), one of the member bodies of the International Union of Geodesy and Geophysics, published a *Bibliography on Tides*,[2] which listed all the known 'publications on the Tides of the Ocean, in a Latin or Teutonic language (from) 1665 (to) 1939'. Some subject categories were excluded, notably 'astronomical, geological and engineering consequences of tidal action', but otherwise the list was comprehensive of all scientific publications on the subject matter from the first published transactions of the Royal Societies of London and Paris. It was later updated to 1969.[2] It is instructive to consider the statistics of publications listed during the three centuries from 1670 to 1969 in 50-year blocks, as follows:

Before 1670	11
1670 – 1719	27
1720 – 1769	29
1770 – 1819	37
1820 – 1869	262
1870 – 1919	1121
1920 – 1969	2100

The marked rise during the 19th century partly reflects the growth of the subject area as described in the last four chapters and also the increased participation of the USA, Canada, Austria, Germany and Scandinavia. The 20th century growth is mostly due to the general spread in all branches of scientific activity as described above and to the introduction of new topics. However, the *rate* of growth as measured by the ratios between consecutive figures has evidently started to diminish, suggesting an approach to a saturation level for the subject matter. It is hard to forecast a probable number for the next 50-year stretch ending 2019, but we shall review the possibility of saturation in Chapter 15.

Because of the increased pace of research, the remaining history will be described in roughly 25-year stretches, some overlapping. The inter-war years 1918–1939, the subject of this chapter, are characterised by a steady development of the ideas laid down in the previous 50 years, without the revolution in instrument-technology and computing which occurred after 1950.

Oceanic tidal friction as earth brake

This subject is a direct sequel to the later sections of Chapter 10, but it is here more closely related to 20th century oceanography than to 19th century astronomy. We have seen that by the end of the first decade of the 20th century most geophysicists accepted tidal friction to be the principal cause of the observed lengthening of the day, which in turn accounted for the discrepancy in the moon's longitude. There was considerable uncertainty, however, whether the frictional dissipation takes place in the ocean or in the solid earth. Tidal currents in the deep ocean were known to be no more than a few centimeters per second and rough estimates of resistance laws showed such speeds to be quite inadequate to account for the astronomical measurements, which required a total dissipation of 1.4 terawatts, according to contemporary calculations.[5] (I shall use the modern SI unit in this context; most of the older literature used the cgs unit of ergs per second $= 10^{-19}$ TW, but some still used the foot-poundal $=$ 0.421×10^{6} ergs.) Oceanic dissipation was clearly concentrated in the marginal shallow seas where tidal speeds of order 50–150 cm s^{-1} or greater were known, albeit over a much restricted area.

In 1916 a geophysicist at Liverpool University named Reginald Street, who had been a graduate-student of Sir Joseph Larmor at Cambridge, decided to take an actual sea area where tidal currents were well documented, and to evaluate its rate of frictional dissipation directly. Street chose the Irish Sea, and used viscous boundary layer theory to construct an elegant mathematical model of tidal flow which gave the dissipation in terms of the recorded surface currents. Applying this model together with surface currents supplied by the Admiralty, he estimated a total rate equivalent to 0.22×10^{-3} TW over an area of 31,700 km^2, about 1/10,000 of the total area of the ocean. Multiplying by 10,000 might suggest that this dissipation meets the required global rate, but of course most ocean currents are much weaker than in the Irish Sea. Besides, as Street pointed out at the start of his 1917 paper,[3] he expected only to provide a lower bound, on account of neglected turbulence in the boundary later.

Street's work on the Irish Sea tends to be underrated because of his use of the laminar boundary layer, but four years later he published another paper on the subject in which he invoked turbulent flow and horizontal flux of energy, obtaining much more realistic results. In his 1921 paper,[3] Street claimed that the calculations for his later approach were already underway in 1916 but that he had been prevented from completing them 'by teaching duties and military service'. In the meantime, his work had become overshadowed by Taylor's celebrated paper of 1919.[4]

As the leading pioneer of the theory of turbulent fluid flow, Geoffrey Ingram Taylor (1886–1975) at once saw that Street's laminar flow model of 1917 would grossly underestimate the drag forces. Using empirical laws for friction in rivers and for air drag over rough surfaces, Taylor applied to the same data formulae for the bottom stress F of the type

$$F = K\,\rho\,V^2, \quad \text{or vectorially,} \quad \mathbf{F} = -\mathbf{K}\,\rho\,\mathbf{V}\,|\mathbf{V}|$$

where ρ is fluid density, V is speed at some distance from the bottom, and K is an empirical constant depending on bottom roughness. The average dissipation rate per unit area over a tidal cycle is then

$$(4/3\pi)\,K\,\rho\,V^3.$$

With K between the admissible limits 0.0016–0.0020 and Street's data for spring tides in the Irish Sea, Taylor obtained dissipation per unit area in the region $(104\text{–}130) \times 10^{-18}\,\mathrm{TW\,cm^{-2}}$. Street's estimate in the same units was only $0.7 \times 10^{-18}\,\mathrm{TW\,cm^{-2}}$, a factor 150 lower than Taylor's lower bound.

Because of this startling discrepancy, Taylor also calculated the dissipation from the mean horizontal flux of tidal energy into the Irish Sea across its north and south entrances. The flux normal to a line element s is approximately

$$E = g\rho \int_0^s D < \zeta . v_n > \mathrm{d}s,$$

where D is depth, $\zeta(t)$ the tidal elevation, $v_n(t)$ the inward normal velocity to ds, and $< >$ denotes a time average over a tidal cycle. (It would appear from Street's second paper that Street had already performed similar calculations for energy flux, but that he had been deterred from publishing by the disparity of the result with his earlier calculation.) There is a small correction for the net rate of working of the moon on the enclosed sea area, which Taylor also calculated. He ignored a small correction for the equilibrium tide which should strictly be included in the flux integral; (ζ in the last formula should strictly be replaced by $(\zeta - \zeta_{\mathrm{eq}})$.)

The final result from flux divergence was a spring-tide dissipation rate of 0.060 TW over an area of $3.9 \times 10^4\,\mathrm{km^2}$, an average of $153 \times 10^{-18}\,\mathrm{TW\,cm^{-2}}$.

Considering the uncertainty in the factor V^3 in the drag formula, the figure 153 was a remarkable confirmation of the upper value of 130 estimated directly from drag. The total average rate over spring and neap tides reduces to 0.0306TW. With the current assumption of a global dissipation rate of 1.4TW from astronomy,[5] this strongly suggested that the latter could be entirely accounted for by tidal friction in shallow seas of total area 50 times that of the Irish Sea, with similar tides. The existence of such seas seemed quite possible.

Taylor's result was quickly seized upon by two men who were to become leading geophysicists of the period, Harold Jeffreys (1891–1989) of Cambridge and Weikko Heiskanen (1895–1971) of Helsinki. Each independently extended the integration of Taylor's drag formula to most of the world's shallow seas for whose tidal currents information was available, and, in brief, concluded that the total dissipation was sufficient to account for the known retardation.[5,6] The information was in many places sparse and had been provided for the guidance of mariners rather than for scientific analysis. Choices of seas and of the dissipation in each sea differed widely. Jeffreys included only semi-enclosed seas, while Heiskanen also included open shelves bordering continents such as the wide shelf sea of Patagonia. The latter sea is now known to be an important energy sink, but Jeffreys considered its data to be too sparse and unreliable for inclusion. On the other hand, he daringly but cautiously attributed two-thirds of his total dissipation to the eastern part of the Bering Sea, on account of some large currents quoted for isolated localities there. Heiskanen also gave the Bering Sea a high figure, but not so high as Jeffreys. (Modern estimates for tidal dissipation in the Bering Sea are much lower than both.[7])

When the total figures were properly reduced by a factor of 0.51 to allow for average as distinct from spring tides (Heiskanen did not at first do this), Jeffreys' calculations gave 1.1TW, Heiskanen's 1.9TW. Despite many questionable points of detail, these estimates were close enough to the apparent astronomical figure of 1.4TW[5] to convince most geophysicists and astronomers that practically all the dissipation in the tides of the geosphere could be attributed to processes in the ocean. Oceanographers showed little interest. Nothing further was done about it until the American oceanographer Walter Munk re-opened the matter in the late 1950's, by which time the astronomical figure had been raised to 2.7TW. We shall return to this subject in Chapter 12.

Research on tidal currents – Sverdrup, Fjeldstad, van Veen

The investigations described in the previous section were the first to make serious scientific use of horizontal as distinct from vertical tidal movements. In fact, quantitative knowledge of tidal currents was fairly rudimentary until oceanographers started to measure them. First to do so was Harald Sverdrup on board Nansen's ship *Maud*, as she drifted in the Arctic ice over the North Siberian shelf in 1922–24. During 14 months of this period, Sverdrup took long series of hourly measurements of tidal current at several depths down to 64 meters below the surface, along some 200 kilometers of drift-track. The vertical range of the tide was also estimated by sounding. He used a self-recording current meter of a type devised by W. Ekman and a similar device of his own manufacture. Sverdrup finally wrote a masterly analysis of all these data and relevant theory while on board the small drifting ship, remote from any library or possibility of scientific discussion.[8]

We saw in Chapter 9 how Sverdrup's tidal maps of the North Siberian shelf revolutionised notions of the Arctic tides. When it came to explaining the observed tidal currents he also had to discard conventional ideas and to devise new theory from basic principles. The usual concept was that tidal currents oscillate linearly in the direction of the waves, which propagate over depth D with speed $\sqrt{(gD)}$. Such a regime is correct for long waves in a non-rotating basin and also for a Kelvin wave along a straight coastline of a rotating sea, (Chapter 7). Sverdrup observed currents whose direction *rotated* and phase speeds greater than $\sqrt{(gD)}$.

He showed theoretically, first, that if one removes the constraint of a nearby straight coast, waves have in general a greater speed

$$c = \sqrt{(gD/(1-s^2))}$$

where $s = (2\Omega \cos\theta/\text{tidal frequency})$, provided $s < 1$. Further, the current normal to the crest has amplitude proportional to $\sqrt{(1/(1-s^2))}$, while a component also exists parallel to the crest, with amplitude proportional to $\sqrt{(s^2/(1-s^2))}$. In other words, the tidal current vector rotates around an ellipse, generally clockwise in the northern hemisphere. All these properties accorded with Sverdrup's observations.

The wave motion just described is often called a Sverdrup wave.[9] G.I. Taylor had also shown that elliptical rotary currents occur when a Kelvin Wave is reflected from the closed end of a rectangular gulf, essentially by the addition of a series of Poincaré waves, (Chapter 7), necessary to satisfy the boundary conditions.[10] However, all wave forms so far mentioned assume currents uniform with depth.

The observed currents were not uniform in depth, and Sverdrup correctly ascribed this property to the effects of friction on the bottom, and in some cases at the undersurface of an ice sheet, transmitted by turbulence to all parts of the water column. With local horizontal rectangular axes x, y, he treated the corresponding components of current u, v as functions of height z above the bottom ($z=0$), and added a term proportional to hypothetical *eddy viscosity coefficient* ν, to the wave equations for horizontal acceleration (u_t, v_t). Thus, for example:

$$u_t(z) - fv(z) = \nu\, u_{zz}(z) - g\, \zeta_x,$$

$$v_t(z) + fu(z) = \nu v_{zz}(z) - g\, \zeta_y,$$

where $f = 2\Omega \cos\theta$ is the Coriolis frequency, ζ is the surface displacement, and suffixes denote differentiation. This form of turbulent shear-stress had been

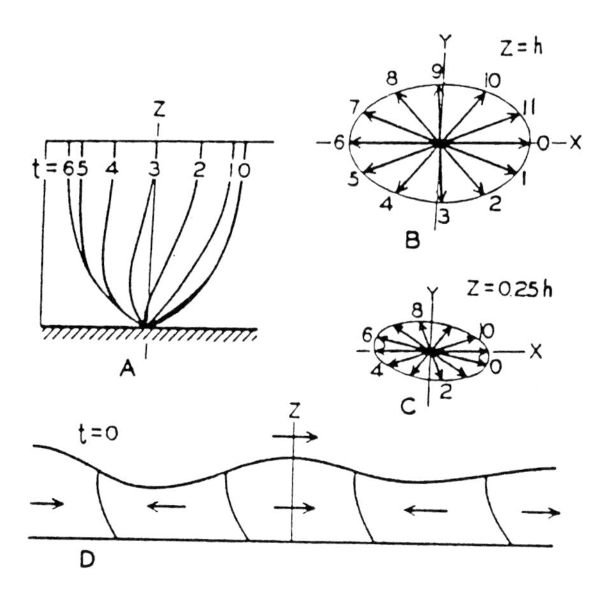

Figure 11.1. Schematic solutions of Sverdrup's semidiurnal wave equations with rotation and friction. (A): The vertical profiles of current $u(z)$ at lunar hours 0–6; note that $u(0)$ is always zero. (B): Hodograph of surface current speed and direction at lunar hours 0–11. (C): As (B) for quarter-depth above bottom. (D): Vertical profile of the wave at hour 0, progressing toward the right; interior arrows denote the direction of current. (From ref. 8.)

applied to wind-driven oceanic circulation by Ekman in 1905. Street had used the same mathematical form in his 1917 paper,[3] but in terms of *molecular* viscosity, as originally postulated by Poisson and Navier in the early 19th century.

Sverdrup chose reasonable values for ν as eddy viscosity, and solved the modified 'Sverdrup wave' equations with simple forms for ν as a function of z. Boundary conditions were $u(0)=0=v(0)$ at the bottom and

$$u_z(h)=0=v_z(D)$$

at the surface $z=D$. As expected, the velocity ellipse varied strongly with height, both in amplitude and orientation. Figures 11.1A–D show a typical solution for $\nu(z)$ uniform with depth. Numbers indicate the time in (lunar) hours relative to the time of maximum wave height. 11.1A shows profiles of water particles which would be aligned vertically in the absence of friction. 11.1B is the velocity ellipse at the surface, maximum at $t=0$ and 6h, but not quite minimum at $t=3$ and 9h. 11.1C shows the ellipse at height $D/4$, much smaller and displaced clockwise (in the northern hemisphere). 11.1D shows the corresponding wave profile and particle displacement beneath the wave. Observations confirmed all these properties.

The results just described are over-simplified by neglecting the real variation of ν with z. Wave equations with variable $\nu(z)$ are harder to solve, but solutions to them were later found by Sverdrup's Norwegian colleague J.E. Fjeldstad.[11]

Density was assumed constant in all this work; vertical variation of density induces *internal tides* below the surface, especially in deeper water. Fjeldstad was also a pioneer in the solution of the internal wave equations for continuous density stratification.(12) However, the importance of internal tides became apparent only later in the century, when current measurements in deep water became technically feasible. We shall return to that subject in Chapter 13.

Finally, mention should be made of the work of the Dutch water engineer, Johann van Veen, architect of the 'Delta Plan' of sea defences across the Rhine delta. In the course of preliminary surveys in the late 1930's van Veen measured the profiles of tidal currents from anchored ships at a large number of stations in the southern North Sea and the Dover Strait. The profiles varied with location, but he found that an empirical law

$$|u(z)| = u(h)(z/h)^{1/k}$$

with k in the region of 5.2, gave a remarkably close fit to most of his profiles.(13) This implies eddy viscosity $\nu(z)$ proportional to

$$(z/h)^{1-1/k},$$

which is not far from the linear form $\nu(z) = Az$ whose exact profile was solved mathematically by Fjeldstad.(11)

Proudman, Doodson, and the Liverpool Tidal Institute

Whereas Harald Sverdrup went on from the physics of tides to become one of the leading physical oceanographers of his time in the USA, the British mathematician Joseph Proudman (1888–1975) stayed close to tidal theory during his whole career. But Proudman also had considerable influence on the development of oceanography between the wars as occupier (from 1933) of the Chair of Physical Oceanography at the University of Liverpool. He also served for many years as Secretary General of the IAPO, and was elected President of IAPO in 1951.

Trained at Cambridge in rigorous disciplines of Applied Mathematics and Mathematical Physics, Proudman later stated that his first inclination had been towards research in the theory of wireless telegraphy, but that he had been persuaded by Professor Lamb of Manchester to work on the 'neglected' field of oceanic tides. Among Proudman's earliest papers to the London Mathematical Society, a series of three papers on the general dynamical theory of tides, presented in 1916 with a fourth part added in 1931,(14) are considered by those competent to judge to be his most outstanding achievement.

It will be recalled from Chapter 7 that Laplace's original problem, on whose solutions S. Hough devoted so much analytical effort, was restricted to an ocean of nearly uniform depth completely covering a sphere. Addition of land boundaries imposes mathematical boundary conditions which transform the problem into a higher stratum of difficulty. Cases of polar basins and oceans bounded by parallels of latitude had indeed been solved by G.R. Goldsbrough,[15] but these are fairly simple extensions of Laplace's problem. Meridional boudaries are much more difficult to cope with. In briefest terms, Proudman showed that the tide generated in an ocean of any analytically defined shape can in principle be expanded in an infinite set of basis-functions, each of which satisfies the boundary conditions and the law of mass conservation. The basis-functions are divided into two classes defining gravity waves and vorticity waves respectively, as Hough and others had found. This division has analogy with Helmholtz's theorem in potential theory, and the functions are now called *Proudman functions* or occasionally *Proudman–Helmholtz functions*. When a series of such functions is summed to satisfy the dynamical relations of Laplace, they yield the normal modes of free oscillation of the defined ocean, which may also be used as a general basis for tidal expansion.

Application of Proudman's theory to even simple geometrically shaped basins was, however, very demanding computationally. In most cases solution was beyond the capability of human computation of the time. Proudman had little patience for computation himself and he spent many years devising simplified but meaningful examples which various students and assistants could reasonably compute. Foremost among his assistants was Arthur T. Doodson (1890–1968), an expert computer who, with Proudman's guidance, became a world authority on many tidal matters.

In 1919, as Professor of Applied Mathematics at Liverpool and a Fellow of Trinity College, Cambridge, Proudman founded a research institute which is unique in the history of our subject. The *Liverpool Tidal Institute* (LTI) was established at the suggestion of Horace Lamb, who chaired a Committee on Tides of the BAAS. The LTI came under the auspices of the University of Liverpool, with funds provided by Alfred and Charles Booth, shipping magnates of the port of Liverpool. Its objects, as stated formally in its first Annual Report of 1920, were:

1. To prosecute continually scientific research into all aspects of knowledge of the tides.
2. To form a training school of applied mathematical research.
3. To form a bureau of organised information concerning the tides.
4. To undertake special pieces of tidal research for commercial or other bodies.

J. Proudman was appointed Honorary Director, A. Doodson Secretary, and at first they had just one junior graduate assistant. Other assistants were added as work developed. The Institute reported annually to a Governing Committee which included the President and the Vice Chancellor of the University, Sir Alfred and Charles Booth, some representatives from the Mersey Docks and Harbour Board, and a number of mathematical professors including Horace Lamb. Later, the LTI became associated with the Harbour Board's moribund time-keeping Observatory and eventually took over the Observatory's premises on Bidston Hill, Birkenhead (of which Figure 10.4 is a reminder.)

During its 50 years of existence, the research programme of LTI covered a range of varied subjects related to tides, including improvements in prediction technique, the preparation of maps of tides and tidal streams in British waters, the design of recording instruments, analysis of storm-surges, and compilation of worldwide data for mean sea level. Underlying this activity was a steady stream of research into Proudman's more esoteric subjects of mathematical theory, for which he had assistance both from formal staff of the Institute such as Doodson and from the Faculty of Mathematics at the University, notably Samuel F. Grace and F. Edith Mercer. When Proudman transferred to the Chair of Physical Oceanography in 1933 he also acquired assistance of a less mathematical nature from that Department, including the use of a small fishing vessel which he employed mainly for measuring turbulent flow in Liverpool Bay.

In 1923, Proudman was awarded the *Adams Prize* of the University of Cambridge for a penetrating essay on the theory of oceanic tides, based on his work of 1916.[14] Harold Jeffreys was a runner-up. The prize entailed a considerable sum of money, which Proudman donated entirely to the new Institute.

I shall review by subject heading those areas of research which advanced significantly under the LTI and its associates, giving due recognition to related work carried out at the same time in other centers in Britain and overseas.

Data analysis and prediction

This was Arthur Doodson's strongest field of interest and one in which he acquired a high degree of expertise. Doodson's methods were of course limited to human computing and the use of the Tide Predicting Machine, (Chapter 8), but he had great skill in devising the most efficient and practical method to use these facilities. Doodson also had an entrepreneurial sense, and could see that after the initial grants of money had been spent the LTI would have to perform a steady paid service to the Community in order to survive. Producing annual tide predictions of good quality was a suitable activity, which Doodson developed into a successful business of national importance.

Since the work of Thomson, Darwin and Roberts (Chapter 8) the 'harmonic method' was considered to provide the most accurate and reliable predictions, with a dubious implication that no further research was required. Edward Roberts had in fact set up one of his sons, H.W.T. Roberts, in a commercial business in Kent, producing tide predictions for all the principal Almanacs and for the Canadian Government, using the 40-component *Universal Tide Predictor* designed by Roberts in 1906. At the LTI, Doodson carefully examined residuals between predictions and observations at the port of Liverpool and at a new tide-gage erected for geodetic use at Newlyn, Cornwall, and found systematic errors. He showed that these were due to neglected harmonic terms representing compound *overtides*, whose parameters he then supplied.

Doodson also had doubts about the accuracy of Darwin's harmonic development of the potential. He computed new tables based upon the latest lunar theory due to E.W. Brown, which included a greater number of terms and improvements on some of Darwin's constituents. He also introduced a new notation for the *argument* of each harmonic term in terms of six small integers, each multiplying one of the standard mean longitudes in the luni-solar theory. The sets of integers, sometimes now known as 'Doodson Numbers', are a more logical notation than the alphabetical characters evolved by Thomson and Darwin (Figure 8.3), but the latter of course remain in use on account of their convenience and familiarity. The tables[16] remained a standard of accuracy for some 50 years, although based on astronomical constants for epoch 1900.

Having a flair for organising large-scale hand-computations, Doodson laid out improved methods for extracting harmonic constituents from long series of hourly tide heights, superseding Darwin's computational method and a popular German method due to W. Börgen (1843–1909) and rivalling one devised for the US Coast and Geodetic Survey by Paul Schureman (1876–1959).[17] He later devised an ingenious method for analysing observations expressed in the form of High and Low Water times and heights, by treating them as low-frequency harmonic series.[18]

To enable LTI to provide predictions based on Doodson's methods, Charles Booth, one of the founders and Chairman of 'Booth Shipping Lines', generously provided funds for the purchase in 1925 of a new 25- to 28-component Tide Predicting Machine for the sole use of LTI. This TPM was constructed to a superior design by the firm of *Kelvin, Bottomley & Baird,* originally founded by William Thomson (Lord Kelvin). Doodson took a great interest in its construction and operation, and used the new TPM to build a growing industry in the production of tide tables. When in 1929 H.W.T. (son of Edward) Roberts unexpectedly died, the LTI took over that business too, together with the 'Roberts TPM' of 1906, which was larger than the new LTI Predictor, but in need of overhaul. (Edward Roberts himself died in 1933, aged 88. Shortly after, his surviving son wrote personally to Doodson:[19] 'He spoke of many of his old associates, and it was a real joy to him to know that his machine was in your hands.' The 'old associates' would have included G.B. Airy, W. Thomson, J.C. Adams and G.H. Darwin, – a distinguished company.)

The culmination of Doodson's association with Tide Predicting Machines was the construction just after World War II of a grand new 42-component TPM by the London firm of Légé & Sons for the LTI. Doodson himself supervised the design, and copies were later made for Argentina, India and Japan. The 'Doodson–Légé TPM', as it was called, was used to provide predictions by contract for up to 180 worldwide ports annually. The machine is now displayed in the foyer of the 'Proudman Oceanographic Laboratory' on Bidston Hill. The original LTI machine and the Roberts TPM may be seen in Liverpool's Maritime Museum.

Only two other major TPMs were constructed after 1935, both in Germany. The first was a unique machine completed in 1938 for the Deutsche

Figure 11.2. The Tide Predicting Machine (*Gezeitenrechenmaschine*) built for the *Deutsche Hydrographische Institut* at Hamburg just before World War II, in its temperature-controlled room. With 62 harmonic components, it was the largest (and heaviest) TPM ever built. The designer, Heinrich Rauschelbach, is standing in front of his machine. (Photograph by courtesy of *Bundesamt für Seeschiffahrt und Hydrographie,* Hamburg.)

Hydrographische Institut at Hamburg. It was constructed by *Mechanoptik, Aude und Reipert* of Potsdam to a design by the hydrographic engineer Heinrich Rauschelbach (1888–1978). With provision for 62 harmonic constituents and weighing 7 tonnes, this was the largest TPM in the world, (Figure 11.2). It is now exhibited at the Deutsches Museum at Munich. A second, somewhat smaller, TPM was made in 1952 for the Deutsche Demokratische Republik. A comprehensive history of all Tide Predicting Machines ever constructed (25 in number) was published by Gunther Sager in 1955.[20]

Fundamentals of Laplacian theory

In 1933 four leading Norwegian and Swedish meteorologists published a treatise on hydrodynamic theory[21] in which they claimed that Laplace's tidal equations (LTE) were inadequate to describe certain types of atmospheric motion, on account of their neglect of vertical acceleration. The examples given concerned only cellular motions of short wavelength, but the manner of writing suggested that LTE were generally unsuitable for application to real dynamic situations.

A large part of Joseph Proudman's work was based on the application of Laplace's equations to the ocean, so he felt compelled to investigate these claims in depth. He finally vindicated the accuracy of LTE in two papers. In the first of these,[22] concerned with fluids of homogeneous density, he compared exact solutions of LTE in seas of geometric shape with the corresponding solutions of the more accurate dynamic equations:

$$u_t - 2\Omega \cos\theta\, v = -(1/r)P\theta,$$

$$v_t + 2\Omega \cos\theta\, u + 2\Omega \sin\theta\, w = -(1/r \sin\theta)P\phi,$$

$$w_t - 2\Omega \sin\theta\, v = -P_r.$$

Here, (u,v,w) are the components of velocity in the directions of increasing (θ,ϕ,r) = (colatitude, east longitude, radial ordinate), and P = (pressure/density − the geopotential). The equation of mass conservation in (u,v,w) takes the usual form. Laplace's equations neglect Ωw compared with v_t and the left side of the third equation compared with g.

Proudman showed that Laplace's original solution for the diurnal tide on a complete sphere (Chapter 7) satisfies the above equation exactly. In bounded seas on a sphere he showed that the differences in the solutions are negligibly small, except for semidiurnal tides near the poles and long-period tides near the equator. The differences are associated only with possible motions whose wavelengths are small compared to the ocean depth. Thus, vindication of LTE was practically complete. At about the same time, Harold Jeffreys[23] verified the above equations in a strictly orthogonal coordinate system on an ellipsoidal earth. Many years later, the whole question was re-examined by J.W. Miles of La Jolla, California.[24]

In his second paper,[25] Proudman gave a simpler analysis for the case of fluids with vertical density stratification. Here even the anomalies near the poles and the equator become negligible in realistic oceanic density structure. The only possible condition for failure of LTE was the internal resonance of a nearly homogeneous fluid; this hardly applies to the ocean.

Tides in mathematical basins on a rotating sphere

The gulf in difficulty between solving for the tides on Laplace's complete sphere and in realistic oceans was so great that it was desirable to find solutions for large bounded oceans of simple geometrical shape. I will not take space to describe the many mathematical studies of tides in flat circular, semicircular, elliptical and rectangular basins by Lamb, Poincaré, Goldstein, Goldsbrough, Taylor and others (including Proudman himself), or in spherical basins with small or zero rotation by Rayleigh, Proudman and Doodson. Such studies gave mathematical insight into the necessary forms of solution but the cases treated were very remote from the conditions of the real ocean. We shall concentrate only on solutions for oceans bounded by complete meridians on a rotating sphere by Goldsbrough, Proudman and Doodson.

Proudman's older friend George R. Goldsbrough (1881–1963) also specialised in tidal theory, but he worked quite independently of Proudman at Armstrong College (later known as King's College), Durham.(26) I have already mentioned Goldsbrough's solutions for tides in oceans bounded by parallels of latitude,(15) which pre-dated Proudman's generalised theorems.(14) That problem is relatively straightforward, since the longitude enters in a simple way and the boundary conditions are easily satisfied by a numerical power series in $\sin\theta$. By contrast, the condition $v=0$ on a meridian leads to a doubly infinite series of functions of θ and ϕ, whose coefficients have to be evaluated from a doubly infinite set of simultaneous equations for any chosen mean depth. The associated computation with contemporary resources was formidable.

From 1927 to 1933 Goldsbrough produced a remarkable series of papers in four parts, exploring the properties of forced tides and free oscillations of oceans on a rotating globe bounded by two meridians, 60° apart. The whole series is well reviewed with full references in the *Biographical Memoir* written by Proudman.(26) (A fifth part followed several years later.) The shape was chosen to simulate roughly the Atlantic Ocean and its possible resonances. He expressed solutions in a double series of 'associated legendre functions', taken for computation up to degree 12. In Part I, Goldsbrough also applied his earlier procedure for a polar ocean up to latitude 45° as a proxy for the Southern Ocean. He showed that its semidiurnal tides were quite small unless the depth is taken to be about 450 meters, which is unreasonably shallow. This result sufficed to discount Whewell's old notion that the tides of the Atlantic originate there.

In order to simplify the algebra, the frequency was chosen to be that of K_2 (sidereal), and the depth of the 60° sectorial ocean was at first assumed proportional to $\sin^2\theta$ (i.e. zero at the poles). This ocean resonated to the K_2 potential when its mean depth was 4725 m, not greatly unrealistic. Alternatively, a sectorial ocean with a more realistic mean depth of 3870 m would resonate if the angle is reduced to 53°. Goldsbrough concluded that the Atlantic Ocean is indeed close to a resonant state for semidiurnal tides, but further calculations for the M_2 tide in uniform depth were needed. These he carried out in Parts III and IV, to great effort, and with comparable results. For example, a 60° ocean of uniform depth 3930 m had a free period of 12 h 33 m, close to that of M_2.

The other great computational effort of the period was that of Proudman and Doodson, who also wrote a sequence of papers on tidal solutions between meridians. In view of Goldsbrough's papers, published first, they concentrated on a hemispherical ocean bounded by a complete meridian. In Part I of their five-part paper, Proudman adapted his general theory of 1916(14) to hemispherical geometry and uniform depth, leading to another type of doubly infinite series of spherical harmonics, for which he computed 716 coefficients common to all solutions. Expansion in terms of this series formed the basis for

Doodson's computations, presented in Parts II and III. In 1968, Proudman wrote in his Biographical Memoir of Doodson for the Royal Society:[27]

> By a feat of computation such as will probably never be repeated without the use of electronic computers, Doodson solved these equations, for diurnal tides in 1935 and for semidiurnal tides in 1937. He drew the resulting cotidal lines for oceans of different depths and determined certain critical depths for resonance.

Figure 11.3 shows examples of the graphical detail to which Doodson went in delineating the contours of both amplitude and phase for the M_2 tide in one quadrant of symmetry of the hemisphere. Values of Lamb's depth parameter $\beta=4(\Omega a)^2/gD$ ranged from 0 to 20. Just as Goldsbrough's sector could be related to the Atlantic Ocean, Doodson's resembled the Pacific, for which the solution for $\beta=18$ shown in Figure 11.3 (bottom right), corresponds roughly to the mean depth. However, the details of the solution evidently vary quite rapidly with β, not to mention other possible variations outside the scope of the model. In Parts IV and V Doodson explored other techniques of computation, including the use of finite differences – a forerunner of later computational methods, (Chapter 12).

Further progress was interrupted by World War II, but we should finally note Goldsbrough's own 'Part V', published in 1950. Goldsbrough here set out a totally new method of solution for the K_2 tide in a sectorial ocean with depth proportional to $\cos^2\theta$. He expressed the solution in terms of a set of orthogonal functions whose coefficients could be expressed as explicit integrals.[26] No computational results were given. By that time, interest in geometrical basins was waning in favor of finite-difference approximations which could in principle be applied to any coastal configuration, given sufficient resources of computation.

Tides in realistic oceans

At the same time as these arcane studies in applied mathematics, progress was being made in semi-empirical mapping of the tides in partially enclosed seas, with considerable realism in some cases. This activity was aided by a steady increase in the number of coastal and island sites where tide-gage records had been harmonically analysed. The International Hydrographic Bureau at Monaco, created in 1921, played a useful part in compiling lists of harmonic constants. (The IHB is a consortium of naval hydrographic authorities, dedicated to improving all aids to navigation. Its sheets of tidal harmonic constants known as 'Special Publication no. 26' were first issued in 1932; the lists have since been computerised.)

Tidal models of the Arctic Ocean by the Austrians, Robert von Sterneck (1871–1928) and Albert Defant (1884–1974), were described at the end of Chapter 9. Both these men paid a lot of attention to the tides of the

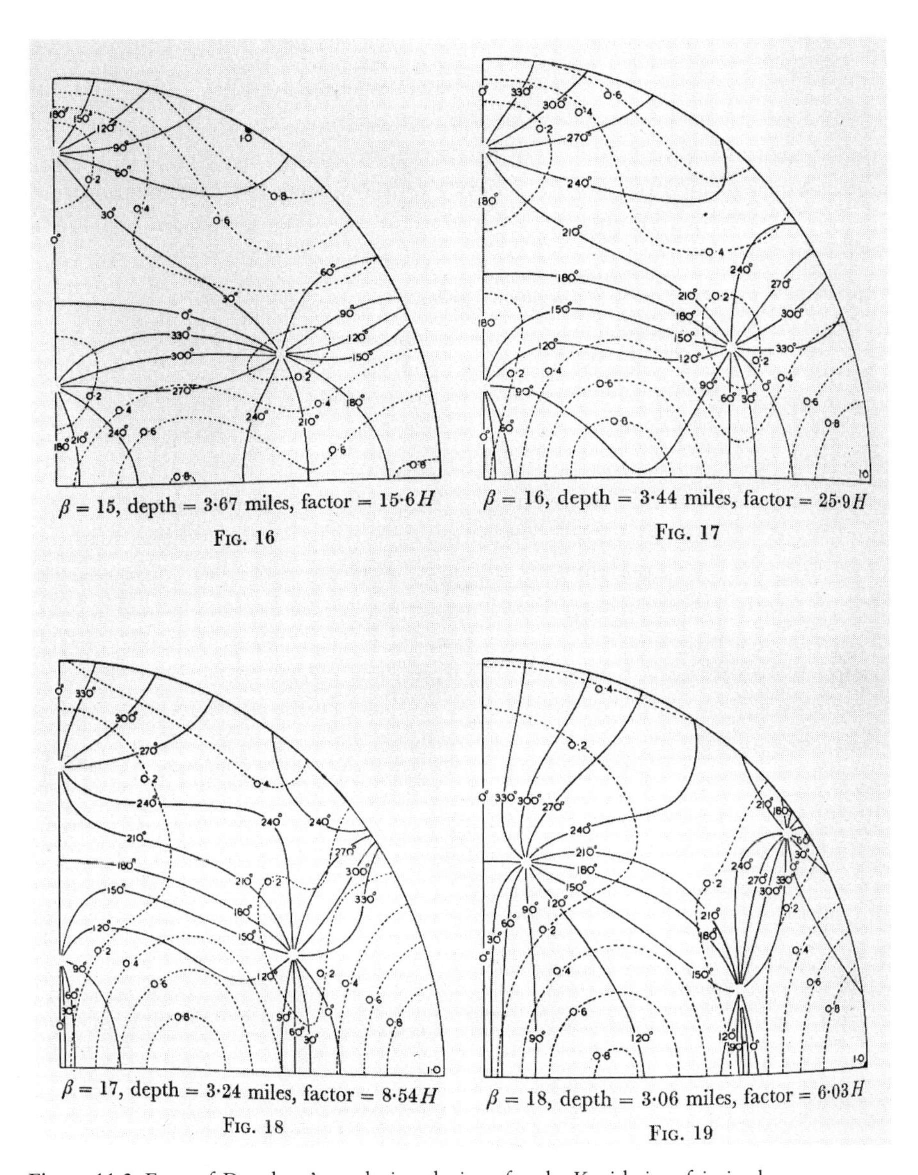

Figure 11.3. Four of Doodson's analytic solutions for the K_2 tide in a frictionless hemispherical ocean with a sequence of constant depths; the solutions for the other three sectors of the hemisphere are symmetrically identical. Full lines show phase lag in degrees with respect to the potential at the central meridian (vertical axis). Dotted lines show amplitudes in arbitrary units; the unit is to be multiplied by the 'Factor' shown below each diagram, where H is the maximum amplitude of the equilibrium tide. The positions of the three amphidromic systems in each diagram are evidently very sensitive to the assumed mean depth. (Courtesy of the Royal Society.)

Mediterranean, Adriatic, Black and Red Seas. While Defant had wider experience of oceanography from his participation in the famous German *Meteor* expeditions, Sterneck was originally a pure mathematician at the University of Graz who turned to the problems of tidal mapping in the last 16 years of his life. In 1920 Sterneck was the first to produce tentative maps of both semidiurnal and diurnal tides in all the world oceans.[28]

The only tractable hydrodynamic theory available for such work was the one-dimensional equation for wave propagation along a canal of variable width $B(x)$ and mean depth $D(x)$:

$$\frac{\partial}{\partial x}\left[B(x)D(x)\,\frac{\partial \zeta}{\partial x}\right]+\frac{\omega^2 B}{g}\,\zeta=0$$

where ζ is wave elevation, and $\zeta' = (\zeta -$ equilibrium tide) as external forcing The Coriolis stress enters only to produce a linear slope normal to x and proportional to the current $u(x)$, which is simply related to $\partial\zeta'/\partial x$. The differential equation can be integrated numerically with suitable conditions at the ends, and obviously works best for long narrow seas such as the Adriatic and Red Seas. The semidiurnal tides in both seas show a simple standing wave about a nodal point in x. Application of the Coriolis correction converts the node to a positive (anticlockwise) amphidromic rotation.[29]

The tides of the Black Sea, though only a few centimeters in amplitude, attracted a surprising number of studies, to account for its unusual *negative* semidiurnal amphidromy. Sterneck[30] gave a reasonable description of semidiurnal and diurnal tides in the Black Sea in terms of a longitudinal oscillation modified by Coriolis stress, together with an equilibrium tide response. Grace[33] got slightly better agreement with the data by using an exact dynamic solution for a rectangular basin of uniform depth.

Defant[29] applied the variable canal theory to the whole Atlantic Ocean between Tristan da Cunha and Iceland. His solution for the semidiurnal tides agreed surprisingly well with measurements at islands along the central axis, and with the phases of currents recorded by Defant himself on the *Meteor* Atlantic Expedition of 1925–27.

Soon after the founding of the LTI, Proudman and Doodson did sterling work in producing the first tidal maps of the North Sea since those of Whewell and Airy (Chapter 9). These were also the first tidal maps of any kind to include isolines of amplitude as well as cotidal lines. (The former are sometimes called 'co-range' lines, which strictly refer to double amplitude, while the latter are now better termed 'co-phase' lines.) Their map is reproduced in Figure 11.4. To produce it, Proudman and Doodson used the novel technique of employing data from *current* measurements to determine the *gradient* of the isolines through Laplace's equations with friction.[31] The same method was employed by Doodson and R.H. Corkan (another member of LTI staff) to produce definitive maps of the tides of the English and Irish Channels and the Irish Sea, thus practically surrounding a good part of the British mainland.[32]

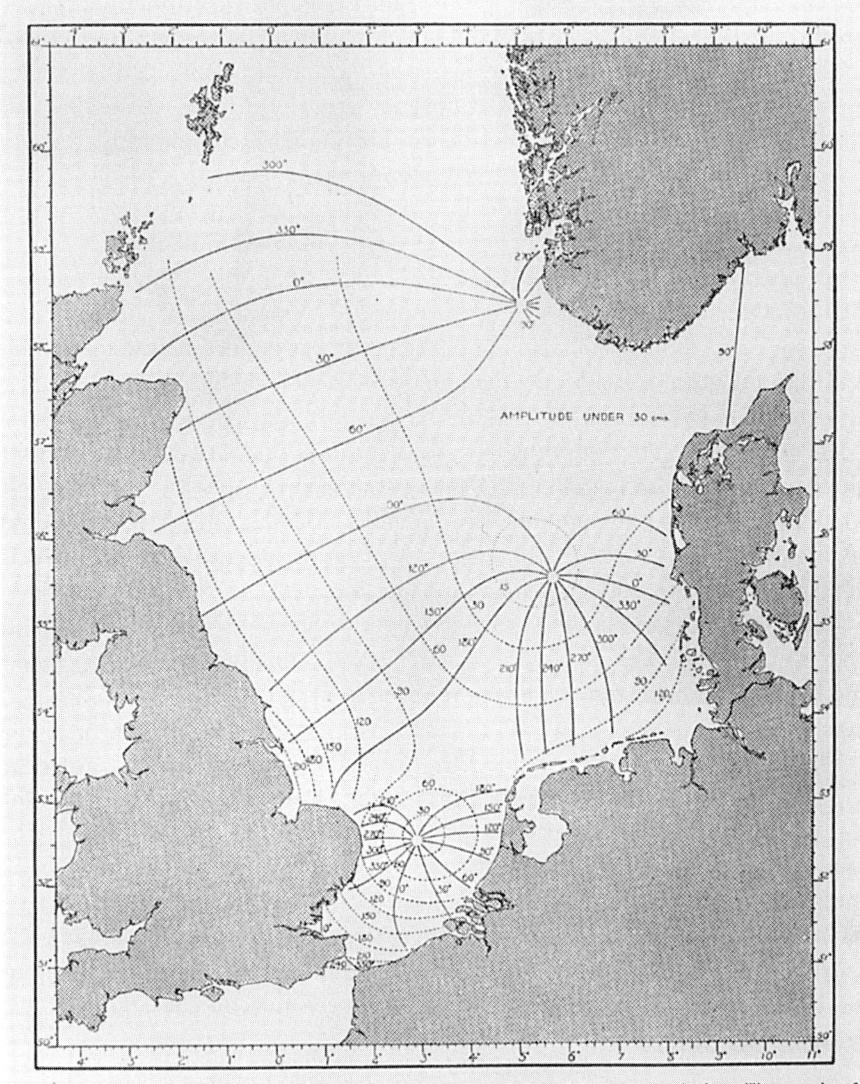

Fig. 17.—Co-tidal lines and co-range lines, denoted by full and dotted lines respectively. The associated numbers give the values of γ in degrees and of H in centimetres.

Figure 11.4. The first definitive map of the M_2 tide in the North Sea, computed by Proudman and Doodson by dynamical theory, making use of all available data for coastal elevations and currents.[31] Full lines are isolines of harmonic phase lag in degrees; dotted lines show harmonic amplitude in centimeters. (cf. Figure 9.2.)

Persisting with his efforts to bridge the gap between geometrical idealism and the complexity of the real ocean, Proudman devised series of 'building blocks' of finite area which might be pieced together to extend tidal solutions out into the deep ocean. One such series consisted of rectangular elements of arbitrary dimensions and depth. With his colleague Samuel F. Grace (1894–1937) he worked out a general dynamic solution for tides in each rectangle with arbitrary boundary parameters. Grace applied sets of these rectangles to the Bay of Biscay and the Gulf of Mexico,[33] but his results had little lasting influence. Grace died prematurely in 1937 from war injuries received in 1918.

Another of Proudman's schemes was more ambitious. He divided the Atlantic Ocean into narrow (5°) strips bounded by parallels of latitude from coast to coast. Along each strip the bathymetry was defined realistically as an arbitrary function of longitude, including continental shelves at both ends. By an ingenious theorem, he devised an explicit formula for the distribution of tidal current along one bounding latitude in terms of the distribution of current along the other. Possible flux of tidal energy out of the strip to the east and west was allowed for. By patching together a number of contiguous strips one could apparently obtain a solution for most of the Atlantic Ocean. The difficulty was in assigning suitable current distributions for the initial northern (or southern) boundary. Proudman accomplished this by a series of arbitrary northbound and southbound Kelvin and Poincaré waves, finally determining their coefficients to provide the best fitting solution to a number of coastal tide elevations.

This time, Proudman performed the computations himself. He presented partial solutions for the M_2 tide of the Atlantic between 50°N and 30°S in his *George Darwin Lecture* to the Royal Astronomical Society in 1944.[34] The results bear some resemblance to the tides as now known, but they are not as complete as Proudman had intended, probably on account of unforseen difficulties. Although his lecture hinted at further calculations, Proudman published no more results for Atlantic tides. In 1945 he resigned his directorship of the LTI in favor of Doodson, in order to concentrate on the affairs of his University Department.

Empirical world cotidal maps by Gunther Dietrich

In practical terms, the most impressive product of the era of tidal research before the automatic computer was the set of cotidal maps for O_1, K_1, M_2, and S_2 in the Atlantic, Indian and Pacific Oceans, constructed by Dietrich during and perhaps a little before World War II.[35] Gunther Dietrich (1911–1972) worked at first under Defant at the Institut für Meereskunde, Berlin; he later became Director of the Institut für Meereskunde at Kiel. Figure 11.5 shows his cotidal maps for K_1 and M_2. They were constructed on the same empirical basis as the much earlier maps of Sterneck,[28] but whereas Sterneck had harmonic constants for only 204 stations at his disposal, Dietrich had direct results from 842 stations, thanks to the compilations of the International Hydrographic Bureau, together with 823 stations whose main constants had been carefully 'inferred' from short series of measurements by the British Admiralty and published just before the war in the *Admiralty Tide Tables* for 1938. Very little hydrodynamic theory was used in the construction, but intelligent use was made of Airy's canal theory and Doodson's calculations for spherical sectors.

As Figure 11.5 shows, only cotidal lines were attempted, following the respectable tradition of Whewell, Airy, Harris and Sterneck, in contrast to the maps of small seas by Proudman and Doodson, (e.g. Figure 11.4). Absence of amplitude contours proved a disadvantage to those wishing to compute energy loss or earth loading, but this was no doubt a symptom of uncertainty about amplitudes in mid-ocean and of difficulty in satisfying mass conservation. Nevertheless, Dietrich did include separate maps in which amplitude was interpolated continuously round the coastlines. Serious attempts to map amplitude across the oceans only began to appear when finite-difference solutions of LTE became possible in the 1950s, but Dietrich's cotidal maps remained the standard source of information for some 25 years. When tides first started to be measured in the deep ocean in the mid-1960s, the phase contours shown in Figure 11.5 were found to be remarkably accurate at the few places where they could be compared with measurements.

Notes and references

1. Buchanan, J.Y., H.N. Moseley, J. Murray, Editor, T.H.Tizard. The Reports of the scientific results of the exploring voyage of HMS *Challenger* during the years 1873–76. 50 volumes, HMSO, London, Edinburgh and Dublin, 1885–1895. (For a non-technical centennial narrative, see: Linklater, E. *The Voyage of the Challenger*. John Murray Pub., London, 288pp., 1972.)
2. Association d'Océanographie Physique (IAPO–IUGG). *Pub. Scientifique no.15*: Bibliography on Tides, 1665–1939, Bergen, 1955. Followed by *Pub. Sci. no. 17*, (1940–1954), Göteborg, 1957; and *Pub. Sci. no. 29*, (1955–1969), Birkenhead, 1971.
3. Street, R.O. The dissipation of energy of the tides in connection with the acceleration of the moon's motion. *Proc. R. Soc. London*, **A, 93**, 348–359, 1917. See also: *Ibid*. The tidal motion in the Irish Sea, its currents and its energy. **A, 98**, 329–344, 1921.
4. Taylor, G.I. Tidal friction in the Irish Sea. *Phil. Trans. R. Soc. London*, **A**, **220**, 1–33, 1919.
5. Jeffreys, H. Tidal friction in shallow seas. *Phil. Trans. R. Soc. London*, **A**, **221**, 239–264, 1920. (Jeffreys' early astronomical estimate of the lunar dissipation is mainly in: The chief cause of the lunar secular acceleration. *Monthly Not. R. astr. Soc.* **53**, (3), 309–317, 1920 – but see Ch. 12, ref. 43.)
6. Heiskanen, W. Über den Einfluss der Gezeiten auf die säkuläre Acceleration des Mondes. *Ann. Acad. Scient. Fennicae*, **A**, **18**, 1–84, 1921.
7. Sündermann, J. The semidiurnal principal lunar tide M_2 in the Bering Sea. *Deutsche Hydrogr. Zeitschrift*, **30**, 91–101, 1977.
8. Sverdrup, H.U. Dynamics of tides on the North Siberian Shelf – Results from the *Maud* expedition. *Geofysiske Publ.*, **4**, (5), Oslo, 75pp., 1927.
9. The Sverdrup wave may also be seen as a special case of a Poincaré wave with one of its two horizontal wavenumbers set to zero, making the wave crests infinitely long and straight.

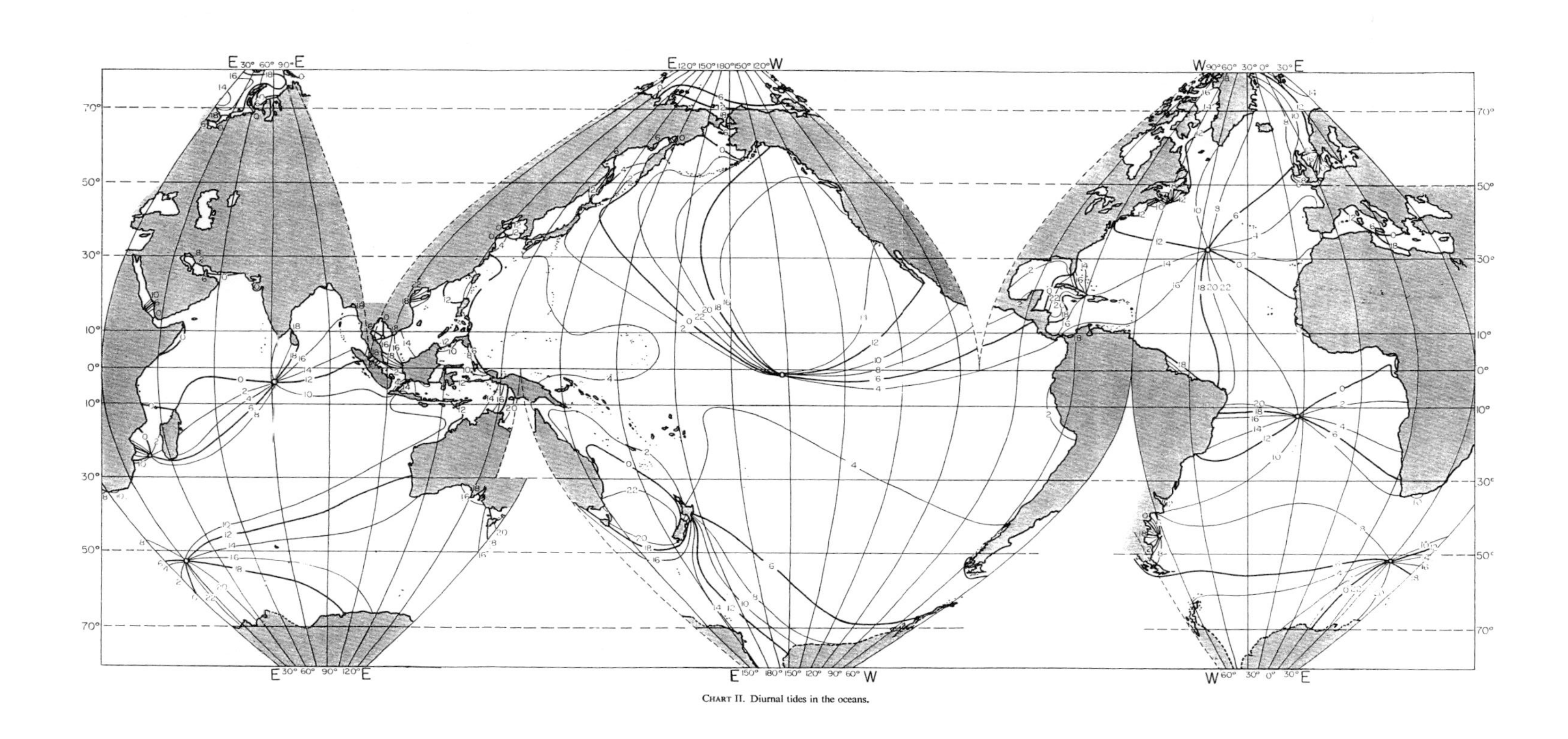

CHART II. Diurnal tides in the oceans.

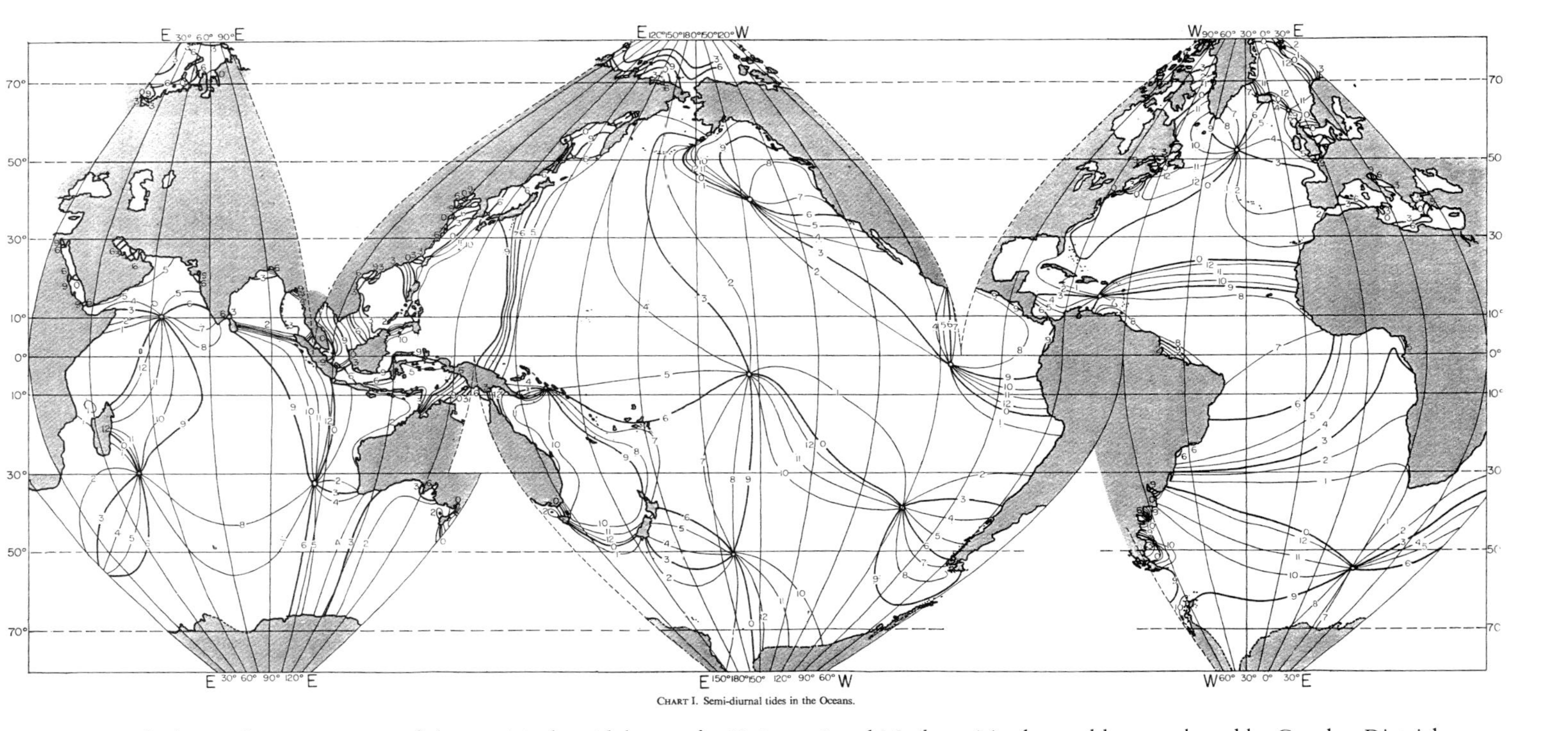

Figure 11.5. The last and most accurate of the empirical cotidal maps for K_1 (upper) and M_2 (lower) in the world ocean, based by Gunther Dietrich on 1665 direct and inferred coastal data stations.[35] Isolines of Greenwich phase lag are denoted by their approximate hours, or more exactly, by $(G/15)$ for K_1, $(G/30)$ for M_2, where G is the harmonic phase lag in degrees.

10. Taylor, G.I. Tidal oscillations in gulfs and rectangular basins. *Proc. London Math. Soc.*, **20**, 148–181, 1921.
11. Fjeldstad, J.E. Contribution to the dynamics of free progressive tidal waves – Norwegian North Polar Expedition with the *Maud*, 1918–1925, *Sci. Results*, **4**, (3), Oslo, 80pp., 1929.
12. Fjeldstad, J.E. *Interne Wellen. Geofysiske. Publ.*, **10**, (6), Oslo, 53pp., 1933.
13. Van Veen, J. Currents in Dover Straits. *J. du Conseil Permt. Intl. pour l'Exploration de la Mer,* **13**, (1), 7–36, 1938.
14. Proudman, J. On the dynamic equations of the tides, Parts 1–3, *Proc. London Math. Soc.*, **18**, 1–68, 1917. (The theory was later extended to an infinity of degrees of freedom in Part 4: *Ibid.* **34**, 293–304, 1931.)
15. Goldsbrough, G.R. The dynamical theory of the tides in a polar basin. *Proc. London Math. Soc.*, **14**, 31–66, 1913. Also, similar title, but for a *zonal* basin: *Ibid.* **14**, 207–229, 1914.
16. Doodson, A.T. Harmonic development of the tide-generating potential. *Proc. R. Soc. London*, **A**, **100**, 305–329, 1921. (Later reprinted with some corrections in: *Internat. Hydrogr. Rev.*, **31**, 11–35, 1954)
17. Schureman, P. Manual of harmonic analysis and prediction of tides. *USCGS Special Pub. no. 98*, Washington, D.C., 1940. (Revised, 1958.)
18. Doodson, A.T. The analysis of high and low water. *Internat. Hydrogr. Rev.*, **28**, 13–77, 1951. (Also issued as *IHB Special Pub. no. 36.*)
19. Letter from E.G.L. Roberts to A.T. Doodson, 23 Aug 1933 – Copy supplied to the writer by Ms. Valerie Doodson (daughter-in-law). During World War II, such was their national importance that the LTI TPM and the Roberts TPM were housed in separate basement vaults at Bidston, in case of bomb damage.
20. Sager, G. *Gezeitenvoraussagen und Gezeitenrechenmaschinen DDR Hydrographic Service*, Warnemünde, 126pp., 1955. (Information about this, and other related publications, and copy of Figure 11.2, kindly supplied by Frau A. Lück, Librarian of the Bundesamt für Seeschiffahrt und Hydrographie, Hamburg.)
21. Bjerknes, V., J. Bjerknes, H. Solberg, and T. Bergeron. *Physikalische Hydrodynamik*, (esp. pp. 450–452), Springer, Berlin, 797pp., 1933.
22. Proudman, J. On Laplace's differential equations for the tides. *Proc. R. Soc. London*, **A**, **179**, 261–288, 1942.
23. Jeffreys, H. A derivation of the tidal equations. *Proc. R. Soc. London*, **A**, **181**, 20–22, 1942.
24. Miles, J.W. On Laplace's tidal equations. *J. Fluid Mech.*, **66**, 241–260, 1974.
25. Proudman, J. The applicability of Laplace's differential equations of the tides. (Paper read at at IAPO General Assembly, Oslo, 1948.) *Internat. Hydrogr. Rev.*, **25**, (2), 112–118, 1948.
26. Proudman, J. George Ridsdale Goldsbrough. *Biogr. Mem. Royal Soc.*, **10**, 107–116, 1964. (Contains full references to Goldsbrough's publications.)
27. Proudman, J. Arthur Thomas Doodson. *Biogr. Mem. Royal Soc.*, **14**, 189–205, 1968. (Contains full references to Doodson's publications.)
28. Sterneck, R. von. Die Gezeiten der Ozeane. *Sitzber. Akad. Wiss. Wien*, **129**, 131–150, 1920. See also *Ibid.* **130**, 363–371, 1921.
29. Defant, A. *Physical Oceanography,* Vol. 2, Pergamon Press, Oxford, 598pp., 1961. (A comprehensive and pedagogic review of most aspects of tidal research in the period 1920–1950.)
30. Sterneck, R. von. Harmonische Analyse und Theorie der Gezeiten des Schwarzen Meeres. *Ann. Hydrogr. Marit. Meteorol.*, **54**, 289–296, 1926.
31. Proudman, J. and A.T. Doodson. The principal constituent of the tides of the North Sea. *Phil. Trans. R. Soc. London*, **A**, **224**, 185–219, 1924.
32. Doodson, A.T. and R.H. Corkan. The principal constituent of the tides in the English and Irish Channels. *Phil. Trans. R. Soc. London*, **A**, **231**, 29–53, 1932.
33. Grace, S.F. Tidal oscillations in rotating rectangular basins of uniform depth. *Monthly Not. R. astr. Soc. – Geophys. Suppl.*, **2**, 385–398, 1931. *Ibid.* The principal diurnal and

semidiurnal tides of the Gulf of Mexico, **3**, 70–83 and 156–162, 1932. *Ibid.* The principal semidiurnal constituent of tidal motion in the Bay of Biscay, **3**, (7), 274–285, 1935.

34. Proudman, J. The tides of the Atlantic Ocean. (George Darwin Lecture). *Monthly Not. R. astr. Soc.* **104**, 244–256, 1944.
35. Dietrich, G. Die Schwingunggssysteme der halb- und eintägigen Tiden in den Ozeanen. *Veröff. Inst. Meereskunde, Berlin*, **A**, **41**, 1–68, 1944.

12

The impact of automatic computers, 1950–1980

Introduction – Some benefits of wartime technology

The major wars of the 20th century interrupted the pursuit of pure research, but after hostilities ceased new scientists with fresh ideas took over senior posts, and new research institutes were formed. New technology devised for advantage in war found commercial applications in peacetime, in some cases beneficial to marine research. World War I had not been noted for useful technology except in air flight, but it appears that several of the warring navies had conducted secret researches on the *sonic echo-sounder* which came into the open in the early 1920s.[1] The echo-sounder transformed understanding of the nature of the ocean floor, with subsequent advantage to tide-modeling. World War II saw more radical innovations in the invention of *radar* and *sonar*, while extensive use of the *bathythermograph* revealed structures of the upper thermocline including internal tides.[2]

Beach landings, atom-bomb tests, and submarine detection showed up the need for better knowledge of waves, currents, temperature structure and tides in arbitrary places. The most notorious case of inadequate tidal knowledge was in the planning of the American invasion of the Pacific atoll of Tarawa in 1943. A rash decision to invade on a neap tide combined with inferior tidal data and ignorance of the vagaries of equatorial sea level caused landing craft loaded with marines to run aground on an offshore coral reef, with consequent heavy loss of life.[3] German U-boats are known to have recorded pressure variations off the west coast of Ireland in 1940, for tidal information in case a beach landing were to be attempted.[4]

However, it was not until the 1950s that wartime technological invention began to impact civilian science. Initial designs had to await improvements brought about by commercial drive and 'Cold War' competition before being accepted as reliable. The electronic computer, developed during World War II for research into cryptology, was applied to a large-scale scientific calculation in 1946–48 to check the accuracy of E.W. Brown's Tables of the Moon,[5] but only in the mid 1950s did electronic computers become available in a convenient form to a wider scientific community. Their applications to tide research will be discussed below.

Digital recording technology grew out of the computer revolution. Its application to open-sea measurements, together with low-powered transistor technology and advances in navigational aids, will form the main subject of Chapter 13. German wartime researches in rocketry led by stages to the first Russian *Sputnik*, launched on 4 October 1957. What at first seemed a daring experiment in military reconaissance soon, with its American competitors, had far-reaching applications to monitoring the earth's gravity field and upper atmosphere, and to precise navigation, time-keeping and geophysical geodesy in general.[6] Satellite geodesy will be the subject of Chapter 14.

The computer revolution

The computer has become such a commonplace piece of the modern scientist's desk furniture that it is hard for members of the younger generation to appreciate how the device has transformed scientific calculations in the 40-odd years since the word 'computer' meant a person. One has to visualise a time (which the present writer remembers well) when, typically in 1954, the only aids to computation were a large sheet of squared paper, a pencil (with eraser!), a slide-rule, a book of tables of standard mathematical functions, and at best a noisy mechanical adding machine which could multipy or divide one laboriously inserted number by another at an average rate of about one decimal digit per second. The earliest accessible electronic computers had slow input/output, were hard to program and liable to break down before the end of a long calculation, but their increase in speed and capacity was greater than that of a jet-plane compared with a slow pedestrian. It was not simply that one could calculate a lot faster; a whole range of bulky computations, previously unthinkable in terms of human effort, suddenly became feasible. And of course, speed, reliability and storage capacity have vastly increased over the years since the mid-1950's.

The principal advances in tide research made possible by electronic computers were the new abilities:

(a) to solve Laplace's differential equations over large seas and oceans of arbitrary shape and bathymetry;
(b) to extract normal modes of oscillation of such basins and their associated resonant frequencies;
(c) to store long series of tidal data and compute their complete spectrum by Fourier transformation;
(d) to compute an adequate lunar/solar ephemeris and hence the 'equilibrium tide' and its component species over a long series of times;

(e) to invert large normal matrices of lagged covariances between sea level and multiple related series.

Additionally, many of the more traditional computations involved in the production of tide-tables could be performed far more quickly and reliably than by the time-honored Tide Predicting Machines (Chapters 8, 11).

The rest of this chapter will trace the progress of such researches, though without pedantic descriptions of the computing procedures themselves. In some cases it will be necessary to go beyond the nominal year limit of 1980 in order to round off a particular line of progress at a suitable resting point. Previously, I have concentrated on the work of outstanding named individuals, but from here onward fewer people will be mentioned by name, partly because many are still alive at the time of writing, and also because most post-war subjects have had so many contributors that it would be tedious to name them all and assign priorities. Important names will be evident from the publications listed at the end of the chapters.

New solutions of tides in seas and oceans

A short technical interlude is necessary by way of introduction to the historical sequence, in order to explain the background to the research and to emphasise how it differed from the classical analytic approaches described in the previous chapter. Most of the techniques are still used in modified form in modern research, but they were relatively new, or had not been applied on a large scale before 1950.

In a typical finite-difference scheme for solving the tidal equations across a chosen sea, the area is divided as finely as possible by a quasi-rectangular grid of the following form:

$$
\begin{array}{cccccc}
 & s-2 & & s & & s+2 \\
r-2 & + & \circ & + & \circ & + \\
 & \circ & + & \circ & + & \circ \\
r & + & \circ & + & \circ & + \\
 & \circ & + & \circ & + & \circ \\
r+2 & + & \circ & + & \circ & +
\end{array}
$$

Here, grid-points p_{rs} marked by crosses are where *current vectors* are evaluated from the momentum equations. Each 'current' point is associated with a known depth D_{rs}. Points p_{rs} marked by '○' arc where *elevations* are evaluated from the equation of mass conservation.

A common approach is to retain only the leading linear terms in the equations and to linearise friction to $-cv_0 u/D_{rs}$, where c is a drag coefficient and v_0 a representative constant speed. The equations for a tidal harmonic of frequency σ are then simplified by removal of a factor $e^{-i\sigma t}$ and expressing the solutions in terms of three complex quantities (u, v, ζ), whose real parts represent the veloc-

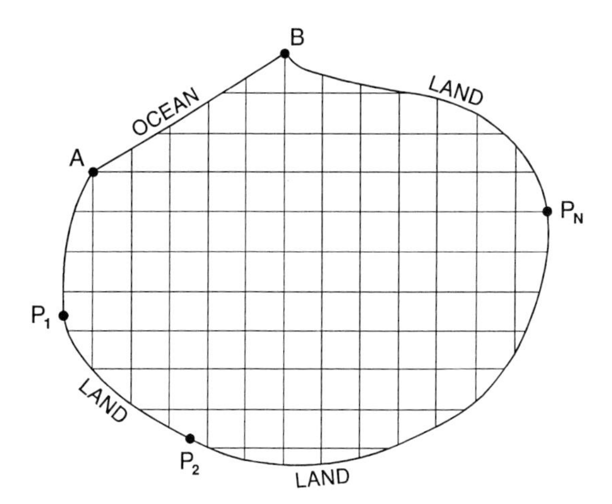

Figure 12.1. Schematic diagram of a semi-enclosed sea exposed to the ocean across a boundary AB, with interior grid of nonuniform depth for solution of the tidal dynamic equations by finite-differences.

ity components and surface elevation in phase with the disturbing potential and whose imaginary parts are lagged in phase by 90°.

The finite-difference approximation for the equation of mass conservation at an 'elevation' point $p_{r,s+1}$ (say) is:

$$2\Delta x.\Delta y.\mathrm{i}\omega\zeta_{r,s+1} = \Delta y[(Du)_{r+1,s+1} - (Du)_{r-1,s+1}] + \Delta x[(Dv)_{r,s+2} - (Dv)_{r,s}],$$

where Δx, Δy are the horizontal distances associated with unit increments in r and s, respectively. At a 'current' point such as p_{rs}, the combinations

$$(cv_0/D - \mathrm{i}\omega)u_{rs} - fv_{rs},\ (cv_0/D - \mathrm{i}\omega)v_{rs} + fu_{rs}$$

(f = Coriolis frequency) are similarly equated to the respective differences

$$g\ (\zeta_{r+1,s} - \zeta_{r-1,s})/2\Delta x,\ g(\zeta_{r,s+1} - \zeta_{r,s-1})/2\Delta y$$

and other terms calculable from the tide potential. The system of equations is closed by equating to zero the components of current (u,v) normal to each land boundary (APB in Figure 12.1), and specifying the elevation ζ at all points along an open-sea boundary such as AB.

Thus, the solution is reduced to a set of simultaneous equations for the unknown parameters u,v,ζ at their allotted points p_{rs}. Assuming the physics is adequately represented, the only fundamental limitations are the size of matrix which the computer can invert and the accuracy of the specified elevations along AB. This is in marked contrast to the methods described in Chapter 11, which aimed at finding an exact mathematical solution for tides in basins of smooth geometrical shape and uniform or gently sloping bathymetry. Here, mathematical elegance is exchanged for greater realism and flexibility. Both amplitude and phase (Arg ζ) are derived, so the traditional 'cotidal maps' were easily enhanced by the addition of co-amplitude contours.

There were variants to the basic approach described above. Some investigators eschewed the condition of zero flow across coastal boundaries in favor of specifying elevations at points such as P1 ... PN in Figure 12.1, which are often readily available and more reliable than an assumed variation along AB. That approach had the disadvantage, realised before electronic computers were available,[7] that coastal elevations are not usually compatible with a null flow condition, owing to small irregularities in the coastal profile. On a global scale, the result of constraining coastal elevations was a small but significant violation of conservation of ocean mass.

Those who were more interested in the tides in shallow seas and estuaries, especially in association with storm surges, found it necessary to include nonlinear terms in the equations, such as the gradient of the *Bernoulli pressure* $(\frac{1}{2})u.\mathrm{grad}(u)$, and a better representation of the friction speed v_0 related to the local ambient speed. This prevented removal of the linear time-variable $e^{-i\sigma t}$; instead, the solution had to be derived by stepping in time as well as in space, with a long 'spin-up' period to reach stability.[8]

As access to and familiarity with the new computers were acquired, solving the tidal equations by finite differences soon became a popular subject for research. Walter Hansen (1909–1991), of the Institut für Meereskunde of the University of Hamburg, was one of the pioneers. Starting with the North Sea[9] in the early 1950's with a mesh of about 1000 points, Hansen expanded into the North Atlantic Ocean and even the whole Atlantic,[10] at the risk of dubious conditions along the southern boundary between South Africa and Argentina. The results of these early experiments were mainly of historic interest. Later followers of Hansen's school of 'Numerical–Hydrodynamical Modeling' at Hamburg made more substantial advances with the aid of larger computers and improved methods.

The Liverpool Tidal Institute (Chapter 11) was slow to enter the computer field because Doodson, who remained as Director until 1960, himself a master computer of the old school, considered the electronic computers of the 1950's unnecessary and untrustworthy. When the LTI started to use the electronic computers of the University of Liverpool under Doodson's successor, Jack Rossiter (1920–1972), their modeling efforts were largely directed towards the prediction of sea floods in the North Sea caused by tides and storm surges. The latter subject had become of national importance since the disastrous floods of 1953 which also stimulated similar effort in Holland and Germany. Rossiter himself had investigated the solution of oceanic tides by finite-difference equations using relaxation methods.[11]

In the 1960's, schools of tidal modeling emerged in the Russian Institutes of Oceanology in Moscow and Leningrad. The Moscow group computed the tides in large ocean areas such as the Pacific, using finite-difference solutions to fit extensive coastal elevation data,[12] but as in Hansen's attempts with the

Atlantic, they were limited by uncertain conditions at open boundaries. Solutions were, in fact, quite sensitive to details along open boundaries; it was not until computers of sufficient size to handle the whole world ocean emerged that they could be dispensed with.

Before considering the early stages of digital modeling for the world ocean, we shall briefly pay some attention to analogue devices which also gave good results for the tides in small seas.

Analogue devices for shallow seas

A new 'National Institute of Oceanography' to study all aspects of the sea was created in Britain in 1949 under the directorship of G.E.R. (later Sir George) Deacon, (1906–1984), who earned a reputation for choosing scientists with unconventional approaches to work at his Institute. In 1959, Deacon invited a Japanese electronic engineer named S. Ishiguro to the NIO to design an 'Electronic Model' for tides and surges in the North Sea. (His son Kazuo Ishiguro is now a highly esteemed English novelist.) S. Ishiguro worked for more than 25 years developing a series of models of increasing complexity, including one for Chesapeake Bay commissioned by an American group. His final model of the North Sea proved to be an accurate tool for predicting coastal floods, given meteorological data in real time, but by 1980 such methods had been entirely superseded in the UK by the digital models developed by a team at LTI under Norman Heaps (1928–1986).

Like the digital models, the Electronic Model divided the sea area by a mesh similar to Figure 12.1. At the center of each mesh an electronic circuit was set as sketched in Figure 12.2, with components representing the physical dimensions of the sea; electromotive force e (measured at the center) represents the sea surface elevation, and currents i_x, i_y the components of horizontal water transport. The capacitor C regulates the continuity equation, inductances L control horizontal acceleration, and resistances R control friction. Increments of emf are injected at the small circles; arrows at the four inner circles indicate the transfer of Coriolis force from the potential differences across resistors at right angles, with the symbol Ω_e standing for f. At the four outer circles, increments e_{wy}, e_{py}, e_{ty} represent external force components in the y-direction due to wind stress, atmospheric pressure and tide potential, respectively, (similarly for components in the x-direction). As such, the electrical equations may be shown to be exactly analogous to the corresponding hydrodynamic equations.[13] Nonlinear friction was also accommodated later, and all time processes were executed continuously without the need for 'time-stepping'.

The circuitry just described is evidently designed for studying the response of the sea to atmospheric forces, but Ishiguro's North Sea model was also shown to give excellent definition of the major harmonic components of the tide.[13]

The other type of analogue model to be used for tides was the 'Coriolis' Rotating Platform ('Plaque tournante') of the University of Grenoble, France,

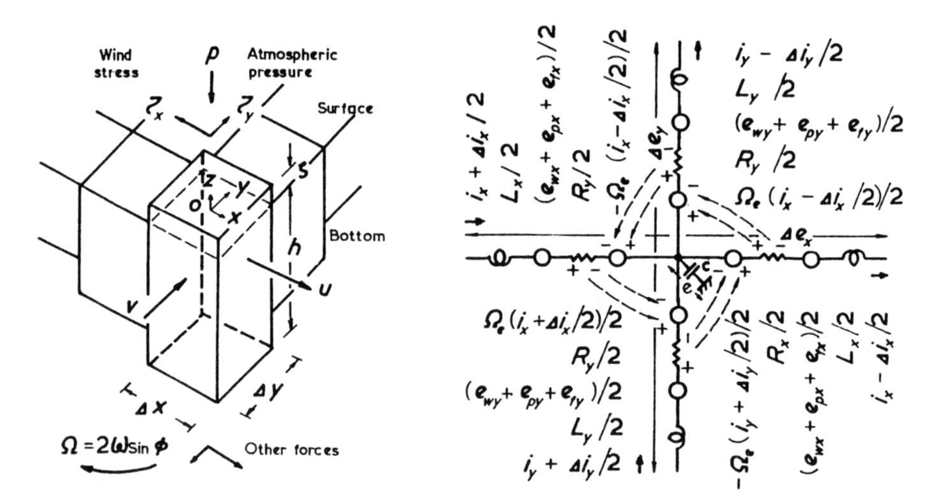

Figure 12.2. Basic principle of analogue representation of the stresses on a small rectangular element of sea of depth h, (left), by a node of electronic circuitry (right). ('Atmospheric pressure' may be replaced by the tide potential, as appropriate.) The capacitor C controls mass conservation; inductances L control horizontal acceleration; resistors R control friction. Transfer of electromotive force across resistors, indicated by broken arrows, satisfies the geostrophic (coriolis) stress relation. The emf at the central point represents surface elevation. See text for further details. (Diagram courtesy of S. Ishiguro.)

(Figure 12.3). This was a linearly scaled model of the English Channel (La Manche), and was similar to other hydraulic models except that the whole sea *rotated* on a turntable to simulate the 'Coriolis' effect. The external tide at the ocean and at Dover Strait were simulated by feeder canals whose water levels were controlled by appropriate harmonic motion; special roughness elements were inserted at all parts of the modeled sea to simulate carefully scaled frictional drag. Small movable probes were used to measure the tidal motion of the surface.

The 'Coriolis' platform was constructed in 1960 for the purpose of studying the effects of a grandiose tidal power plant projected by *L'Electricité de France,* to enclose the whole of the Baie de Mont St. Michel as far as the Iles Chausey with a barrage some 35 kilometers long. The well known and successful tidal power plant in the River Rance above St. Malo was originally intended as a pilot for the Chausey project. The larger project was ultimately shelved in favor of nuclear power, but the rotating model was put to good use for intensive studies of the tides, (and later for other hydrodynamic studies). Figure 12.4, for example, shows the amplitude and phase contours for the nonlinearly generated term M_4, taken from the model.[14] This wave is largely responsible (with M_6) for the notorious 'Double High Water' around Southampton and the long stand of High Water at Le Havre, which had not previously been satisfactorily explained. The two amphidromic systems centered near Newhaven and west of Cherbourg were unexpected.

However, the 'Coriolis' Model of La Manche suffered the same fate as the

Figure 12.3. The 'Coriolis' Rotating Platform of the University of Grenoble, as used with a hydraulic model of the tides in La Manche (English Channel). With a diameter of 14 meters, the model had a horizontal scale of 1/50, 000 and a vertical scale of 1/500, corresponding to a time scale of 1/2, 250. Rotation at a stable 50.4 seconds period, representing (2π/Coriolis frequency), simulated the Coriolis stress at the appropriate latitude. The scaled period of the M_2 tide was 20 seconds. Since 1987, the platform has carried a cylindrical tank of 13 meters diameter, used to simulate internal waves with rotation. (Photograph courtesy of Gabriel Chabert-d'Hières.)

electronic models through being overtaken by the digital methods, which produced more accurate results with greater flexibility and less trouble. Despite their ingenuity, their interest is now mainly historical.

Tide models for the world ocean

It was evident that a 'breakthrough' in digital tide modeling would come with the appearance of a computer with enough capacity to solve the equations on a grid covering the world's oceans, thus avoiding the open boundary problem. In principle, the only boundary conditions required are either null flow along land boundaries or constraints to the tidal elevation at a sequence of points along the boundaries. The latter constraints may be relaxed if the physics is perfectly represented, but this is hard to achieve. There were adherents of both null-flow and data-constrained models, but the models constrained by data gave more accurate results in most respects, despite the difficulty with mass conservation mentioned above.

First to present a global solution was Professor Chaim L. Pekeris (1908–1993) of Israel, who is also reputed for his distinguished contributions to many branches of mathematical geophysics and meteorology. His first global solution for M_2 (of the unconstrained null-flow type) was produced by use of

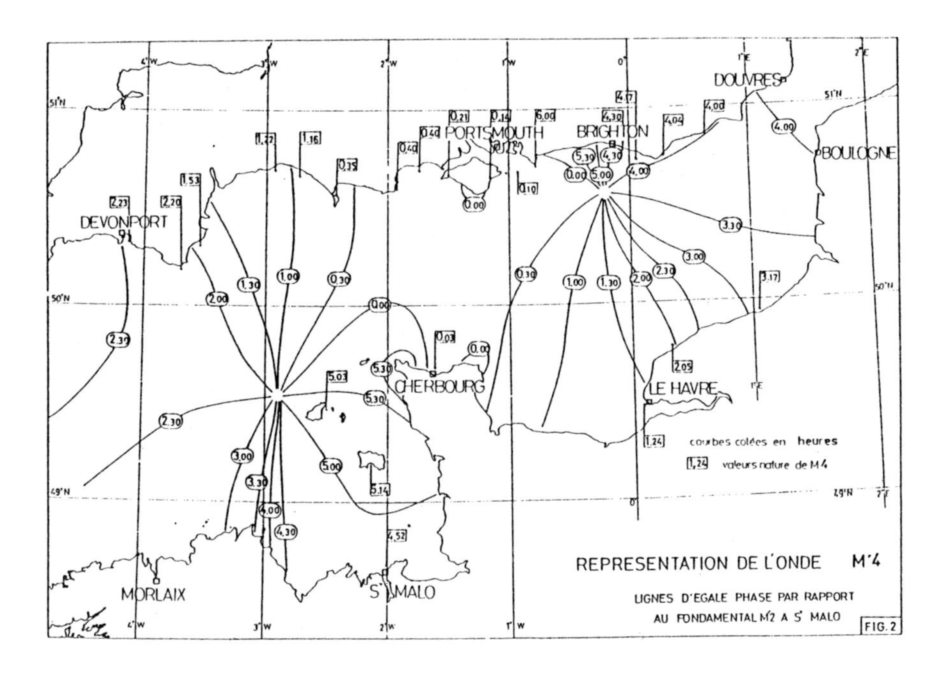

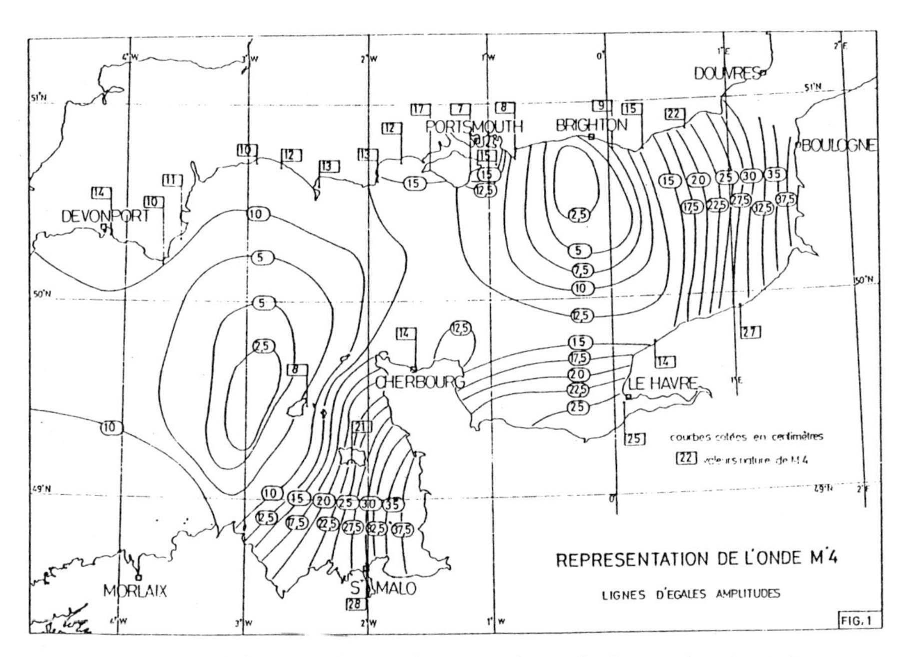

Figure 12.4. Maps of the 'overtide' M_4, showing isolines of (above): phase lag in hours; (below): amplitude in centimeters, derived from measurements with the 'Coriolis' Rotating Platform (Figure 12.3). Many other leading overtides and linear constituents were produced.[13] Representation of overtides by mathematical models would require nonlinear terms to be added to Laplace's equations.

the *WEIZAC* computer which had been built to Pekeris's own design in the Weizmann Institute at Rehovot. Pekeris presented a preliminary tidal map – the first global map ever to show amplitude contours – at the IUGG General Assembly at Helsinki in 1960, but he did not see fit to publish his results in the literature until 1969, after some revision.[15]

Joseph Proudman, present at the Assembly, rose to congratulate Pekeris, adding that he had not expected to see this problem solved during his lifetime. (In retrospect, the solution was far from perfect, but one can understand Proudman's reaction, having spent most of his professional career searching for such solutions by analytical mathematics.) Between 1960 and 1969, Pekeris and his colleagues revised their linear representation of friction, tested their finite-difference computational procedure by comparison with an analytical solution for a geometrical sea of uniform depth, and (later) even verified Laplace's Tidal Equations by an unusual method.[16]

By 1973, at least six other global tide models had been constructed in different countries, using different computers and methods. A global model computed at the Japan Meteorological Agency was actually the first to appear in the literature,[17] but it had no pretence to accuracy. Figure 12.5 shows six other solutions (a–f) for the M_2 tide in the Atlantic Ocean, extracted from global maps. (f) is from the work of C.L. Pekeris, mentioned above,[15] and (e) is the pre-1950 empirical map of G. Dietrich described in Chapter 11. The origins of the first four maps are:

(a) Institute of Oceanology (Moscow),[18]
(b) University of California, La Jolla,[19]
(c) Institut für Meereskunde, Hamburg,[20]
(d) Institute of Oceanology (Leningrad).[21]

See the Notes and References for names of individual authors.

Maps (a) and (b), and of course (e), are constrained to coastal data, and so rely on the latter to embody the net effects of friction from shallow seas. Maps (c), (d) and (f) use only the null-flow condition, but they include frictional terms with adjustment of parameters to provide an optimal but rough average fit to known elevations.

Without going into too much detail, one sees that all models agree in placing a cyclonic amphidromic system centered in the northwest Atlantic, but they differ radically in the South Atlantic. The null-flow models (c,d,f) all place an anticyclonic system between South Africa and South America. This was at first claimed as an important new discovery, but later more accurate models (e.g. Figure 12.6) show that the area is in reality occupied by an elongated trough of low amplitude, terminating in a *pair* of oppositely rotating amphidromies off Brazil, a situation to which the old empirical map of Dietrich (e) perhaps comes closest.

Amphidromic systems have more visual impact on maps than their importance deserves. More important are the regions of maximum amplitude off

Figure 12.5. Six early computations for the M_2 tide in the world ocean, compared here in the Atlantic Ocean, as grouped by M.C. Hendershott from diverse sources. Above, left to right; (a), (b), (c); below; (d), (e), (f). – see text for details and references. (Reproduced from *The Sea* (edited by E.D. Goldberg et al.), Vol. 6, Chapter 2, Wiley-Interscience, New York, 1977.)

western Europe and at the mouth of the River Amazon at the equator. All maps shown, excepting (d), agree on these zones, at least qualitatively. None adequately describes the complex system with large amplitudes known to exist on the wide shelf east of Patagonia.

Ocean loading and self-attraction

Disagreements in the Atlantic as shown in Figure 12.5 and worse discrepancies in other oceans suggested deficiencies in physical representation. A major deficiency, soon recognised, was the neglect of earth loading and self-attraction of the ocean tide itself.[22] From the geophysical discussions of Chapter 10 it was well understood that these two effects distort the primary tidal potential U of the moon and sun as felt by the ocean by the factor

$$\gamma_2 = 1 + k_2 - h_2 = 0.693,$$

involving the Love Numbers k, h of degree 2. This implies that the spatial gradient on the right hand side of LTE, normally written

$$(g/a)\delta/\delta\theta(\zeta - \zeta_{\text{equilibrium}})$$

is in effect

$$(g/a)\delta/\delta\theta(\zeta - \gamma_2 U/g).$$

It was now realised that for accuracy the final bracket should be further expanded to read

$$\zeta - \gamma_2 U/g - \sum_n \gamma'_n \alpha_n \zeta_n,$$

where $\gamma'_n = 1 + k'_n - h'_n$ are the *loading* Love numbers associated with the nth degree spherical harmonics ζ_n of ζ itself, and

$$\alpha_n = [3/(2n+1)] \times (\text{sea density/earth density}) = 0.563/(2n+1).$$

Neglect of this term in all models up to 1972 was causing serious errors. It will be readily appreciated that n has to be taken to high values in the summation to yield an adequate spherical harmonic expansion of ζ. Numerical values of k'_n, h'_n were becoming known to high order from models of the earth's interior.[22]

Formally, this is a considerable complication, because it implies an integral of ζ itself over the whole globe, making the equations to be solved of *integro-differential* type. It took several years for those concerned to find ways of making the necessary adjustment. Eventually, three methods were employed, with varying degrees of approximation:

(1) By summing a special series of *Basis Functions* to give least-squares fit to known tidal elevations at mid-oceanic islands (constraints for the coastal elevations being guaranteed),[19A]

(2) By an approximation of type

$$\sum_n k'_n \alpha_n \zeta_n \approx K\zeta, \sum_n h'_n \alpha_n \zeta_n \approx H\zeta,$$

where K and H are roughly appropriate constants,[15A,20A,21A,23]

(3) By evaluating all ζ_n from another global solution for ζ, known to be reasonably accurate.

('A' has been appended to reference numbers when one of the authors of solutions (a–f) shown in Figure 12.5 was also involved in producing an improved model with the above corrections.)

The new solutions appeared in the years 1978–81, and represented a 'new wave' of tidal maps of much more mutual consistency than before. It would be wasteful of space to show them all, but Figure 12.6 shows the complete map for

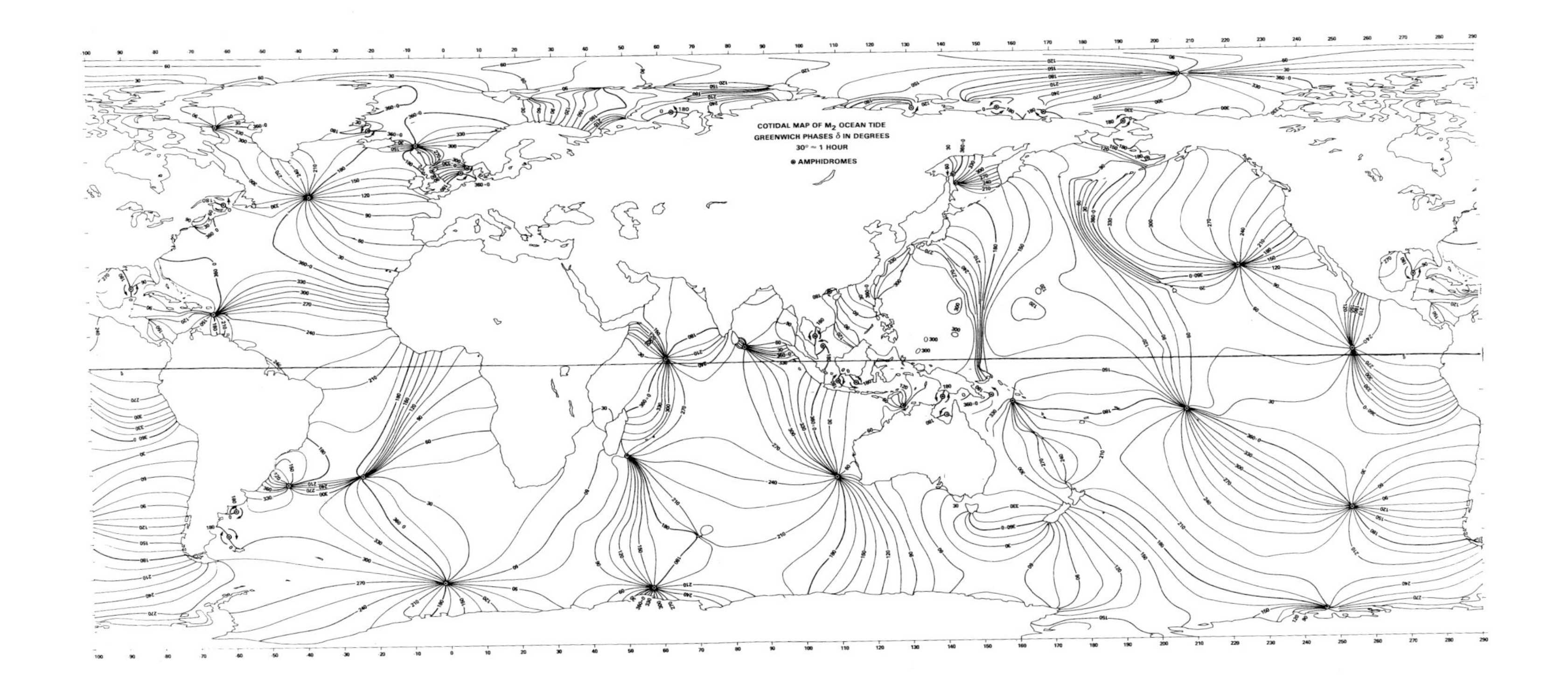
COTIDAL MAP OF M_2 OCEAN TIDE
GREENWICH PHASES δ IN DEGREES
$30^\circ \approx 1$ HOUR
⊛ AMPHIDROMES

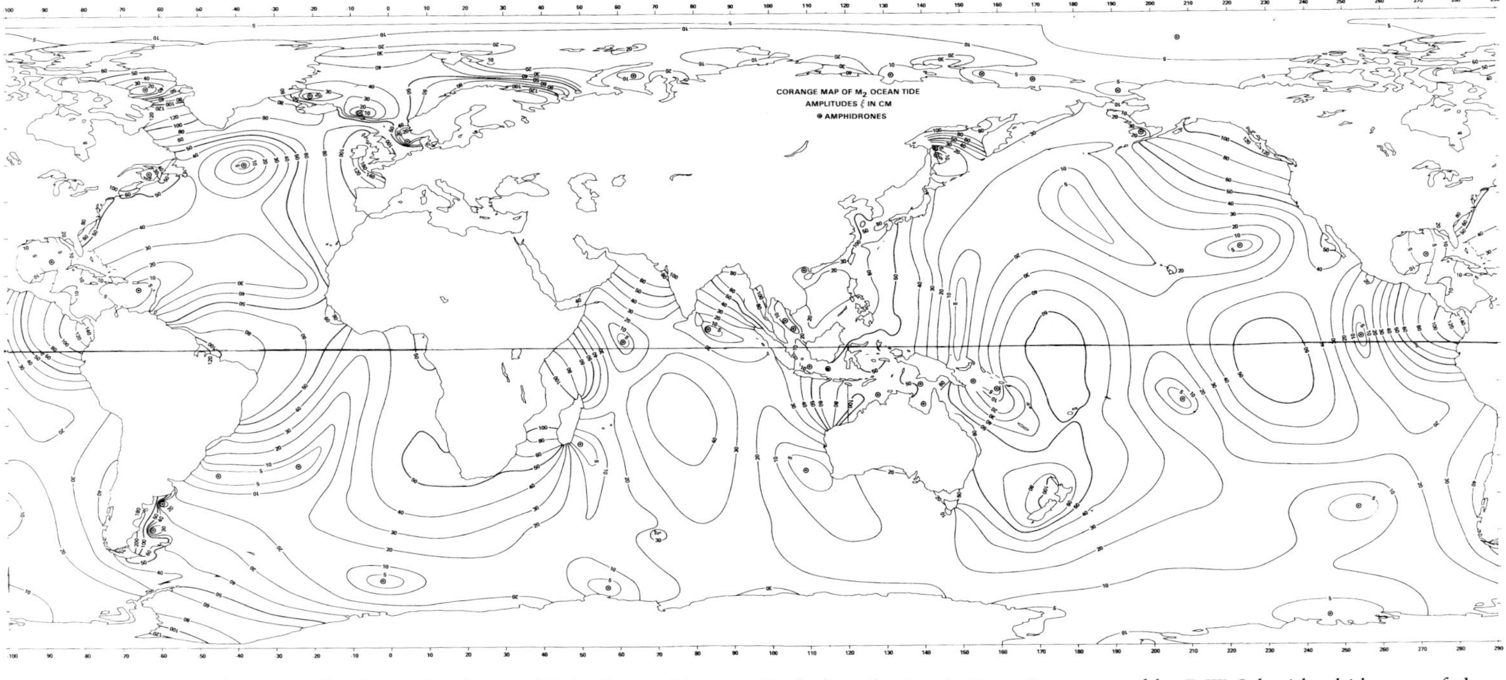

Figure 12.6. Numerical-empirical solution for the M_2 tide in the world ocean (including the Arctic Ocean), computed by E.W. Schwiderski by use of the methods described in ref. 23. Above: Greenwich phase lag; below: amplitude in centimeters. The map projection is linear in latitude and longitude. (Courtesy, E.W. Schwiderski, private communication.)

M_2 by Schwiderski,[24] which was in some respects the best of that generation of global solutions. Ernst W. Schwiderski deserves mentioning by name, because his tidal models for $(Q,O,P,K)_1$, $(N,M,S,K)_2$, and later Mf and Ssa, though not beyond criticism, were so carefully worked out that his maps, produced for the US Naval Surface Weapons Center around 1980–83, were accepted as a standard of excellence for over 12 years. The maps, in digital arrays of 1° resolution, were used in many applications and so have earned a special place in the history of the subject, though superseded later. Schwiderski's method[23] was the solution of elaborate physical formulations, constrained to fit a large body of coastal and island data. He used a nonlinear form for friction, including a term of debatable validity to represent horizontal eddy viscosity, and an approximation to ocean loading of type (2), with $1+K-H=0.10$.

Schwiderski's maps for M_2 in Figure 12.6 have an accuracy better than 10cm in amplitude and 10° in phase over much of the deep ocean, but they are less accurate over continental shelf seas. They are still used by some as the basis for the higher approximation to ocean loading (type 3). The latest generation of tidal models, based on satellite altimetry, will be discussed in Chapter 14.

Normal modes of the ocean

The earliest computations of the global tides, though of no great accuracy, revealed that solutions were sensitive to whatever parameters were used to represent friction and bathymetry, a symptom of closeness to resonance. This begs the questions: 'how close are the ocean's resonant frequencies to the astronomical tidal frequencies, and how is the ocean's response to tidal forcing affected by friction?' Both questions are of fundamental importance.

Calculating the ocean's resonances (that is, its *normal modes*) amounts, in brief, to finding the eigenfrequencies σ_n and their associated eigenfunctions $E_n(\theta,\phi)$ in spherical coordinates θ,ϕ such that

$$E_n(\theta,\phi)\exp(-\mathrm{i}\sigma_n t)$$

satisfies the free-wave equations, (that is, LTE without the external forcing), and all null-flow boundary conditions necessary to describe the configuration of the ocean. The formal mathematical framework for this had been laid out by Proudman (Chapter 11) in 1917, long before there was any hope of being able to compute normal modes for a realistic ocean. The first complete set (E_n, σ_n) for Laplace's ocean (or atmosphere) of uniform depth covering a sphere was carried out in the late 1960's at the British NIO;[25] a complete sequence of the modes and their resonant frequencies were elegantly plotted against a parameter $\varepsilon^{-\frac{1}{2}}$ where

$$\varepsilon = 4a^2\Omega^2/gD$$

is the same as Lamb's parameter β. The waves of class 1 (gravity waves) and class 2 (vorticity waves) were found to be further divided into two types, in the first of which σ_n is asymptotic to $\varepsilon^{-\frac{1}{4}}$ for large ε, and in the second, σ_n is asymptotic to $\varepsilon^{-\frac{1}{2}}$. Note that for realistic ocean depths, ε is about 10–100, that is 'large'. The concept of eigenfunctions for numerically *negative* equivalent depths D ($\varepsilon < 0$), which had been discovered a little earlier in the context of atmospheric tides (Chapter 10), was re-confirmed and seen to be necessary for a complete description of the dynamical response to forcing.

In the same computation, wave modes $E_n(\theta, \phi)$ traveled both eastward and westward, and were infinite in number, as postulated by Proudman. Their resonant frequencies were clustered around Ω, 2Ω, 3Ω, . . for realistic values of ε, that is fortuitously in the region of the tidal frequencies. Thus, one may expect tides in a spherical ocean to be near resonance, as Laplace and Hough had suggested. Similar, but more complicated, results were also computed at NIO for a hemispherical ocean bounded by meridians.[26]

Computing the normal modes of a *realistic* ocean is less elegant mathematically than in the cases just described, but the results are more interesting in a geophysical context. In the years around 1980 independent solutions were published from the Institute of Oceanology, Leningrad[27] and from the University of Chicago,[28] both preceded by pilot calculations with smaller areas of ocean. Their results are similar, but not identical. Reference (27) used a rather coarse grid of 5° in latitude and longitude while (28) used an elaborate mesh of spherical triangular elements, specially adjusted to take account of large islands like Madagascar and New Zealand. These islands produce important diffraction patterns in the tides.

As in the geometric oceans, resonances σ_n are densely clustered about the natural tide frequencies, and their associated wave forms $E_n(\theta, \phi)$ qualitatively resemble tidal maps[28, Pt. 2]. In theory, all normal modes will be generated by a given gravitational potential, for example of form

$$P_2^{\,1}(\theta, \phi)\exp[\mathrm{i}(\sigma t + \phi)]$$

(ϕ is east longitude), with amplifications depending not only on the smallness of $|\sigma_n - \sigma|$ but also on the spatial convolution of E_n with $P_2^{\,1}$. Figure 12.7 shows the set of spatial amplifications excited by the $P_2^{\,1}$ harmonic (lower panel) and by $P_2^{\,2}$ (upper panel), at all resonant frequencies $\sigma_n/2\pi$. Each involves the same set of modes but the amplifications (termed in the Figure 'spectral densities') differ for the two spatial harmonics. The heavy black arrows show the frequency limits of those modes close enough to typical tidal frequencies to form an important contribution to the resonance, according to the linear response function

$$\Phi(\sigma, \sigma_n) = [(1 - \sigma/\sigma_n)^2 + (1/2Q)^2]^{-\frac{1}{2}},$$

where Q = *Quality Factor* = $\sigma \times$ energy/dissipation rate. (Φ has a maximum value $2Q$ at $\sigma_n = \sigma$.) By matching summations

$$\sum_n A_n E_n \Phi_n$$

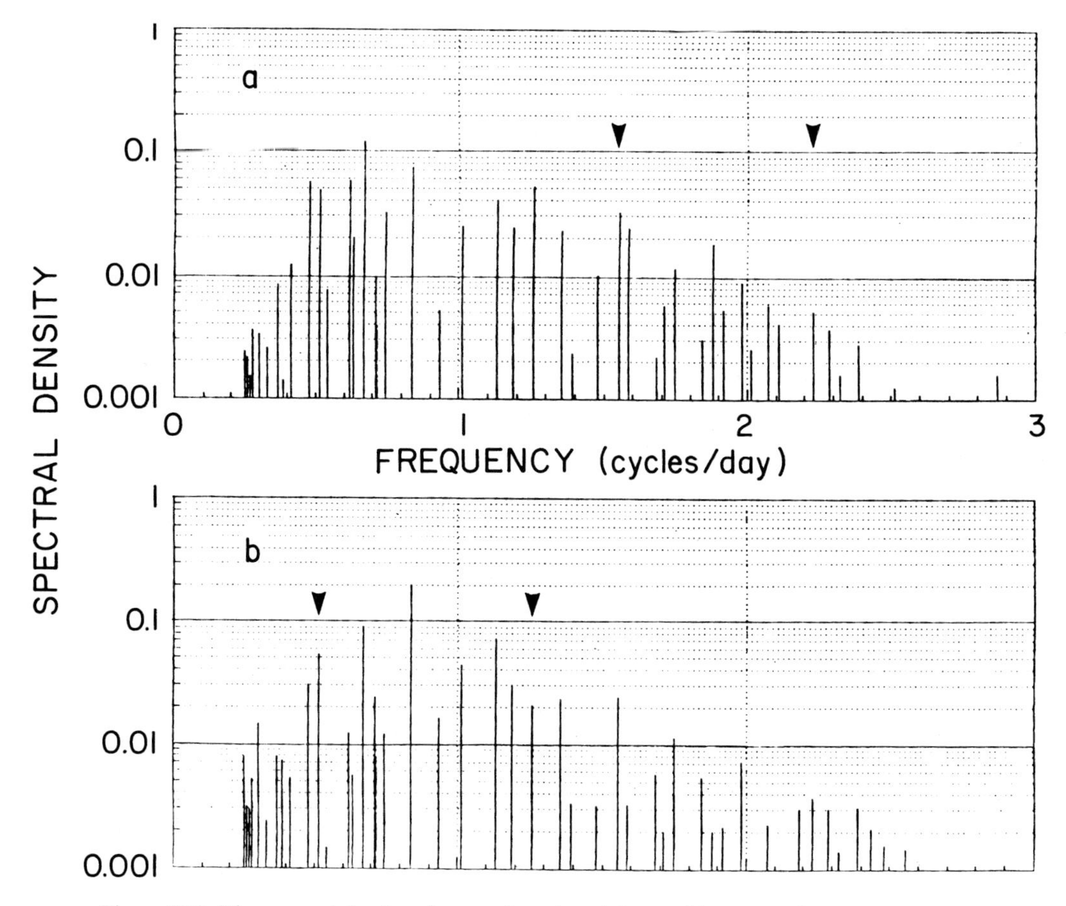

Figure 12.7. The spectral density of normal modes of the world ocean at frequencies σ_n, convoluted with the principal tide potentials proportional to P_2^2 (semidiurnal – above) and P_2^1 (diurnal – below). Vertical arrows show the effective frequency limits of the admittance $\Phi(\sigma, \sigma_n)$ over which these modes contribute to the tides of the ocean – see text.

with observed tidal data, it was deduced that the diurnal tides have an effective *Q* of about 7 and the semidiurnals have *Q* about 15.[28, Pt. 4] Thus, the diurnal tides are more strongly affected by friction than are the semidiurnals. This was the first discovery of such facts; the *Q* values just quoted were considerably lower than those deduced from a previous discussion of tidal *Age*,[29] but they were confirmed by later investigations based on satellites, (Chapter 14).

Another interesting fact observable in Figure 12.7 is that, whereas the M_2 tide (1.932c/d) has three or four modes which contribute to it in similar proportion, the diurnal tide O_1 (0.930c/d) appears to be dominated by a single mode of frequency 0.84c/d. The wave form E_n for this mode has strong Kelvin-wave trapping round Antarctica and a half-wave resonance across the Pacific Ocean, features which are indeed observed in the known diurnal tides. There is insufficient space here to discuss these and other features revealed by normal mode analysis, but we shall return to the subject in Chapter 15.

Spectral analysis of data – noise and coherence

Computations with time-series such as the analysis of power spectra, autocovariances and cross-covariances, quickly became popular in the 1950's in the fields of meteorology and oceanography, especially in application to waves. One might have thought that tidalists, with their long series of hourly data from tide-gages, would soon take advantage of the available computer programs, but they were preoccupied with the extraction of harmonic constants by traditional, well understood methods. Even when conversion to the electronic computer became imperative in the early 1960's most tidal authorities concentrated at first on automating the old paper-and-pencil methods.

At this point it is hard to find circumlocution to avoid mentioning the name of Walter H. Munk of Scripps Institution of Oceanography of the University of California at La Jolla. Munk did more than anybody to re-vitalise all branches of tidal science during the 1960's and early 1970's. Introduced to oceanography in the war years by his friend and early mentor Harald Sverdrup, whose work on tidal currents was described in Chapter 11, Munk brought fresh scientific approaches to several branches of geophysics, especially those involving wave motion. His interest in tides stemmed from a ten-year study of variable earth rotation,[30] from which he perceived that the problem of tidal increase in the length of day was far from solved, and that none of the recognised tidal authorities showed much interest in trying to solve it.

As Director of the Institute of Geophysics and Planetary Physics (IGPP) of the University of California at La Jolla, Munk set up a Tidal Study Group (TSG), consisting of a succession of oceanographers, technicians and computer specialists whom he led in a coordinated research programme. Some aspects of this programme will be discussed in Chapter 13; at present we are concerned with computer-orientated activity. As a start, many of the long series of hourly tide-gage data from diverse parts of the world were assembled in computer-accessible format at IGPP, with careful scrutiny for transcription errors. (Such errors had frequently been overlooked in data series which had been assembled before automatic computers were available.)

The TSG paid much attention to the *power spectrum* of sea level, a hitherto neglected topic. The traditional view was that, at frequencies below about 3 cycles per day (cpd) the variations of sea level could be considered to be similar to the generating potential, with a slight disturbance by weather. Variability at frequencies not present in the tide potential, or indeed at the tidal frequencies themselves, had been ignored for convenience. Such attitudes were pushed to absurdity when attempts were made to 'improve' tidal predictions by proliferating the number of small harmonic terms to be independently extracted from the data, without considering the structure of the ocean's *admittance* (linear and nonlinear) which produced these terms.

By computing the power spectrum of many years of hourly sea levels, the TSG found that the spectrum had a non-tidal *continuum* of characteristic shape. Spectral density rose more or less monotonically from low levels above about 6 cpd to very high levels at the lowest frequencies which could be resolved. The tides appeared as strong quasi-line spectra embedded in the continuum, but the smooth shape of the continuum could be traced on either side of the principal groups of tidal lines.[31] On closer inspection it was found that the continuum rose sharply very close to the tidal lines, effectively converting the lines into narrow bands of cusp-like shape.[32] These *tidal cusps*, as they were called, could be interpreted as slow modulations of the tides on a climatic time scale. Their precise cause was hard to identify, but the most promising suggestion was that they were due to the surface manifestation of internal tides. In any case, the presence of cusps made the direct extraction of Darwinian 'Harmonic Constants' much less reliable than was previously thought.

The same spectral filters which were applied to the sea level data were also applied to simultaneous series of the tide potential itself, now routinely computed to a useful accuracy. This parallel process was in effect a *cross-spectral* analysis of the two series, allowing a separation of the spectrum of sea level into two parts, one part coherent with the generating potential, the other part consisting of 'noise'. The coherent part defined reliable estimates of the oceanic admittance, expressible as an amplitude ratio and a phase lag. The spectral noise ratio defined the confidence limits of such estimates.

All the above characteristics are illustrated in Figure 12.8, which shows the cross-power spectra of 19 years of tide-gage data from Honolulu, as analysed by Munk's TSG in 1965. The most important spectral range, 0–2.5 cpd, is shown in relation to the tide-generating potential. The spectrum of the potential itself is shown in the top panel; below it the spectrum of sea level is divided into its tidally coherent part (filled columns) and its noise content (unfilled columns). The two bottom panels show the corresponding tidal admittances and their confidence limits. The admittance is seen, typically, to be a relatively smooth but not a linear function of frequency.

A great deal of work of this sort by the TSG cast the tides in a new analytical light. Various technical refinements were devised to allow for an anomaly in the solar tides due indirectly to radiation through interaction with the surface atmospheric tide. Allowances were also made to accommodate the small element in the lunar tide potential of degree 3, and for weak nonlinear effects.[33] The same techniques, with some revision, were applied at the UK–NIO to tides and surges and their interactions in the context of coastal flood forecasts for the North Sea.[34]

Data analysis and prediction by the Response Method

The concept of *admittance* so introduced suggested a new formalism for predicting tides. Instead of summing an indefinitely large number of harmonic constituents in the tradition of Darwin with further extension by Doodson, the Munk TSG proposed computing tides as an integrated product of the gener-

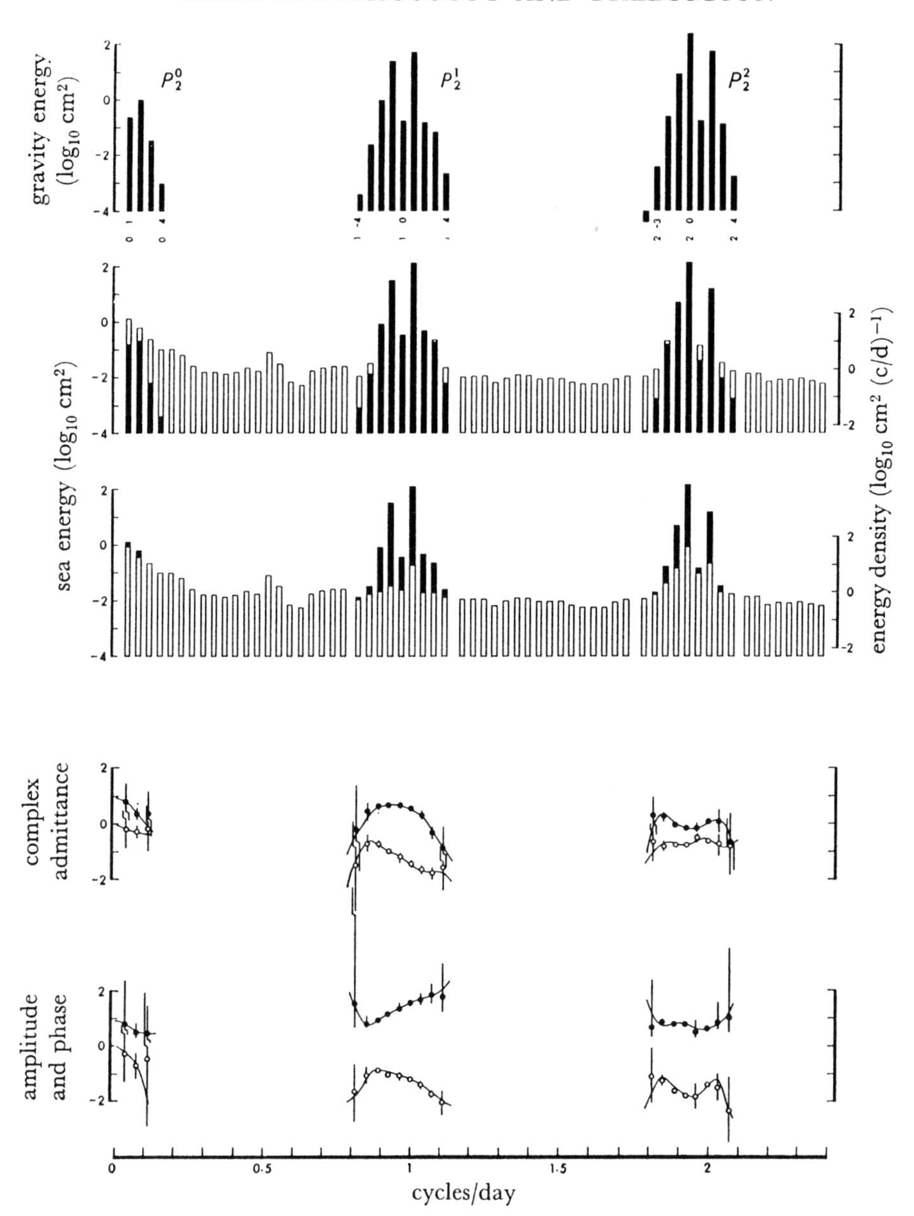

Figure 12.8. Cross-spectral analysis of 19 years of tide-gage data from Honolulu with differentiation between tidally coherent signal and noise, and hence estimation of the station admittance. Top panel: spectrum of tide potential; 2nd panel: spectrum of total sea level and its coherent part (black columns); 3rd panel: As in panel 2 but with non-coherent part (noise) as unfilled columns. All spectra are plotted on a logarithmic scale. Panels 4 and 5 show the admittance components; real and imaginary parts (above); amplitude and phase (below), with statistical estimates of confidence limits – see text. (Courtesy, The Royal Society.)

ating potential with a *response function*. The response function could be derived from the admittance, or from a direct analysis of the data in relation to the potential. All the harmonic structure was fully built into the computed potential, including small terms which would be hard to evaluate independently by classical harmonic analysis. Weakly nonlinear terms due to shallow water hydrodynamics and friction were also included. Anomalies due to solar radiation were accommodated through a separately computed *radiational potential*. This method of analysing and predicting tides was termed the 'Response Method'.[33]

Several careful, unbiassed, comparisons were made between predictions by the Response Method and by Harmonic Methods with the latest improvements. All demonstrated that the former method gives a more accurate prediction of $\zeta(t)$ with the use of fewer arbitrary constants, even when shallow-water terms are included.[35] However, the improvement in predictable variance is numerically small compared with the natural noise in sea level. Because of this, and the fact that the Response Method is harder for a routine operator to grasp, it has never been adopted for ordinary tide-table production. It remains essentially a research tool for specialists.

More of the pole tide

With a few unimportant exceptions, the 14-month pole tide in sea level had not been revisited since the time of Sir George Darwin (Chapter 10). In 1933 the International Association of Physical Oceanography, IAPO, (later IAPSO) set up a *Permanent Service for Mean Sea Level* (PSMSL) to collect and disseminate monthly mean data from tide-gages all over the world. By 1958, the PSMSL data bank contained many thousand station-months, with at least 16 stations having more than 42 years' data and some more than a century. Walter Munk saw that these data warranted a new attack on the problems of the pole tide. He and a colleague selected the eleven longest of the PSMSL records, including an average of six similar records from the coast of Holland, and subjected these to the *Blackman-Tukey* power-spectrum estimation process which involves computing a series of lagged auto- and cross-covariances.[36]

The computations were lightweight by modern standards, and the proud 1959–ish phrase 'performed with the aid of high speed digital computers' may now raise a smile, but this was the first time that such an analysis had been applied to tide-gage data. Power spectra over the whole frequency range showed that the pole tide signal was only just detectable above noise level, if at all. In fact, only three data sets gave significant results: The combined Dutch stations, one station (Swinemunde) in the Baltic Sea, and Marseille in the Mediterranean Sea.

There was an advanced discussion of the theoretical relation of an 'equilibrium' pole tide to the Love numbers, but the only deducible results from the above stations were that their signal appeared to have an amplitude about twice the equilibrium value and a phase lag indistinguishable from zero.

Later results based on Fourier analysis of more extensive data confirmed that the large amplitudes were confined to the east coast of the North Sea, the western Baltic Sea and the Gulf of Bothnia;[37] elsewhere the pole tide, if detectable, was close to 'equilibrium'. There were also speculations about a possible dynamic resonance in the eastern North Sea related to depth topography,[38] or whether the 14-month peak there could be driven by wind.[39]

Globally, the matter was fairly well clinched by a new analysis in 1989 by A. Trupin and J. Wahr at the University of Colorado, who used the *stacking* technique, whereby the records are weighted according to geographical position to match a P_2^1 spatial harmonic.[40] Their analysis showed a clear global pole tide signal, with enhanced amplitudes only in the northern sea areas mentioned. When these were omitted, the pole tide was virtually in equilibrium with the Chandler excitation. Further, the enhanced stations proved to be significantly coherent with barometric pressure and hence, by implication, with the wind stress. These results seem to dispose of the old suggestion that the oceanic pole tide may be damping the Chandler wobble.

Most recently, a group at the *Proudman Oceanographic Laboratory* (a successor, after other changes of name, to the LTI at Bidston), reproduced the observed North Sea pole tide by running a storm-surge model with 30 years of real weather data. Their results confirm the hypothesis of weather excitation.[41] However, the physical reality of a finely-tuned peak at the Chandler frequency in the spectrum of wind-stress is controversial. Amplification of weather noise by a local hydrodynamic resonance is still a plausible, if imperfectly understood, hypothesis.[38]

Quantifying global tidal dissipation

To close this chapter, we shall consider progress in an important research area, which, though not specifically computer-driven, fits most conveniently in the current time-frame. In Chapter 10 I brought the saga of the so-called 'lunar acceleration' to the point about 1920 when it was first convincingly shown that tidal braking of the earth's rotation (with coupled *deceleration* of the moon's longitude) was sufficient to account for the astronomical observations. Nearly half a century later, Walter Munk introduced his *Harold Jeffreys Lecture* to the Royal Astronomical Society in London[42] with the words: 'In 1920 it appeared that Jeffreys had solved the problem of tidal dissipation. We have gone backwards ever since.'

The retrogression referred to consisted partly in the growing incredibility of the Bering Sea as a sink for two-thirds of the world's tidal dissipation, and more importantly, that intensive analysis of the mean longitudes of the moon, sun and Mercury recorded since 1680 had practically doubled Jeffreys' estimate of the overall rate of global dissipation.[43] This was scientific *progress* of course, but the apparent agreement of 1920 between oceanographic and astronomic theory had been seriously undermined.

The astronomical method is straightforward in principle. The apparent mean longitude of the moon, (that is the observed longitude relative to the equinox, minus the orbital motion and the Adams-Delaunay planetary term) can be written

$$L = a + bT + (q + s)T^2 + B(T),$$

where T is the time in centuries, $2q$ (<0) is the true acceleration of the moon induced by tides, and $2s$ (>0) is the apparent acceleration of all other bodies which reflects the deceleration of the earth's rotation by tidal braking. $B(T)$ is the observed short-term (decadal) fluctuation due to exchanges of momentum and inertia within the geosphere, of great geophysical interest in itself but irrelevant to tides. The sun's mean longitude is similarly

$$L' = a' + b'T + (n'/n)[sT^2 + B(T)],$$

where $n'/n = 1/13.37$ is the ratio of mean motions, sun:moon. The quantity $sT^2 + B(T)$ may be easily eliminated from these two equations, and q evaluated by a least-squares parabolic fit to observations.

Applying conservation of angular momentum and Kepler's 3rd Law for orbital radius and period, the torque Γ about the earth's axis induced by the lunar tides is then expressed in terms of known quantities:

$$\Gamma = -\frac{2}{3}\frac{Mq}{1+r}(a_m)^2,$$

where M is the moon's mass, r the moon:earth mass ratio 1/81.3, and a_m is the semi major axis of the moon's orbit. Finally, the rate of energy dissipation by lunar tides is seen to be:

$$-\mathrm{d}E/\mathrm{d}t = \Gamma(\Omega - n).$$

The pioneering calculation along these lines was by Harold Spencer Jones (1890–1960), the tenth Astronomer Royal.[44] He obtained for the true lunar tidal acceleration of L,

$$2q = -22''.44 \pm 1.06 \text{ century}^{-2},$$

$$\text{from which } -\mathrm{d}E/\mathrm{d}t = (2.7 \pm 0.1) \times 10^{12} \text{ watts},$$

compared with the 1920 estimate 1.4 TW of Jeffreys.[43]

We shall not delve into why Jeffreys' first estimate was so low. It was based on current estimates of the accelerations $(q+s)$ and s from eclipse data recorded in antiquity, most of which have uncertain authenticity, debated to the present day. Jeffreys accepted the higher figure for dissipation in the later editions of *The Earth*,[43] and suggested, with foresight, that there may be other sinks of tidal energy additional to the shallow seas.

Munk was concerned at the gap which had opened up between Spencer Jones's astronomical estimate of 2.7 TW and the rough estimate of about 1.1–1.9 TW for the dissipation in shallow seas which had not been re-examined since 1920. In an attempt to verify one or other of the extremes, he integrated the rate of working of the moon on the tides of the *deep* ocean from the formula

$$W = \rho \int < \gamma_2 U \; \delta\zeta / \delta t > \mathrm{d}A$$

(U is primary tide potential, ζ is surface elevation, $<>$ implies average over a cycle). The best available data for surface amplitude and phase at the time (1960) were Dietrich's empirical maps (Chapter 11). Figure 11.5 shows only the phase lags, but other maps in Dietrich's publication gave rough indications of coastal amplitude. The integrand varies between positive and negative values over the oceans,[30, Table 11.3] but Munk's total integral came out to be +3.2TW (lunar), which considering the rudimentary nature of the data was commendably close to the astronomical figure.

In 1965, a student of Munk named Gaylord Miller (1931–1976) re-evaluated the dissipation in shallow seas, using better data and methods than had been available to Jeffreys and Heiskanen.[45] Miller brought Jeffreys' estimate of 0.75TW for the Bering Sea down to 0.24TW, but he increased the estimates for some other seas and added sea areas not considered by Jeffreys. His resulting total for M_2 was 1.7TW, which he considered an upper bound since only currents at the surface were reported. This was significantly greater than Jeffreys' 1.1TW but still far short of the total from astronomy. Here, it should be added that the first realistic numerical model of the M_2 tide in the Bering Sea, from the Institut für Meereskunde, Hamburg,[Chapter 11, ref. 7], reduced the dissipation rate in that sea still further to about 0.03TW. If Miller had used this value his total would have been only 1.5TW.

Munk[42] examined the possibilities for other sinks of energy, in the solid earth, and in the conversion of barotropic tidal currents to internal (baroclinic) tides, which tend to dissipate themselves by generating turbulence and vertical mixing. It appeared to be hard to account for more than 0.5TW by these mechanisms, so the gap in the energy budget remained. We shall hear of still further efforts to remedy this lacuna of physical understanding in the following two chapters.

Notes and references

1. Anonymous. Echo Sounders. *Internat. Hydrogr. Rev.*, **1**, (2), 39–49, 1924. Extensions in: *ibid.*, 2, (1), 53–121, 1924
2. LaFond, E.C. and B.K. Couper. The Bathythermograph. 2nd Internat. Conf. on History of Oceanography, 1972. *Proc. R. Soc. Edinburgh*, 72, B, 433–448, 1972.
3. Zetler, B.D. Military tide predictions in World War-2. Pp. 791–797 in: Parker, B.B. (ed.) *Tidal Hydrodynamics*, J. Wiley & Sons, New York, 1991.
4. Private communication (1975) from Walter Habich, tidal officer at the *Deutsche Hydrographische Institut*, Hamburg.
5. Eckert, W.J., Rebecca Jones, and H.K. Clark. Construction of the lunar ephemeris. Pp. 283–363 in: *Improved lunar ephemeris, 1952–1959*. Nautical Almanac Offices of USA and UK, U.S. Govt. Printing Office, Washington, D.C., 422pp., 1954.
6. King-Hele, D.G. – see Chapter 14, ref. 1.

7. Doodson, A.T., J.R. Rossiter, and R.H. Corkan. Tidal charts based on coastal data: Irish Sea. *Proc. R. Soc. Edinburgh*, **64**, (1), 90–111, 1954.
8. Banks, J.E. A mathematical model of a river–shallow-sea system used to investigate tide, surge, and their interaction in the Thames–North Sea region. *Phil. Trans. R. Soc. London*, **A**, **275**, 567–609, 1974. (This is by no means the first tidal model to use time-stepping, but an early and explicit example of the use of finite differences to solve non-linear tide equations in a two-dimensional sea.)
9. Hansen, W. Theorie zur Errechnung des Wasserstandes und der Strömungen in Randmeeren nebst Anwendungen. *Tellus*, **8**, (3), 287–300, 1956.
10. Hansen, W. The reproduction of the motion in the sea by means of hydrodynamical-numerical methods. Mitteilung Inst. Meereskünde, Univ. Hamburg, Pub. no. 5, 1–57, 1966. (Hansen's solution for the Atlantic tides, c.1952, is reproduced and described in pp. 502–503 of: Defant, A., *Physical Oceanography*, Vol. 2, Pergamon Press, Oxford, 598pp., 1961.)
11. Rossiter, J.R. On the application of relaxation methods to oceanic tides. Proc. R. Soc. London, **A**, **248**, 482–498, 1958.
12. Bogdanov, K.T., K.V. Kim, and V.A. Magarik. Numerical solution of the tidal hydrodynamic equations for the Pacific Ocean by the electronic computing machine BECM-2. *Trudy Inst. Okeanol.* **75**, 73–98, 1964.
13. Ishiguro, S. Electronic analogues in oceanography. *Oceanogr. & Mar. Biol. Annual Rev.*, **10**, 27–96, 1972. (Much of Ishiguro's work in this field was presented in unpublished Technical Reports; the above survey is his most substantial published review.)
14. Chabert-d'Hières, G. and C. Le Provost. Détermination des charactéristiques des ondes M4 et M6 dans la Manche sur modèle reduit hydraulique. *C.R. Acad. Sci., Paris*, **270**, 1703–1706, 1970. (Several similar papers by the same authors describe other tide constituents of 'La Manche'.)
15. Pekeris, C.L. and Y. Accad. Solution of Laplace's equations for the M_2 tide in the World Ocean. *Phil. Trans. R. Soc. London*, **A**, **265**, 413–426, 1969.
15A. Accad, Y. and C.L. Pekeris. Solution of the tidal equations for M_2 and S_2 in the World Ocean from a knowledge of the tidal potential alone. *Phil. Trans. R. Soc. London*, **A**, **290**, 235–266, 1978.
16. Accad, Y. and C.L. Pekeris. The K_2 tide in oceans bounded by meridians and parallels. *Proc. R. Soc. London*, **A**, **278**, 110–128, 1964. Also: Pekeris, C.L. A derivation of Laplace's tidal equation from the theory of inertial oscillations. *Ibid.* **344**, 81–86, 1975.
17. Ueno, T. Theoretical studies on tidal waves travelling over the rotating globe, Parts 1 & 2. *Oceanogr. Magazine*, (Japan), **15**, (2), 99–101, and **16**, (3), 47–54, 1964.
18. Bogdanov, K.T. and V.A. Magarik. A numerical solution of the problem of tidal wave propagation in the world ocean. *Izv. Atmos. Oceanic Physics*, **5**, (12), 1309–1317, 1969. (English transl.: *Bull. Acad. Sci. USSR, Atmos. Oceanic Physics*, **5**, 757–761, 1969.)
19. Hendershott, M.C. Ocean tides. *Eos*, (Amer. Geophys. Union), **54**, 76–86, 1973.
19A. Parke, M.E. and M.C. Hendershott. M2, S2, K1 models of the global ocean tide on an elastic earth. *Marine Geodesy*, **3**, 379–408, 1981.
20. Zahel, W. Die Reproduktion Gezeitenbedingter Bewegungsvorgänge in Weltozean Mittels des Hydrodynamisch-numerischen Verfahrens. Mitteilung Inst. Meereskünde Univ. Hamburg, no. 17, 1970. (Similar solution for the K_1 tide in: Zahel, W. *Pure and Appl. Geophysics*, **109**, (8), 1819–1825, 1973.)
20A. Zahel, W. The influence of solid earth deformations on semidiurnal and diurnal tides. Pp. 98–124 in: *Tidal friction and the Earth's rotation*, (Ed. J. Broche & J. Sündermann), Springer-Verlag, Berlin, 243pp., 1978.
21. Gordeyev, R.G., B.A. Kagan, and V.Y. Rivkind. Numerical solution of the equations of tidal dynamics in the world ocean. *Dokl. Akad. Nauk (SSSR)*, **209**, (2), 340–343, 1973.
21A. Marchuk, G.I. and B.A. Kagan, *Dynamics of Ocean Tides*. Kluwer Acad. Pub., Dordrecht, 327pp., 1989. (Transl. of Russian text published 1983.)
22. Hendershott, M.C. The effects of solid earth deformation on global ocean tides. *Geophys. J. R. astr. Soc.*, **29**, 389–402, 1972.
23. Schwiderski, E.W. Ocean tides; 1 – Global ocean tide equations; 2 – A hydrodynamic interpolation model. *Marine Geodesy*, **3**, 161–217 and 219–255, 1980.

24. Schwiderski, E.W. Atlas of ocean tidal charts and maps; 1 – The semidiurnal principal lunar tide M_2. *Marine Geodesy*, **6**, (3–4), 219–255, 1983. (Maps for eleven leading tidal constituents were published by *Naval Surface Weapons Center*, Dahlgren, Virginia.)
25. Longuet-Higgins, M.S. The eigenfunctions of Laplace's tidal equations over a sphere. *Phil. Trans. R. Soc. London*, **A**, **262**, 511–607, 1968.
26. Longuet-Higgins, M.S. and G.S. Pond. The free oscillations of a fluid on a hemisphere bounded by meridians of longitude. *Phil. Trans. R. Soc. London*, **A**, **266**, 193–223, 1970.
27. Gotlib, V. Yu. and B.A. Kagan. Resonance periods of the World Ocean. *Dokl. Akad. Nauk, SSSR*, **252**, 725–728, 1980. Also: *idem*. Numerical solution of tides of the world ocean – 3 – A solution to the spectral problem. *Deutsche Hydrogr. Zeitschrift*, **35**, 45–58, 1982. (See refs. 21, 21A for other related papers by B.A. Kagan.)
28. Platzman, G.W. Normal modes of the world ocean. Part 1: Design of a finite-element barotropic model. *J. Phys. Oceanogr.*, **8**, (3), 323–343, 1978. Part 2 (with G.A. Curtis, K.S. Hansen, and R.D. Slater): Description of modes in the period range 8 to 80 hours. *Ibid.* **11**, (5), 579–603, 1981. Parts 3 and 4: A procedure for tidal synthesis; synthesis of diurnal and semidiurnal tides. *Ibid.* **14**, (10), 1521–1550, 1984. Also: Platzman, G.W. World ocean tides synthesised by normal modes. *Science*, **220**, 602–604, 1983.
29. Garrett, C.J.R. and W.H. Munk. The age of the tides and the *Q* of the oceans. *Deep-Sea Res.*, **18**, 493–503, 1971.
30. Munk, W.H. and G.J.F. MacDonald. *The Rotation of the Earth: a Geophysical Discussion*. Cambridge Univ. Press, 323pp., 1960.
31. Munk, W.H. and E.C. Bullard. Patching the long-wave spectrum across the tides. *J. Geophys. Res.*, **68**, (12), 3627–3634, 1963.
32. Munk, W.H., B.D. Zetler and G.W. Groves. Tidal cusps. *Geophys. J. R. astr. Soc.* **10**, 211–219, 1965.
33. Munk, W.H. and D.E. Cartwright. Tidal spectroscopy and prediction. *Phil. Trans. R. Soc. London*, **A**, **259**, 533–581, 1966.
34. Cartwright, D.E. A unified analysis of tides and surges round north and east Britain. *Phil. Trans. R. Soc. London*, **A**, **263**, 1–55, 1968.
35. Zetler, B., D. Cartwright and S. Berkman. Some comparisons of Response and Harmonic tide predictions. *Internat. Hydrogr. Rev.*, **56**, (2), 105–115, 1979.
36. Haubrich, R. and W.H. Munk. The pole tide. *J. Geophys. Res.*, **64**, (12), 2373–2388, 1959.
37. Miller, S.P. and C. Wunsch. The pole tide. *Nature*, Phys. Sci. **246**, 98–102, 1973. Also: Maksimov, I.V. and V.P. Karklin. The Baltic polar tide. *Dokl. Akad. Nauk, SSSR*, **161**, 580–582, 1965.
38. Wunsch, C. Dynamics of the pole tide and damping of the Chandler Wobble. Geophys. *J. R. astr. Soc.*, **39**, 539–550, 1974. Also, *ibid.* **40**, 311, 1975; *ibid.* **87**, 869–884, 1984; and Xie, L. & S.R. Dickman, North Sea pole tide dynamics. *Geophys. J. Internat.*, **121**, 117–135, 1995.
39. O'Connor, W.P. The 14-month wind-stressed circulation (pole tide) in the North Sea. Natl. Aeronautics & Space Admin. (NASA) Tech Memo. no. 87800, 1986.
40. Trupin, A. and J. Wahr, Spectroscopic analysis of global tide-gauge sea level data. *Geophys. J. Internat.*, **100**, 441–453, 1990.
41. Tsimplis, M.N., R.A. Flather and J.M. Vassie. The North Sea pole tide described through a tide-surge numerical model. *Geophys. Research Letters*, **21**, (6), 449–452, 1994.
42. Munk, W.H. Once again – Tidal Friction. (Harold Jeffreys Lecture, 1968). *Q. J. R. astr. Soc.*, **9**, 352–375, 1968.
43. Jeffreys' first estimate of 1920 (Chapter 11, ref. 5), was revised several times. Later summaries of his views are in Ch. 8 of: *The Earth*, 6th Edition, 1976, and in: Jeffreys, H. Tidal friction; the core, mountain and continent formation. *Geophys. J. R. astr. Soc.*, **71**, 555–566, 1982.
44. Spencer Jones, H. The rotation of the Earth, and the secular accelerations of the Sun, Moon and planets. *Monthly Not. R. astr. Soc.*, **99**, 541–558, 1939.
45. Miller, G.R. The flux of tidal energy out of the deep oceans. *J. Geophys. Res.*, **71**, (10), 2485–2489, 1966.

13

The impact of instrument technology, 1960–1990

Early postwar advances in marine instrumentation

The equipment used for an oceanographic expedition in 1955 differed little from that of 1925. Apart from the Bathythermograph, mentioned in the opening paragraph of Chapter 12, (whose recording principle was a trace scratched by a stylus on a smoked glass plate), the basic tool was the 'Nansen' water-sampling bottle with reversing thermometers, evolved from earlier designs around the turn of the century by Fridthof Nansen (1861–1930), the Arctic explorer. A string of about twelve Nansen bottles lowered on a vertical wire in the deep ocean would, after interpretation of their thermometer readings and skilful chemical analysis of their samples, give the broad features of the vertical stratification in salinity, temperature, and density with respect to depth below a stationary ship. Comparison of the stratification in at least three well-spaced stations determined the horizontal pressure gradients at any depth, from which the mean geostrophic flow could be estimated. The whole measuring procedure was far slower than the tides.

Form and composition of the sea floor were monitored by echo-sounder and occasional dredges of rock samples. These were soon to be supplemented by the towed *proton magnetometer* and seismic sounding of controlled explosions by hydrophone. Following a long-established tradition, the trawling of nets for biological samples was always an important (for some the most important) part of every oceanographic expedition.

I will not attempt to trace every facet of the rapid development in marine recording equipment from the late 1950's onwards, but I will highlight those features which eventually found application to the recording of *tides* in the open sea. Probably the first important innovation was the neutrally buoyant float invented around 1955 by John Swallow (1923–1994) at the UK National Institute of Oceanography. *Swallow floats*, as they were usually called, were thick metal tubes of some 2–3 meters length, whose weight and compressibility were so adjusted to reach neutral buoyancy at a pre-determined depth. Acoustic 'pinks' emitted at roughly one second intervals from each float enabled a ship to fix the float's geographical position every half-hour or so by a form of Doppler navigation, and hence to determine the water transport at the chosen depth by direct measurement. Early results of tracking Swallow floats in the western North Atlantic revolutionised oceanographers' understanding of the physics of deep currents.(2)

The operations just described involved two important techniques, (a) the use of acoustic signals to locate an underwater object, and (b) accurate positional navigation of the ship itself. Technique (a), derived from wartime *sonar* research, was later to be used to retrieve instruments moored to the sea bed. The original tool for (b) was radar distance and bearing of a moored floating target known as a 'dan buoy'. (Radar was another wartime invention; fixes by sextant would have been far too infrequent and inaccurate for the purpose.) Facilities for rapid and accurate navigation steadily improved over the years, with networks of transmitting beacons such as the *Loran* system, and later by systematic satellite transmissions. All this development proved of great value for the location and recovery of moored instruments, including tide-recording devices.

The ability to deposit recording instruments on or attached to the ocean floor, and later to find and recover them, was a great step forward, dating from the mid-1960s. It enabled the variations in time of physical quantities such as current velocity and temperature at chosen locations to be recorded without a ship in attendance, (and, to quote a witticism of the 1960s, 'without interference from technicians'). Strings of current meters, each counting rotations of a vertically mounted rotor and recording the magnetic orientation of a vane and a themistor signal, were the most common type of moored assembly. The recorded currents are *Eulerian* (motion past a fixed location) as distinct from *Lagrangian* (motion following the water mass) which is measured by Swallow floats; the two types are complementary. The whole assembly is necessarily buoyant, and held to the bottom by a heavy anchor weight. A releasing mechanism severs a link above the anchor weight on reception of a coded acoustic signal from the parent ship. The assembly of instruments and buoyant spheres then floats to the surface and is hauled on board the ship.

A great deal of technical research was necessary to make such systems practicable. It required the design of robust recording devices of low weight and power consumption, the provision of a stable time-base, and reliable response to acoustic command signals. The research was progressed in several marine research institutions, principally in the USA, Britain and Norway. As oceanography became 'big business', with the growth of offshore mineral prospecting and the search for safe repositories of nuclear waste, commercial instrument manufacturers took over the routine production of the best-proven underwater systems of the sort described.

Applications to tidal research

Tidal variations were clearly discernible in long-term current-meter and thermistor records from the deep ocean. Their amplitudes were found to be much larger than predicted by barotropic tide theory and their phase varied with depth.[3] Such records revived interest in the idea of *baroclinic* or *internal tides*, which had been recognised in the Norwegian Sea early in the century by Helland-Hansen and Nansen, and developed in theory by Zeilon,[22] who called them *tidal boundary waves*, implying a boundary between two layers of different density. Internal tides had also been encountered as an essential ingredient of air tide theory (Chapter 10 and Appendix D). Their ocean manifestation will be discussed in a later section of the present chapter.

Baroclinic effects are less obvious in current-meter records from shallow seas where the barotropic tide usually dominates, modified by turbulent friction in the manner discussed by Sverdrup, (Chapter 11). Horizontal arrays of current-meters near the ocean-shelf-break showed up interesting wave behavior in the *diurnal* tidal currents.[4,40] These invoked the theory of *second-class waves* in variable depth, which had first been identified theoretically in the 19th century by Lamb,[5] (Chapter 7). They will be discussed in a later section under the heading 'Continental Shelf waves'.

Bottom pressure recorders

But the most exciting application of the new ocean technology to tidal science was the ability to record variations of *pressure* on the sea-bed. To high approximation, the pressure at a fixed depth D below the mean surface may be expressed as:

$$p(t) = p_a(t) + \rho g(D + \zeta(t)),$$

where ρ is the mean density of the water column, g is the local gravity, and ζ is the variable elevation of the surface relative to that of the sea-bed, that is, 'surface tide minus earth tide'. $p_a(t)$ is the surface atmospheric pressure, which imposes a small perturbation due to the solar air tide. The factor ρg is close to 1.005 millibars ($1\,\text{mb} = 10^{-5}$ pascals) per centimeter of water elevation z in most parts of the ocean. The surface tide relative to the earth's center (as sensed by satellite altimetry – Chapter 14) is roughly 0.94ζ in mid-ocean. Thus, measurements of $p(t)$ give a reliable record of $\zeta(t)$ remote from dry land. Such measurements first became possible in oceanic depths from about 1964.

The reader may wonder why, during more than a century of theoretical speculation about the form of the tide in the open sea, nobody had made direct measurements of this fundamental geophysical variation. Harbor records from small islands were too few and widely spaced for simple interpolation. In fact, several mechanical devices for recording pressure variation in shallow seas were invented from the 1890's onwards but all had serious technical limitations, and none gave reliable results, even in shelf seas. As an unsatisfactory alternative, one may recall Captain Hewett's laboriously gathered series of soundings in the southern North Sea in 1840 at the behest of William Whewell (Chapter 9).

The principal difficulties were the extremely high ambient pressure which any sea-bed instrument has to withstand, and the smallness of the ratio ζ/D, which, though greater than 10^{-1} in coastal waters, is more typically 10^{-4} in deep ocean locations. The recording accuracy of z must be smaller still by 10^{-2} to 10^{-3}, thus requiring an accuracy of 10^{-6} to 10^{-7}.

Early mechanical devices

Initial researches had been pursued mainly by the Hydrographic Services of the Navies of France, Germany and Britain,[6] with the object of making tidal corrections for offshore soundings. The basic principle of operation was summarised by Darwin[1] in describing a device by Captain Adolf Mensing (1845–1929) of the Imperial German Navy, patented in 1904:

> [Mensing's] instrument is completely enclosed in a massive airtight vessel, which is let down to the bottom of the sea and is attached to a buoy to mark its position... . The air in the airtight vessel is first pumped up until it attains [the mean pressure at the bottom] ... After the instrument has been in position about half an hour an ingenious arrangement comes into play whereby the previously airtight vessel opens automatically [to the sea. An aneroid barometer then records the variations in pressure on a clockwork disc.]

From contemporary reports Darwin believed Mensing's gage to be a viable instrument, but later German assessments concluded that the mechanism was too delicate for use at sea.[6] An earlier French device by Louis Favé (1853–1922) known as the 'marégraphe plongeur'[7] (not mentioned by Darwin) achieved some success, and with later modifications such as the use of a bourdon tube for pressure sensing, survived until World War II. However, few results of scientific value were reported. Its main disadvantages were a very insensitive reading scale per meter change in z requiring a special micrometer to examine the data, and uncertain corrections for variations in temperature.

A more direct measure of changes in the *total* depth of water was patented in Britain in 1908 by the Navy Hydrographer Admiral Mostyn-Field, with the help of the Cambridge Scientific Instrument Company. The open end of a long thin flexible tube was fixed to a framework lowered to rest on the bottom. The other end was connected to a large cylinder of compressed air with pressure-reducing valve, on land or on an attendant ship. Pressure and air flow were adjusted so that air continuously bubbled slowly from the open end. The pressure in the tube was then the same as that of the total head of water and was measured by a recording barometer. Darwin[1] reported credible measurements of the tide from three stations in mid English Channel recorded this way in 1909 by a naval party on board HMS *Triton*. Handling Mostyn-Field's gage from a ship at sea presented obvious difficulties, and the gage was soon deemed impractical. However, with modern instrumentation the principle of the *bubbler-gage*, as it is now called, is still used in harbors and rivers.

Finally among pre-1950 devices for measuring tides in shallow seas, mention should be made of the instrument invented around 1928 by Heinrich Rauschelbach (1888–1978), the designer of the remarkable *Gezeitenrechenmaschinen* (or TPM's) for the *Deutsche Seewarte* (Figure 11.2). Rauschelbach's tide-gage purported to measure absolute pressure by photographically recording the interface of water forced into a chamber of compressed air.[8] Thermographs were also included. Mechanical design was fairly sophisticated, and the gage was said to have given some good results in the German Bight, but it too was clumsy to handle at sea; interest in using it eventually faded.

Thus, none of the mechanical instruments designed for measuring tides offshore before 1960 had any lasting success, and none was ever used in depths of more than 100 meters. The above account is given mainly for the historical record and to serve as background for the modern deep-sea tide-gages developed between 1960 and 1990.

The modern era of pressure recording

First to show a continuing interest in recording tides from the deep ocean was the 'Association francaise pour l'exploitation des grands profonds de l'océan' (AFEGPO), working under the auspices of the 'Service hydrographique de la Marine' (SHOM) in Paris and later in Brest.[9] The AFEGPO/SHOM gage was based on principles similar to the 'marégraphe plongeur' of Favé. Recorded pressure was not total pressure but relative to that of a volume of compressed nitrogen, released from a cylinder on reaching the bottom then sealed on command after isothermal adjustment. (Nitrogen was chosen for its low thermal coefficient of expansion.) The principal innovation in measuring the pressure differential between the compressed gas and the ambient sea water was the use of a taut wire vibrating in a magnetic field, mounted in a sealed Invar cylinder. The natural frequency of vibration depends on the pressure on the cylinder; the vibrations were counted electromagnetically and registered every 6 minutes on a mechanical counter with photographic recording. The temperature of the nitrogen was sensed by a separate vibrating wire system, monitored in the same photographic frame (Figure 13.1 – top).

The first AFEGPO instrument required direct electrical connection with an attendant ship, through a cable. This imposed serious limitations in several ways, not least being the limit in duration of each record. Nevertheless, the first recordings, though limited to 38–73 hours, were evidently of excellent quality (Figure 13.1 – lowest panel). The record shown here is the first of a group of four taken in depths of 150 m to 470 m at the shelf-break north of Biscay Bay in July

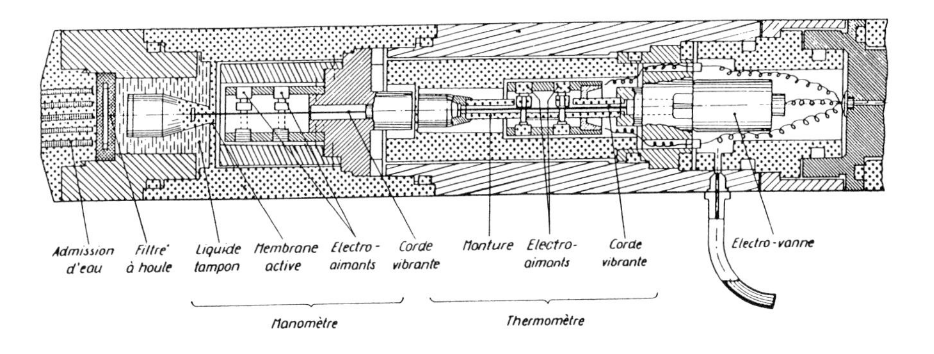

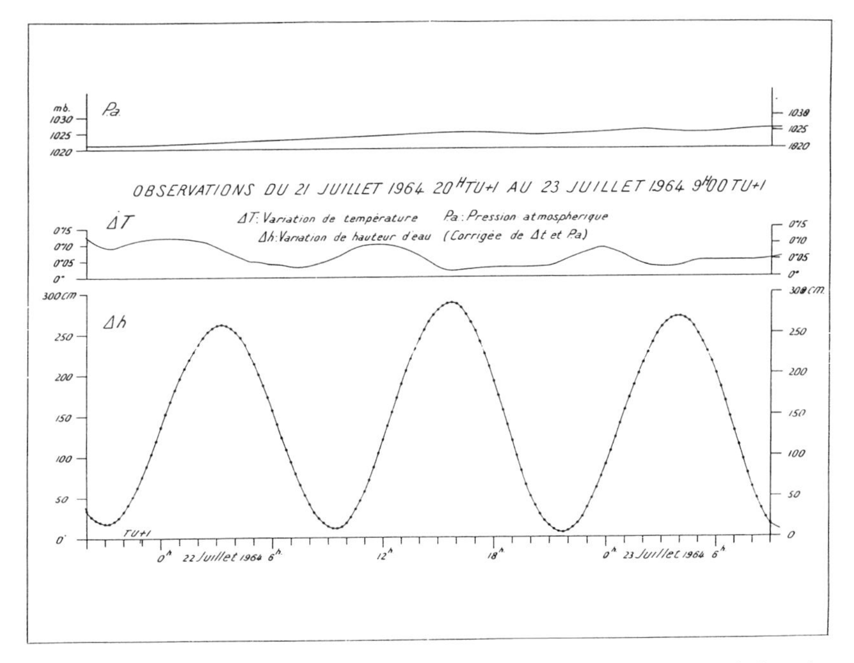

Figure 13.1. (Above): Sketch of instrumental details of the AFEGPO gage, including the vibrating cords and electromagnets for pressure (left) and temperature (right). (Below): The first of four pioneering records of tidal pressure and temperature variations from the shelf-break south of Brest taken with the French AFEGPO gage[9a] in 1964. The pressure variation (bottom) has been converted to equivalent change in water level (cm), after correction for the variations in temperature (middle) and in surface air pressure (top). Times are marked in French Standard Time = UT + 1.

1964. They were the first ever *bona fide* recordings of the tide in what may be termed the 'open ocean'.[9a]

An autonomous version of the AFEGPO/SHOM gage which could be left on the bottom for much longer periods without a ship in attendance was designed at the 'Centre Océanologique de Brest' (COB) around 1972, but after one or two losses of equipment at sea the French research programme in this area was abandoned for several years.

Pelagic tide recording in USA and Britain

Shortly after the pioneering French records in 1964, the United States Coast & Geodetic Survey (USCGS) obtained pressure records in similar depths at the shelf-break east of New York, using a strain-gage which recorded total pressure.(10) The USCGS subsequently adopted a more accurate and reliable absolute pressure-gage based on a skilfully engineered bourdon-tube sensor with optical read-out, designed at Gulf General Atomic Inc. of San Diego.(11) The bourdon-tube device was further developed by its designer Jean Filloux into a versatile digitally recording pressure recorder with a wide frequency range, and used by him to provide a large number of tidal records from various parts of the Pacific Ocean.(12) The Filloux gage has also been extensively used by the US National Oceanographic and Atmospheric Administration (NOAA).

As evidence of breadth of interest in recording ocean tides at this period, an unique record was made in 1967 of both ocean (pressure and current) and solid earth (gravity) tides at a location 3900m deep in the Pacific Ocean west of San Francisco by the Lamont Geological Observatory of Columbia University. The pressure sensor was a vibrating wire manufactured in the USA under the trade name 'Vibrotron'. All electrical signals were connected to a recording station on the coast through a cable 160 km long – a very expensive item for a single recording.(13)

The most sophisticated deep-sea instrument capsule developed in the USA to measure pressure, temperature and currents, was that of Frank E. Snodgrass (1920–1985), who worked in Walter Munk's Institute (IGPP) at La Jolla, California. Snodgrass's capsules(14) were the ultimate extension of a series of pressure recorders constructed by him in the early 1960s to record surface waves in various frequency bands. The prototype tidal capsule used a 'Vibrotron' wire to sense pressure with a resolution of 10^{-7}, quartz crystals for temperature to 10^{-6} degrees C, and heated probes to sense current speed to 1mm/s with some directional resolution. All signals were processed by internal computer and stored digitally on magnetic tape; the computing and recording equipment was housed in two 1.2m diameter aluminium spheres which also provided the necessary buoyancy for recovery. A heavy battery pack served for anchor weight. In later versions, the 'Vibrotron' was replaced by a quartz crystal oscillator free from temperature sensitivity, the tape-deck was replaced by a solid-state digital store, and the battery pack was greatly lightened and housed internally. Figure 13.2 shows an early version of the assembly being deployed in the Pacific Ocean in 1967.

A set of three one-month records from Snodgrass capsules off California in 1967–1968, together with the the Lamont GO record, one record from a 'Filloux' bourdon tube, data from some current meters and from several shore gages, were analysed by Munk's 'Tidal Study Group' to produce a realistic synthesis of the diurnal and semidiurnal tide maps off California. The synthesis was made from a mix of suitably chosen Kelvin waves and Poincaré waves, and

Figure 13.2. An early version of the deep-sea tide capsule designed by Frank Snodgrass[14] after recovery from the Pacific Ocean off California in 1967. The ballast which held the capsule to the sea bed, (at that time a stack of car batteries), had been released some 45 minutes earlier by an electromagnetic device activated by an acoustic signal. The upper (left) sphere was a buoyancy chamber; the lower sphere contained computing circuits, acoustic controls, and a digital tape deck. The upper masts were antennae for acoustic and radio location. The pressure and current sensors (not seen here) were suspended from the lower sphere. The persons in the photo are: (left) Frank Snodgrass, (right) Walter Munk. (Author's photograph.)

the results included the first direct identification of an oceanic amphidromic system – that for M_2 centered near 27° N, 135° W.[15]

Munk and Snodgrass went on to take some high precision tide records between Australia and Antarctica[16] and in the western North Atlantic;[17] they then turned their attention to other projects. Some groups from American universities continued to take bottom pressure records occasionally, usually in conjunction with other oceanographic experiments.

Meanwhile, a more persistent campaign of 'pelagic' tide recording had been started in the British *Institute of Oceanographic Sciences* (IOS), an amalgamation of the marine laboratories originally known as LTI and NIO (Chapters 11, 12). The word *pelagic*, pertaining to the open sea, was borrowed from the field of marine biology. A dedicated team of technicians at IOS developed a series of their own bottom pressure recorders, starting with modest instruments for the shelf-edge, going on to seamounts and ocean ridges, and finally achieving first class recording equipment for all oceanic depths. Specialised calibration facilities for pressure and temperature sensors and a well tried system for acoustic recovery were essential accompaniments to the recording equipment. A long series of sea-going campaigns from 1970 to 1985 spread tidal data stations from the entire northwest European shelf-edge to the mid-Atlantic Ridge and along lines from the Azores to French Guiana and from Brazil to the Gulf of Guinea.[4,18] After 1985, IOS occupied several very longterm bottom pressure recorders in the Southern Ocean for purposes other than tidal research.

Harmonic tidal constants resulting from all these exercises together with pelagic records taken by Canada, Brazil and Japan were collated from the many scientific publications where they first appeared by the 'International Association for the Physical Sciences of the Ocean' (IAPSO, formerly known as IAPO) for the ready access of analysts. Figure 13.3a shows a world map of pelagic station positions as of 1992, published by IAPSO.[19] The map shows 348 stations, of which 142 were taken by the UK–IOS, 117 by organisations in USA, and 89 by other nations, mostly Canada. Not surprisingly, the geographical distribution is heavily concentrated in the North Atlantic and the eastern North Pacific Oceans, with only about 3 percent of all stations in the southern hemisphere.

An international Working Group on ocean tides

The activity based on the use of Snodgrass pressure capsules was an important part of the contribution of Walter Munk's Institute to the new wave of tidal research which Munk instigated. The ultimate goal was objective definition of the tides in the world ocean, as discussed in Chapter 12, but this could hardly be achieved without international cooperation. Starting with informal discussions at international assemblies such as those of IAPO/IAPSO, Munk and various colleagues formulated a programme involving a coordinated attack on the problem by computer modellists and oceanographers.[20] This included a campaign to record pressure variations at a wide array of ocean stations, tentatively 300 stations on a global 10°×10° grid. The programme attracted a number of enthusiastic meetings under IAPO/IAPSO in 1965–1967 which led to the formation of a 'Working Group on deep-sea Tides', aided by funds from the Scientific Committee on Ocean Research of ICSU. ('Deep-sea tides' was later modified to 'Tides of the open sea', in recognition of the increasing number of measurements at the shelf-edge.) Munk was founder-chairman of the Working Group, and meetings were attended by representatives from eleven countries.[21a,21b]

The IAPO/IAPSO/SCOR Working Group (WG-27) remained active until 1975; its meetings included three sea-going expeditions to compare techniques of handling the various recording equipments at sea.[21c] WG-27 served a useful purpose in focussing attention on the possibilities and difficulties of operations at sea with their associated problems of calibration and data analysis, but it ultimately failed to attract the worldwide cooperation which Munk had hoped for. The instrumentation was necessarily expensive and it required dedicated technical specialists and extensive use of research ships. Willing individuals failed to get sufficient financial backing from their national ocean research organisations to make an effective contribution to the programme. The research institutes had other pressing problems on which to spend their resources.

The computer modellers, while expressing detached interest in the field measurements, did not interact with the sea-going fraternity and, perhaps unwittingly, gave an impression that the problem of the ocean tides would ultimately be solved by computers alone. At best, there was a suggestion that measurements should be made at certain positions of extreme amplitude near the equator, in order to provide critical tests for the computer models. However, the practitioners had difficulty obtaining ships to travel the long distances required to reach these places, and mostly had to cooperate with oceanographic programmes in other parts of the ocean. Nevertheless, some of the initial enthusiasm of WG-27 remained after the Working Group was disbanded, as testified by Figure 13.3a and by the advances in modelling technique discussed in Chapter 12.

Eventually the problem of global measurement was solved in the 1990s by the analysis of satellite altimetry. This will be the subject of Chapter 14, where it will be seen that the bottom pressure records also played an important part.

Tidal currents and internal tides

Longterm recording current meters which could be moored in the deep ocean were developed about the same time as bottom pressure recorders and they used similar technology. Their primary object was to monitor steady currents and their variations at all depths, but records always showed diurnal and semidiurnal fluctuations of tidal character, albeit much less regular than in the surface tide or bottom pressure. Figure 15 of the paper by Munk and others on the tides off California[15] illustrates the nature of such fluctuations in one vector

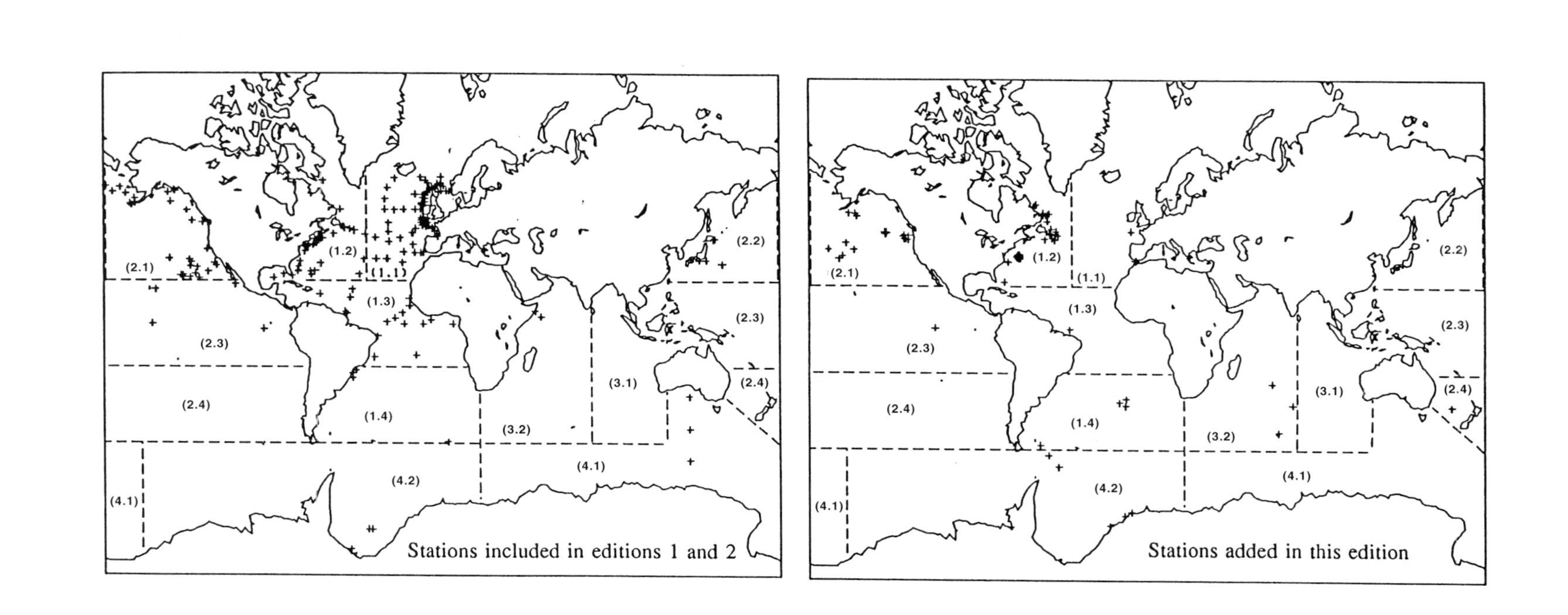
(1.1)
(1.2)
(1.3)
(1.4)
(2.1)
(2.2)
(2.3)
(2.4)
(3.1)
(3.2)
(4.1)
(4.2)
Stations included in editions 1 and 2
Stations added in this edition

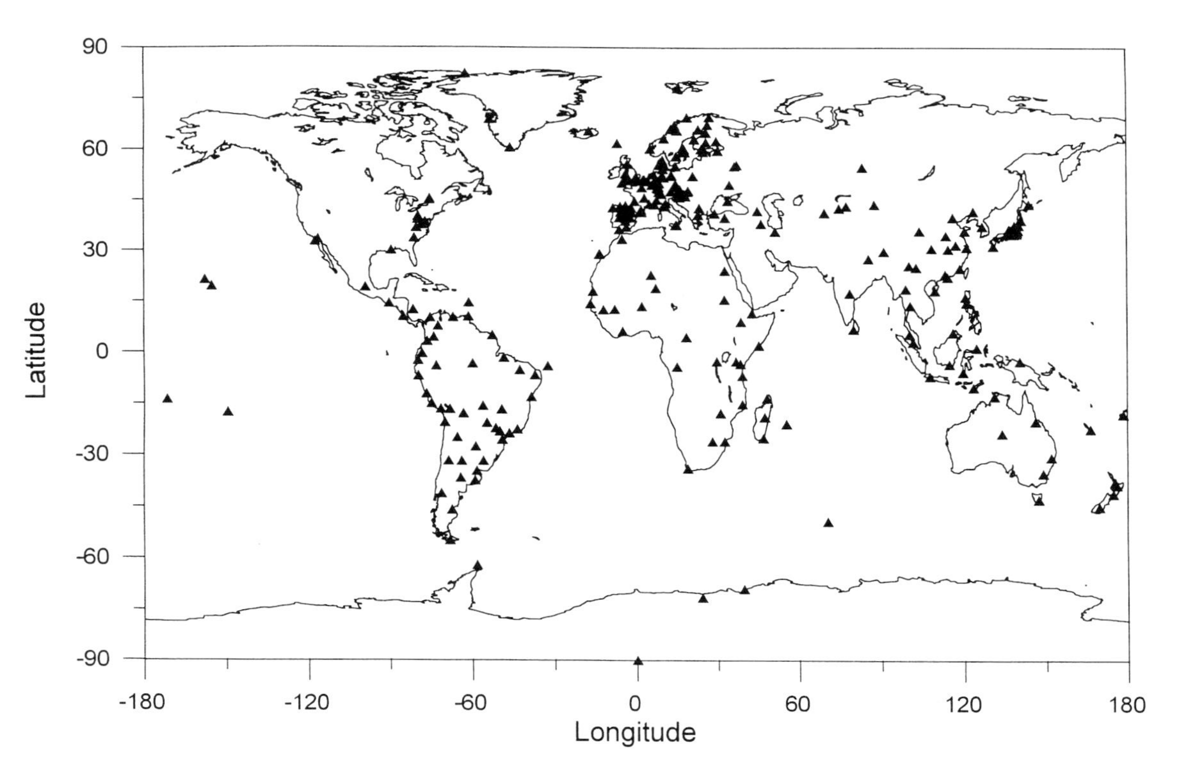

Figure 13.3. (a – upper): The positions of 348 pelagic tidal stations occupied from 1964 to 1991, whose parameters were compiled and published by IAPSO.[19] Left-hand panel shows compilation as of 1985; right-hand panel, additional stations 1986–1991. Bottom pressure records from very shallow locations close to land are not shown. (b – lower): The positions of 289 good quality gravity-tide stations, compiled in the data bank of the ICET of the Royal Observatory of Belgium in 1994. A gravity-tide station at the South Pole is included. (Lower diagram by courtesy of Professor Paul Melchior.)

component near the bottom. Irregular structure and vertical changes of amplitude and phase were also observed.[23a,23b,26]

Most current meters included a thermistor, which also showed irregular tide-like variations at depth. Since mean temperature in the ocean generally decreases with depth, these variations were manifestations of vertical motions of isopycnal (i.e. constant density) surfaces. Deduced amplitudes were commonly as large as 30 meters, at least one order of magnitude greater than the surface tide but with much shorter wavelength and carrying much less energy. Coherence between adjacent moorings about 100km apart was very small.

All these properties accorded qualitatively with the known theory of internal waves,[24] and pointed to a new subject of research – *internal tides*. Two comprehensive reviews of knowledge of internal tides were published in 1975 and 1981, respectively.[3,25] The first review lists 130 papers on the subject, of which more than 100 have dates later than 1960. A hundred papers in 15 years reflects a very different pattern of scientific research from anything yet encountered in this book. The surge of activity was as much due to the novelty and sudden accessibility of measurable data as to the realisation of a relatively unexplored field. It is not possible to describe the highlights of each study here. An outline of the principal trends of thought is all that can be attempted in a context of recent scientific history. Although the emphasis of this chapter is on progress driven by technology, understanding the measurements required a parallel development of theoretical research; again some preliminary discussion of this theory is necessary in the following few paragraphs.

The theory of freely propagating internal waves in a flat rotating sea of uniform depth is outlined in Appendix D, where it served as an introduction to the theory of air tides (Chapter 10) which is closely related but physically more complex. In brief, the internal wave equations depend on the vertical distribution of a parameter $N(z)$ known as the *buoyancy frequency*, defined in terms of the mean vertical density-gradient at height z. The three-dimensional forms of Laplace's equations in vertical and two horizontal components of the velocity support motions which are wavelike in both vertical and horizontal directions. The requirement that the vertical motion be zero at the surface $z=0$ and at the bottom, $z=-D$, is met by a sequence of eigen-parameters D_n, known as *equivalent depths*. These depend on D, $N(z)$, the wave frequency and the Coriolis frequency. D_n determines the character of the various possible wave-modes, with $(n-1)$ zeros of velocity between $z=0$ and D. The horizontal wave equations are then identical in form to the classic LTE in a sea of depth D_n and constant density.

With typical values of $N(z)$ and D in the ocean, the depth of the lowest mode, D_1, which predominates, is less than 1m, so with phase-velocity of order $\sqrt{(gD_n)}$ it is clear that the horizontal wavelength of the internal tidal waves of lowest mode is very much shorter than those of the surface tide, whose horizontal velocity is uniform from top to bottom. A typical wavelength of internal mode 1 is about 130km, compared with that of the surface tide, several thousand kilometers. Higher order modes have even shorter wavelengths. Because of their very short natural wavelength, there is no possibility of simple coupling of the internal tides with the surface tide, or indeed with the tide-generating potential.

Generating mechanisms

The next question, an important one, was: 'how are the internal tides generated?' The horizontal wavelength becomes larger near the 'critical latitudes' where $f = \sigma$, and it was once thought that the external potential might feed energy into the internal tides near these latitudes.[27] However, closer examination showed this to be theoretically unsound;[25] in any case there is no evidence for enhanced internal motion of diurnal character near the critical latitudes of 30°.[3]

It thus became clear that the internal tides draw their energy from the surface tides themselves. This has already been mentioned as a possible increment to global tide dissipation (Chapter 12). The energy level of internal tides was found to be comparable with surface tides,[26] although they are unable to advect that energy away fast enough to provide a major sink. The mechanism for energy transfer has been the subject of much research.

A theoretical coupling mechanism involving the (usually neglected) horizontal component of the Coriolis stress[27] has not been investigated quantitatively, but it is thought to be very small in magnitude. A more promising mechanism is the interaction of barotropic currents with the irregular bathymetry of the ocean. The bottom undulations provoke vertical velocity over a wide spectrum of wavelengths, generating internal motions at the appropriate eigen-modes which advect from their place of origin. Such a mechanism had been explored as long ago as 1912 by Nils Zeilon[22] in the context of two-layer tidal flow in fjords and in a wave tank (Figure 13.4) but it was only seriously applied to the deep ocean in the 1960's.[28] The internal wave energy which is generated is eventually dissipated by instability and nonlinear scattering into turbulent motion.

Calculation of bathymetric generation of internal tides was at first considered[28] as a linear perturbation in vertical velocity U.gradε, where U is the horizontal barotropic tidal velocity and the bathymetry is expressed as $D + \varepsilon(x,y)$. Such a representation is conveniently tractable but it is only strictly valid for unrealistically small values of ε/D. For the abrupt changes of bathymetry at the edges of continental shelves an altogether different technique was required. It was found that the *characteristic paths* of the internal wave equations describe discrete packets of energy concentrated in rays which travel at angles to the horizontal of order 10^{-1}, and which are reflected from the bottom and from the surface in sawtooth fashion. (The ray representation is equivalent to a mix of high-order wave modes.) Where the ray slope coincides with the slope of the continental shelf, as happens at many sites, there is strong energy conversion into the internal-wave ray.[29]

Elaborate techniques have been devised to compute the rate of energy conver-

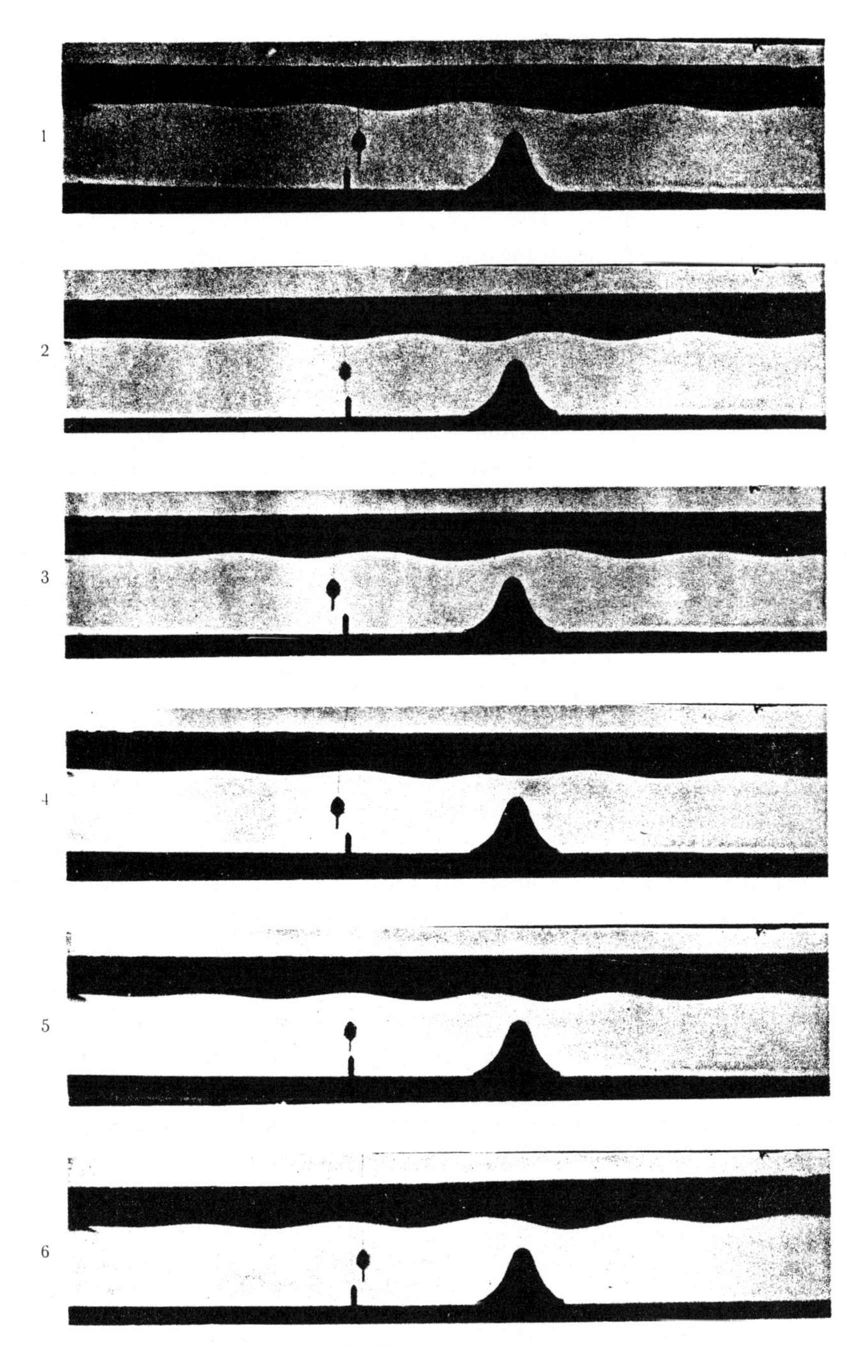

Figure 13.4. Photograph by Nils Zeilon[22] of internal waves generated at an interface of two fluids of different density in a long tank by slowly varying flow past a ridge (center right). The directions of the main flow at six different times are indicated by the positions of a small pendulum with respect to a fixed marker (center left). The waves propagate from the ridge in both directions.

sion at shelf-edges, but they are difficult to apply as a worldwide integral. Rough estimates[30] of the global rate are however only about 0.015 terawatts for M_2, compared with estimates of order 0.1 TW for the scattering over the entire ocean floor by the linear method.[28] These figures are hardly significant compared with the known global tidal dissipation of 2.50 TW. The smallness of the shelf-boundary contribution was also confirmed by field measurements in the North Atlantic.[31]

Quite recently, another form of calculation of energy transfer from bottom scattering was carried out, in which all parts of the ocean floor are approximated by a massive array of vertical steps. This gave the highest figures yet encountered, exceeding 1 TW.[32] The integrand is far from uniform and appears to be concentrated near shelf boundaries and mid-ocean ridges. However, experimental verification of the theoretical results is as yet inadequate. Some of the best measurements of internal tidal structure near the shelf-edge itself have indeed shown unusually high levels.[23a,23b,26] Experiments in the North Pacific using the new technique of *acoustic tomography* have suggested a coherent internal tide emanating from the Hawaiian ridge, later confirmed by satellite altimetry.[33]

Barotropic waves of second class – Rossby waves and Continental Shelf waves

The discovery by the analytical theory of second-class waves in fluid of uniform density by Lamb, Margules and Hough at the end of the 19th century was described in Chapter 7. It was not until the time-frame of the present chapter that measurements existed in sufficient detail for such theory to have direct application to ocean tides. Second-class waves, it will be recalled, are barotropic (i.e. vertically uniform) waves of low frequency with a high proportion of kinetic:potential energy, in theory tending to steady currents at zero frequency. In practice they may be divided into waves which depend on the variation of the Coriolis parameter f with latitude, and those which depend on variations of depth.

Once again, in order to avoid encumbering the text with mathematics, I have described the basic theory of such waves separately, in Appendix E. Surface elevation does not appear in the dynamic equations in an important way, and the salient properties may be deduced from simplified relations in which the horizontal divergence is set to zero. The resulting wave motions are defined essentially in terms of the *vorticity* of horizontal current, but the associated surface elevations may be recovered from the pressure gradients.

In the first situation outlined in Appendix E, the depth D is assumed uniform but the Coriolis frequency f is allowed to vary linearly with north latitude, y/a; the gradient df/dy is positive at all latitudes. Such conditions give rise to a class of long waves whose phase velocity always has a westward (never eastward) component. These are the *Rossby* or *Planetary* waves, first identified by the meteorologist Carl-Gustav Rossby (1898–1957),[34] and later studied in relation to ocean current.[35] It has been suggested that irregularities in the long-period tides Mf, Mm observed in the central Pacific Ocean are due to Rossby-wave structures at these frequencies.[36] Most recently, satellite altimetry has shown Rossby waves to be common in the ocean at very low frequencies, not specifically tidal.

Assuming the depth D to vary in the direction y (which may now have any geographical orientation) but with f assumed uniform over the assumed scale of motion, another type of wave motion is considered in Appendix E. Here, the dynamics are determined by the change of vorticity, not with latitude but with the compression of a vortex column as it moves into shallower water. The effect is most pronounced along the steep gradients of a continental shelf, and is characterised by waves of short length whose phase progresses in the direction parallel to the shelf- break with shallow water to the right in the northern hemisphere, left in the southern hemisphere. Amplitude of the variation in current is greatest at the shelf-break (maximum depth gradient), while surface elevations are greatest in the shallow water of the shelf itself. The wave amplitude decays rapidly in the direction of the deep ocean, so these waves are 'trapped' along the shelf edge. This property is analogous to the classic Kelvin wave in a sea of uniform depth, but the wavelengths are much shorter than those of Kelvin waves. A sequence of modes may be identified, the nth mode having n reversals of phase normal to the shelf. Such waves are known as *Continental Shelf waves* or *Topographical Rossby waves*.

Continental Shelf waves were first recognised in the ocean as low-frequency waves observed in 1962 propagating northwards along the east coast of Australia.[37] Similar observations in other parts of the world soon followed, and were supported by an evolving literature of theoretical studies.[38] Discovery of their manifestation in tides, however, has an interesting history which goes back to the 17th century.

Sir Robert Moray's 'extraordinary tydes'

Shortly before the Restoration of King Charles II and the foundation of the Royal Society of London, Sir Robert Moray, friend of the king and a founder-member of the Royal Society, stayed for some weeks at the remote island of Berneray between Uist and Harris in the Outer Hebrides west of Scotland (possibly in refuge from the anti-royalist faction). There he observed that the tidal streams behaved in a most unusual manner. Whereas the rise and fall of the local tide is predominantly semidiurnal as elsewhere in Britain, the tidal streams between the islands sometimes exhibited a strong diurnal motion. At semidiurnal neaps in summer, the currents through the Sounds of Berneray and Harris flow northwestwards (towards the open ocean) all night without reversal, then flow in the opposite direction all day. The local residents further told Moray that the currents behaved similarly in winter but with the night and day directions reversed.

Moray reported these 'extraordinary tydes', as he called them, to the Society in 1665; his paper appeared in the first volume of the *Philosophical Transactions* and is partly reproduced in Figure 13.5. At that time, scientists concerned with tides were too occupied with trying to understand ordinary tidal behavior to pay much attention to such irregularities. Sir Robert Moray himself was busy encouraging proper tidal observations in all parts of the country (Chapter 6). In fact, nearly 200 years passed before a decent set of hydrographic measurements in the Sound of Harris, confirming Moray's description, were reported by the Navy.[39] An attempted explanation of the observed phenomenon in terms of changes of the sun's declination accompanying the report is unconvincing and without interest.

Another century elapsed before the National Institute of Oceanography set up an array of recording current meters over an area of some 500km^2 of shelf sea, centered on the island of St. Kilda which is about 65km west of the Outer Hebrides. Analysis of a month's records in 1968 revealed that strong diurnal currents were prevalent over the entire shelf, exceeding the semidiurnal currents in certain directions.[40] The currents observed by Moray had evidently been moving in sympathy with the wider shelf regime out to the deep ocean.

By 1968 the theory of Continental Shelf waves was wellknown, and it was possible to show from calculations based on the shelf bathymetry that all the required properties of a first-mode wave, in direction, speed of propagation, sense of current rotation, and cross-shelf pressure gradients were satisfied by the diurnal tides of the Hebridean shelf.[4] The semidiurnal tides had more of the characteristics of Kelvin waves, with a much greater phase velocity. This was understandable, because a general requirement of all second-class waves is that their frequency should be substantially less than the inertial frequency f_0, which at 57° latitude is about 1.7 cycles per day. So Moray's 'extraordinary tydes' were at last put into a rational framework as a result of late 20th century instrumentation and theoretical technique.

Once diagnosed on the Hebridean shelf, Continental Shelf wave-trapping of diurnal tides was recognised in other high-latitude shelf seas, from observations of currents, cross-shelf pressure differences, and in fine-scale modelling.[4,41]

Advances in understanding earth tides

We last encountered the tidal deformations of the solid earth at or shortly after the time of Sir George Darwin (Chapter 10). Progress in this field was slow until after World War II, when it was stimulated, like ocean research, by the revolution in scientific technology. Seismic research associated with the detection of underground explosions also led to refined instrumentation and new models of the earth's interior structure, both highly relevant. New geodetic techniques for precise fixing of earth positions, such as VLBI and laser ranging, needed tidal corrections. All of these developments lent a sense of modernity to the subject, somehow lacking on the oceanic side.

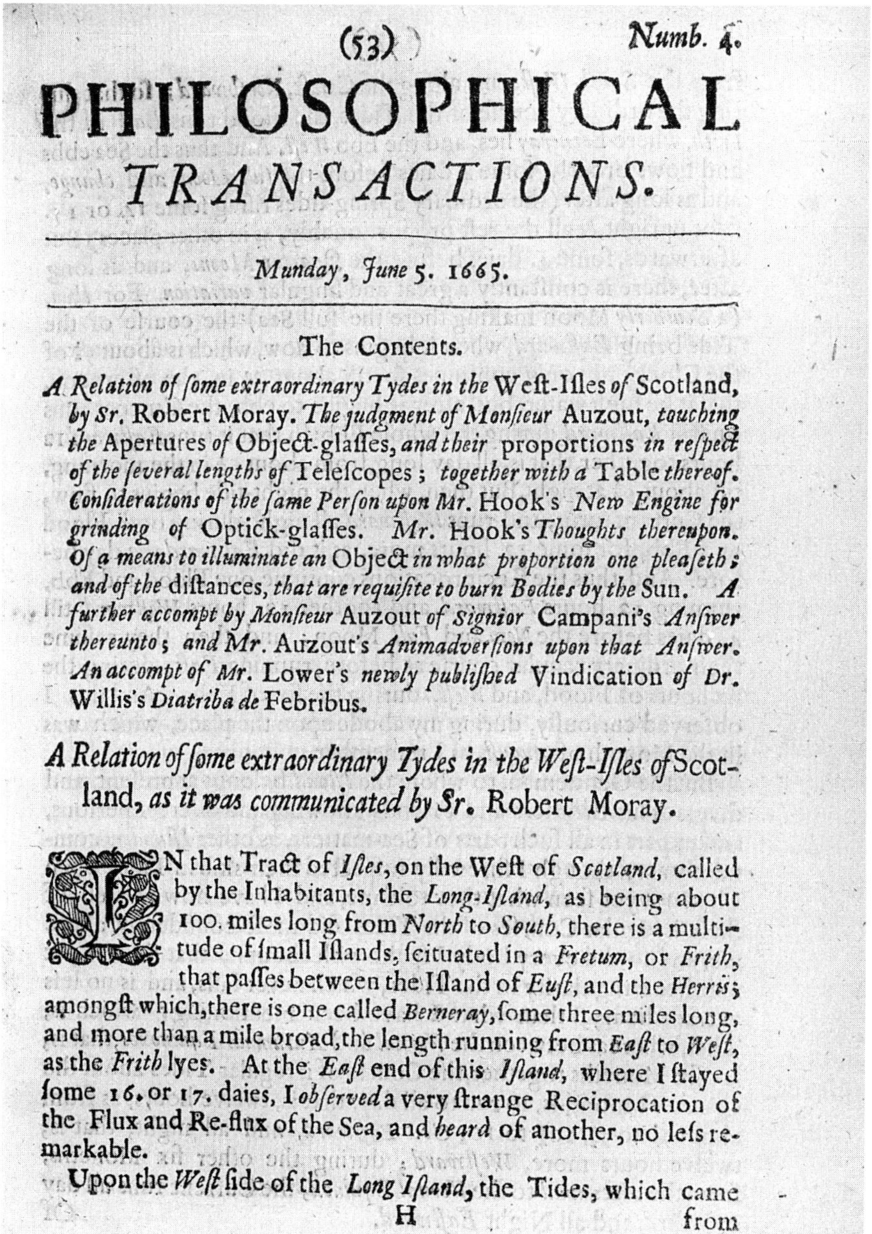

(53) **Numb. 4.**

PHILOSOPHICAL *TRANSACTIONS.*

Munday, June 5. 1665.

The Contents.

A Relation of ſome extraordinary Tydes in the Weſt-Iſles *of* Scotland, *by Sr.* Robert Moray. *The judgment of Monſieur* Auzout, *touching the* Apertures *of* Object-glaſſes, *and their* proportions *in reſpect of the ſeveral lengths of* Teleſcopes; *together with a* Table *thereof. Conſiderations of the ſame Perſon upon Mr.* Hook's *New Engine for grinding of* Optick-glaſſes. *Mr.* Hook's *Thoughts thereupon. Of a means to illuminate an* Object *in what proportion one pleaſeth; and of the* diſtances, *that are requiſite to burn Bodies by the* Sun. *A further accompt by Monſieur* Auzout *of Signior* Campani's *Anſwer thereunto; and Mr.* Auzout's *Animadverſions upon that Anſwer. An accompt of Mr.* Lower's *newly publiſhed* Vindication *of Dr.* Willis's *Diatriba de* Febribus.

A Relation of ſome extraordinary Tydes in the Weſt-Iſles of Scotland, *as it was communicated by Sr.* Robert Moray.

IN that Tract of *Iſles*, on the Weſt of *Scotland*, called by the Inhabitants, the *Long-Iſland*, as being about 100. miles long from *North* to *South*, there is a multitude of ſmall Iſlands, ſcituated in a *Fretum*, or *Frith*, that paſſes between the Iſland of *Euſt*, and the *Herris*; amongſt which, there is one called *Berneray*, ſome three miles long, and more than a mile broad, the length running from *Eaſt* to *Weſt*, as the *Frith* lyes. At the *Eaſt* end of this *Iſland*, where I ſtayed ſome 16. or 17. daies, I *obſerved* a very ſtrange Reciprocation of the Flux and Re-flux of the Sea, and *heard* of another, no leſs remarkable.

Upon the *Weſt* ſide of the *Long Iſland*, the Tides, which came

H from

Figure 13.5. The first two pages of Sir Robert Moray's paper on the tides in the region of Harris and Berneray (Outer Hebrides), copied from the first published edition of the *Phil. Trans. R. Soc. London* (1, 53–55, 1665). The language is of the pre-Newtonian period described in Chapter 4, but the observed phenomena were not satisfactorily explained for 300 years. (Photograph courtesy of the Royal Society.)

from the *South-Weſt*, run along the Coaſt, **Northward**; **ſo that du-**ring the ordinary courſe of the Tides, the Flood runs *Eaſt* in the *Frith*, where *Berneray* lies, and the Ebb *Weſt*. And thus the Sea ebbs and flows orderly, ſome 4 daies before the *full Moon*, and *change*, and as long after (the ordinary Spring-tides riſing ſome 14. or 15. foot upright, & all the reſt proportionably, as in other places) But afterwards, ſome 4. daies before the *Quarter Moons*, and as long after, there is conſtantly a great and ſingular *variation*. For *then*, (a *Southerly* Moon making there the full Sea) the courſe of the Tide being *Eaſtward*, when it begins to flow, which is about 9½ of the Clock, not only continues ſo till about 3½ in the afternoon, that it be high water, but, after it begins to ebb, the Current runs on ſtill *Eaſtward*, during the whole Ebb; ſo that it runs *Eaſtward* 12 hours together, that is, all day long, from about 9½ in the morning, till about 9½ at night. But then, when the night-tide begins to flow, the Current turns, and runs *Weſtward* all night, during both Flood and Ebb, for ſome 12. hours more, as it did *Eaſtward* the day before. And thus the Reciprocations continue, one Flood and Ebb, running 12. hours *Eaſtward*, and another 12. hours *Weſtward*, till 4. daies before the *New* and *Full* Moon; and then they reſume their ordinary regular courſe as before, running *Eaſt*, during the ſix hours of Flood, and *Weſt*, during the ſix of Ebb. And this I obſerved curiouſly, during my abode upon the place, which was in the Moneth of *Auguſt*, as I remember.

But the Gentleman, to whom the *Iſland* belongs at preſent, and divers of his Brothers and Friends, knowing and diſcreet perſons, and expert in all ſuch parts of Sea-matters, as other *Iſlanders* commonly are, though I ſhrewdly ſuſpected their skill in Tides, when I had not yet ſeen what they told me, and I have now related of theſe irregular Courſes of the Tides, did moſt confidently aſſure me, and ſo did every body I ſpake with about it, that there is yet another irregularity in the Tides, which never fails, and is no leſs extraordinary, than what I have been mentioning; which is, That, whereas between the *Vernal* and *Autumnal Equinoxes*, that is, for ſix Moneths together, the Courſe of irregular Tides about the Quarter Moons, is, to run to all day, that is, twelve hours, as from about 9½ to 9½, 10¼ to 10½, &c. *Eaſtward*, and all night, that is, twelve hours more, *Weſtward*; during the other ſix Moneths, from the *Autumnal* to the *Vernal Equinox*, the Current runs all day *Weſtward*, and all Night *Eaſtward*.

Of

Figure 13.5 (*cont.*)

Attention was still focussed on the *tilt factor*

$$\gamma = 1 + k - h$$

and the *gravimetric factor*

$$\delta = 1 + h - (3/2)k,$$

(suffix 2 understood), which for a spherically symmetric earth determined the ratio of the amplitudes of tidal variations of tilt and gravity to their corresponding values in the primary tide potential. However, the main requirement now was not merely a rigidity modulus for the earth, but criteria for choice of the best model of the earth's interior. Measurements of γ and δ had to be better than 1 percent. Such precision was now becoming possible from modern tiltmeters and gravimeters with nominal precisions of order 10^{-9} radians and 10^{-6}Gals respectively – (1 Gal = 10^{-2}ms^{-2}). Having obtained the best theoretical earth model, one could then use it to deduce the individual Love numbers h (vertical displacement) and l (horizontal displacement) which are also useful to quantify but difficult to measure directly.

The theoretical models became more complex when ellipticity of the figure and rotation were allowed for.[42] The Love number notation used above is then no longer appropriate and the gravimetric factor varies with latitude (by about 3 percent it was at first thought, but much less according to revised calculations at the time of writing), and with tidal frequency. The ellipsoidal models also showed a curious resonance close to but slightly faster than a sidereal day, due to a coupling between the liquid core and the mantle, first predicted by Jeffreys and Vicente.[43] Being sharply tuned, this resonance scarcely affects the harmonic constituent O_1 and is only just detectable at K_1. In theory the peak resonance is near a very minor constituent known as ψ_1, but searches for an anomaly at ψ_1 in gravity or tilt or in oceanic records have so far failed to yield a convincing result.[44]

There was interest too in whether a phase lag exists in the earth's response to the potential. A phase lag of as little as 0°.1 would indicate tidal energy dissipation within the solid earth as well as in the ocean, whose relevance was discussed in Chapters 10 and 12. Theoretical studies based on the *Q-factor* of seismic disturbances showed that the phase lag was more probably of order 10^{-3} degrees and therefore negligible.[45] but experimental evidence for this was desirable.

Instruments and results

The reader is referred to a modern treatise on earth tides for details of the new instrumentation.[46] As a rough summary, the two components of tidal *tilt* were measured, either by the minute changes in fluid level at the ends of two long perpendicular horizontal tubes, or by the deflections of the free end of a rigidly suspended pendulum. Very long (80m) pendula to produce maximum deflection were used at one exceptional site,[47] but later technology permitted much shorter lengths (0.6m) to be used at a greater variety of locations. In all cases, extremely sensitive transducers (interferometer or capacitance plates) were necessary to detect the minute deflections. Protection from diurnal variations in temperature and other environmental factors was found to be essential. This was usually achieved by mounting instruments in laboratory basements, disused railway tunnels and mines. Boreholes 30 meters deep were eventually found to give best results.

As detectors of the global elastic properties, tiltmeters were found to have two disadvantages. Firstly, in a large irregular cavity like a tunnel the tidal signal was found to vary with position within the cavity by several percent, due to the straining of the cavity itself. It was for this reason that purpose-drilled boreholes were preferred. The other disadvantage is that the signal is strongly affected by the attraction of sea tides within 50 kilometers of the cavity (as in Figure 10.4), and these effects are not negligible even at distances of order 1000km. Corrections for sea tide loading were always carried out of course, but marine tidal maps were often deficient in coastal detail. There was also sensitivity to local irregularities in crustal structure. Similar difficulties were found in the measurement of tidal *strain*. Nevertheless, the tidal signal in tilt has been found useful in detecting horizontal changes in local geological structure and in assessing liability to earthquakes.[(48)]

With the introduction of *spring gravimeters* in the 1950s, the tidal variations in gravity proved to be the most valuable in testing the global structure of the earth. Gravity is less critically site-dependent than tilt or strain, and ocean loading outside about 40km can be more reliably calculated. Spring gravimeters use a mass suspended by a carefully designed spring system, well insulated from changes in temperature and atmospheric pressure. In the earlier models, the vertical movement of the mass was measured directly, but it was found to be more accurate to 'null' the movement by an electronic feedback mechanism. The change in gravity is then measured by the amount of feedback. Calibration is achieved either by raising the instrument in a known gravity gradient or by comparing the tidal signal with that from a known instrument at a standardising laboratory.

By far the most sensitive and noise-free tidal gravimeter, developed in the late 1970's, was the *superconducting gravimeter* (SCG). Here, a superconducting mass is supported by the force gradient of a magnetic field generated by a pair of superconducting current-carrying coils. Spectroscopy of an 18-month record from a SCG showed signal:noise ratios of 70 decibels, compared with about 40–50 decibels found in the best ocean pressure records. Detailed correlation with the harmonic expansion of the tide potential and removal of loading effects showed a gravitational factor of $\delta = 1.160 \pm 0.002$ with less than 0.1° phase shift at latitude 34° North.[(49)] However, the SCG is expensive and delicate, and so it is unsuitable for transportation to a variety of sites.

Determining gravity tides at globally distributed sites was a prominent activity of the 'International Geophysical Year' (1957–58),[(50)] and as with ocean tides it gained international support from the ICSU, principally focussed on the foundation of a 'Permanent Commission' (PCET) and an 'International Centre for Earth Tides' (ICET) at Brussels.[46, pp.471–472] From its inception in 1954, the ICET has taken and collected tidal gravity records from all continents. The earliest record kept in their data bank is one taken in 1954 with a spring gravimeter by the Austrian pioneer in gravimetry and first president of the PCET, Rudolf Tomaschek (1895–1966). (The record, historically interesting in its own right, was taken in the extreme north of the Shetland Isles as a by-product of an unsuccessful attempt to measure a possible absorption of the sun's gravity during a total eclipse.[51]) Another record comes from a permanent observatory at the South Pole,[52] The gravity stations number 382 to date, 289 of whose records are deemed worthy of detailed analysis, (Figure 13.3b).

A plot of gravimetric factors derived from 34 of the best ICET stations as of 1983 showed a scatter of some ±2 percent, mostly higher than the best theoretical curve, and a spread of about ±0°.5 in phase. These results were at first taken to reflect local geological anomalies, but a team from the 'Proudman Oceanographic Laboratory' at Bidston later obtained factors from five carefully calibrated European stations which differ from the theoretical value by only a tenth of 1 percent with phases within 0°.1.[53] The latter result strongly suggested that the theoretical model is correct to at least 0.1 percent, and caused the calibration standards of the ICET to be adjusted in 1992. Another potential source of error was the limitations of current oceanic tide maps, a subject now under careful investigation.

Finally, mention should be made of the *inverse approach* to worldwide gravity data. If one assumes the latest earth models to be correct, including the Green's functions for ocean loading, then one may in principle solve for the ocean tide itself as a problem in inverse theory. With the gravity data supplemented by ocean pressure data, this approach has been shown to produce plausible maps of the ocean tide.[54] However, the results to date cannot compare in accuracy to the most recent generation of tide maps based on satellite altimetry, to be considered in the next chapter.

Notes and references

1. Darwin, G.H. *The Tides* ... See Ch.1, ref. 2.
2. Swallow, J.C. and L.V. Worthington. An observation of a deep counter-current in the western North Atlantic. *Deep-Sea Res.*, **8**, 1–19, 1961.
3. Wunsch, C. Internal Tides in the Ocean. *Rev. Geophys. & Space Physics*, **13**, (1), 167–182, 1975.
4. Cartwright, D.E., A.C. Edden, R. Spencer, and J.M. Vassie. The tides of the northeast Atlantic Ocean. *Phil. Trans. R. Soc. London*, **A**, **298**, 87–139, 1980.
5. Lamb, H. See Ch. 7, ref. 14.
6. Matthäus, W. Contribution à l'Histoire du Marégraphe de Haute Mer. (Translated from a German original dated 1968). *Cahiers Océanographiques*, **22**, (2), Service Hydrogr. de la Marine, Paris, 1970.
7. Favé, L. Marégraphe Plongeur. *J. de Physique*, **10**, 404–414, 1891.
8. Rauschelbach, H. Zur Geschichte des Hochseepegels; Grundlagen für einem neuen Hochseepegel. *Ann. Hydrogr. und Mar. Meteorol.* **60**, 73–76 and 129–136, 1932.

9. Eyriés, M. Marégraphes de grandes profondeurs. *Cahiers Océanogr.* **20**, 355–368, Service Hydrogr. de la Marine, Paris, 1968. **9a.** Eyriés, M., M. Dars, and L. Erdelyi. *Ibid.*, **16**, (9), 781–798, 1964.
10. Hicks, S.D., A.J. Goodheart, and C.W. Iseley. Observations of the tide on the Atlantic continental shelf. *J. Geophys. Res.*, **70**, (8), 1827–1830, 1965.
11. Filloux, J.H. Deep Sea tide gauge with optical read-out of bourdon tube rotations. *Nature*, **226**, 935–937, 1970.
12. Filloux, J.H., D.S. Luther, and A.D. Chave. Update on seafloor pressure and electric field observations from the north Central and NE Pacific: Tides, infratidal fluctuations and barotropic flow. Pp. 617–639 in: Parker, B.B. (ed.) *Tidal Hydrodynamics*, John Wiley, New York, 883pp., 1991.
13. Nowroozi, A., J.T. Kuo, and M. Ewing. Solid earth and oceanic tides recorded on the ocean floor off the coast of North California. *J. Geophys. Res.*, **74**, 605–614, 1969.
14. Snodgrass, F.E. Deep Sea Instrument Capsule. *Science*, **162**, 78–87, 1968.
15. Munk, W., F. Snodgrass, and M. Wimbush. Tides offshore: transition from Californian coastal to deep-sea waters. *Geophys. Fl. Dyn.*, **1**, 161–235, 1970.
16. Irish, J.D. and F.E. Snodgrass. Australian–Antarctic tides. *Amer. Geophys. Union – Antarctic Research series*, **19**, 101–116, 1972.
17. Zetler, B., W. Munk, H. Mofjeld, W. Brown, and F. Dormer. MODE tides. *J. Phys. Oceanogr.*, **5**, (3), 430–441, 1975.
18. Cartwright, D.E., R. Spencer, J.M. Vassie, and P.L. Woodworth. The tides of the Atlantic Ocean, 60° N to 30° S. *Phil. Trans. R. Soc. London*, **A**, **324**, 513–563, 1988.
19. Smithson, M.J. (Ed.) Pelagic Tidal Constants – 3. IAPSO (IUGG) *Pub. Sci.* 35, 191pp., 1992.
20. Munk, W.H. and B.D. Zetler. Deep-sea Tides: A Program. *Science*, **158**, 884–886, 1967.
21. Reports in *Proceedings of the Sci. Committee on Oceanic Research*, (SCOR–IAPSO); (**a**) **1**, (2), 98–101, 1965; (**b**) **3**, (1), 56–64, 1967; (**c**) **11**, 22–24, 1976.
22. Zeilon, N. On tidal boundary-waves and related hydrodynamical problems. *Kung. Svenska Vetenskapakad. Hand.* **47**, (4), 1–46, 1912.
23. (**a**) Gould, W.J., and W.D. Mc.Kee. Vertical structure of semidiurnal tidal currents in the Bay of Biscay. *Nature*, **244**, 88–91, 1973. (**b**) Magaard, L. and W.D. McKee. Semidiurnal currents at Site D. *Deep-Sea Res.*, **20**, 997–1009, 1973.
24. Lamb, H. *Hydrodynamics*. Art 235, Cambridge Univ. Press, 6th Edn. 1932.
25. Hendershott, M.C. Long waves and ocean tides – Internal Tides. Pp. 329–339 in: *Evolution of Physical Oceanography*, (Scientific surveys in honor of Henry Stommel). M.I.T. Press, Cambridge, Mass., 623pp., 1981.
26. Hendry, R. Observations of the semidiurnal internal tides in the western North Atlantic Ocean. *Phil. Trans. R. Soc. London*, **A**, **286**, 1–24, 1977.
27. Miles, J.W. On Laplace's Tidal Equations. *J. Fluid Mech.*, **66**, 241–260, 1974.
28. Cox, C.S. and H. Sandstrom. Coupling of internal and surface waves in water of variable depth. *J. Oceanogr. Soc. Japan*, 20th Anniv. Vol., 499–513, 1962.
29. Rattray, M., J.G. Dworski, and P.E. Kovala. Generation of long internal waves at the continental slope. *Deep-Sea Res.*, **16**, 179–195, 1969.
30. Baines, P.G. On internal tide generation models. *Deep-Sea Res.* **29**, (3A), 307–338, 1982.
31. Schott, F. On the energetics of baroclinic tides in the North Atlantic. *Annales Geophys.*, **33**, (1/2), 41–62, 1977.
32. Sjöberg, B. and A. Stigebrand. Computations of the geographical distribution of the energy flux to mixing processes via internal tides and the associated vertical acceleration in the ocean. *Deep-Sea Res.*, **39**, (2), 269–291, 1992.
33. Papers by Dushaw *et al.* (1995) and by Ray & Mitchum (1996) are referenced in Chapter 1, ref. 7.
34. Rossby, C.-G. and Collaborators. Relation between variations in the intensity of the zonal circulation of the atmosphere and the displacements of the semi-permanent centers of action. *J. Marine Res.* **2**, 38–55, 1939.
35. Longuet-Higgins, M.S. Planetary waves on a rotating sphere (I): *Proc. R. Soc. London*, **A**, **279**, 446–473, 1964; (II): *Ibid.* **284**, 40–68, 1965.
36. Wunsch, C. The long-period tides. *Reviews Geophys.*, **5**, 447–475, 1967.

37. Hamon, B.V. Continental shelf waves and the effects of atmospheric pressure and wind stress on sea level. *J Geophys. Res.*, **71**, 2883–93, 1966.
38. Mysak, L.A. Recent advances in Shelf Wave dynamics. *Reviews Geophys. & Space Phys.*, **18**, (1), 211–241, 1980.
39. Otter, H.C. (Captain, RN) On the tides in the Sound of Harris. *Proc. R. Soc. Edinburgh*, **4**, 89–90, 1858. (Includes a brief theoretical note by 'Dr. Stark', who communicated the paper.)
40. Cartwright, D.E. Extraordinary tides near St. Kilda, *Nature*, **223**, 928–932, 1969. Sequel: Cartwright, D.E., J.M. Huthnance, R. Spencer, and J.M. Vassie. On the St. Kilda shelf tidal regime. *Deep-Sea Res.*, **27A**, 61–70, 1980.
41. Crawford, W.R., R.E.Thomson, and W.S. Huggett. Shelf waves of diurnal period along Vancouver Island. Pp. 225–235 in: *Coastal Oceanography*, (H. Gade, A. Edwards, H. Svenson, Editors), Plenum Press, New York, 582pp., 1983.
42. Wahr, J.M. Body tides on an elliptical, rotating, elastic and oceanless earth. *Geophys. J. R. astr. Soc.*, **64**, 677–703, 1981.
43. Jeffreys, H. and R.O. Vicente. The theory of nutation and the variation of latitude. *Monthly Not. R. astr. Soc.*, **117**, 142–161 and 162–173, 1957.
44. Wahr, J.M. and T. Sasao. A diurnal resonance in the ocean tide and in the earth's load response due to the resonant free 'core nutation'. *Geophys. J. R. astr. Soc.*, **64**, 747–765, 1981.
45. Zschau, J. Tidal friction in the solid earth: loading tides versus body tides. Pp. 62–94 in: *Tidal Friction and the Earth's Rotation.* (P. Brosche and J. Sündermann, Editors), Springer-Verlag, Berlin, 243pp., 1978.
46. Melchior, P. *The Tides of the Planet Earth.* 2nd Edn., Pergamon Press, Oxford, 641pp., 1983.
47. Bolt, B.A. and A. Marussi. Eigenvibrations of the Earth observed at Trieste. *Geophys. J. R astr. Soc.*, **6**, 299–311, 1961.
48. Gerstenecker, C., J. Zschau, and M. Bonatz. Finite element modelling of the Hunsrück tilt anomalies: a model comparison. Pp. 797–803 in: *Proc. 10th Internat. Symposium on Earth Tides*, 1985. Consejo superior de investigaciones Cientificas, Madrid, 1986.
49. Warburton, R.J., C. Beaumont, and J.M. Goodkind. The effect of ocean tide loading on tides of the solid earth observed with the superconducting gravimeter. *Geophys. J. R. astr. Soc.*, **43**, 707–720, 1975.
50. Harrison, J.C. plus 5 others. Earth-tide observations made during the 'International Geophysical Year'. *J. Geophys. Res.*, **68**, (5), 1497–1516, 1963.
51. Tomaschek, R. Tidal gravity measurements in the Shetlands. *Nature*, **175**, 937–939, 1954.
52. Knopoff, L., P.A. Rydelek, W. Zürn, and D.C. Agnew. Observations of load tides at the South Pole. *Physics Earth & Planetary Interiors*, **54**, 33–37, 1989.
53. Baker, T.F., R.J. Edge, and G. Jeffries. European tidal gravity: an improved agreement between observations and models. *Geophys. Res. Letters*, **16**, (10), 1109–1112, 1989.
54. Jourdin, F., O. Francis, P. Vincent, and P. Mazzega. Some results of heterogeneous data inversions for oceanic tides. *J. Geophys. Res.*, **96**, (B12), 20, 267–288, 1991.

14

The impact of satellite technology, 1970–1995

Prelude: 1957–1969

The successful launch by the USSR of the earth's first artificial satellite on 4 October 1957 took western scientists by surprise. Within a few days thousands of people including the writer turned out on clear evenings to watch *Sputnik-1* (or rather its rocket which was brighter) travel across the background of the stars, obeying Newton's laws of motion and gravitation. Apart from the impressive technological feat, its intended purpose was not at first clear, whether for military reconnaissance, weather surveillance or telecommunication. When its successor, *Sputnik-2*, launched in December, contained a dog named 'Laika', it was realised that the *Sputniks* were the prelude to man's first venture into space, achieved in due course in April 1961 by the astronaut Yuri Gagarin.

Few can have envisaged in 1957 that earth satellites would one day revolutionise synoptic oceanography. Such a revolution dawned gradually, starting with infrared imagery in the late 1960s and leading by stages to the precise radar altimetry of the TOPEX/POSEIDON satellite, designed in the USA and France and launched in 1992. Radar altimetry was the ultimate tool for global measurement of the ocean tides as well as many other dynamic features of the ocean surface. However, exciting new information about the tides was also provided in the 1970's and 1980's through the analysis of satellite orbits, before altimetry had reached a useful level of precision. For the historical record it is important to trace the development of orbital analysis from *Sputnik-1* onwards. While the tides were hardly relevant in the early stages of satellite research, the orbits traced out by the satellites revealed astonishing properties of the earth's Figure and of its gravity field, hardly dreamed in the pre-satellite days of geophysical and geodetic research.

A small group of British mathematicians working at the Royal Aircraft Establishment (RAE) was the first to achieve scientific results.[1] In 1957 this group had already devised a dynamic theory for the motion of an arbitrary body orbiting round an oblate earth and subject to atmospheric drag, as an extension to research on high-trajectory missiles; they had even designed a launching system for a hypothetical satellite. The combination of this expertise with the recent acquisition of an electronic computer, and a Radio Department at RAE keen to track the emitted 'beeps', enabled *Sputnik-1* to be regularly tracked from a few days after launch. Its initial orbital parameters were determined within two weeks. Little more than a month after launch, the RAE group had a paper published in *Nature* which included a new estimate of air density at a height of about 241 km at a time of high solar activity, deduced from the daily rate of decrease of the orbital period.[(2b)] (The Mullard Radio Observatory at Cambridge were doing similar, equally timely, work.[(2a)])

The property of the shape and gravity field of the solid earth most directly deducible from a satellite's orbit is its *oblateness*, usually expressed in terms of a *flattening* factor f equal to the eccentricity of the earth's Figure. Flattening causes the orbit's *nodes* (equatorial crossing points) to regress round the equator at an observable rate. This is a suitable point to introduce the principal symbols associated with satellite orbits, which will be used throughout this chapter.

The gravitational potential U of the earth at a distance r from its centre may be expressed, in zonal average:

$$U = (GM/r)\left[1 - \sum_{n=2}^{\infty} (a_e/r)^n)\, J_n\, P_n(\cos\theta)\right]$$

where GM is the fundamental astronomical constant related to earth mass, a_e is the equatorial radius, and the P_n are Legendre polynomials of colatitude θ. The J_n are a series of coefficients of flattening which would all be zero for a spherical earth. In reality, J_2 is by far the greatest coefficient; it is related to f by known formulae.[(3)]

Figure 14.1 shows the basic geometry of a satellite orbit as far as it concerns us here. N denotes the *ascending node* and I the *angle of inclination* of the orbit, which has semi-major axis 'a' and eccentricity 'e'. The rate of progression of N round the equator (positive eastwards) is given by

$$\frac{\mathrm{dO}}{\mathrm{d}t} = -\left(\frac{GM}{a^3}\right)^{\frac{1}{2}} \left(\frac{a_e}{a(1-e^2)}\right)^2 [1.5J_2 - F_4J_4 - F_6J_6 - \cdots]\cos I$$
$$+ \text{minor perturbations,}$$

where F_n are numerically small, known functions of I and e. (The usual symbol for O is Ω, but I prefer to continue to use Ω here for the angular velocity of the earth.) The above expression is nearly a simple rule for determining J_2 from measurement of $\mathrm{dO}/\mathrm{d}t$, but for accuracy J_4 may not be neglected. With observations of several satellites of different inclination, a fair number of even coefficients J_2, J_4, J_6 ..., could be calculated independently by least squares. The

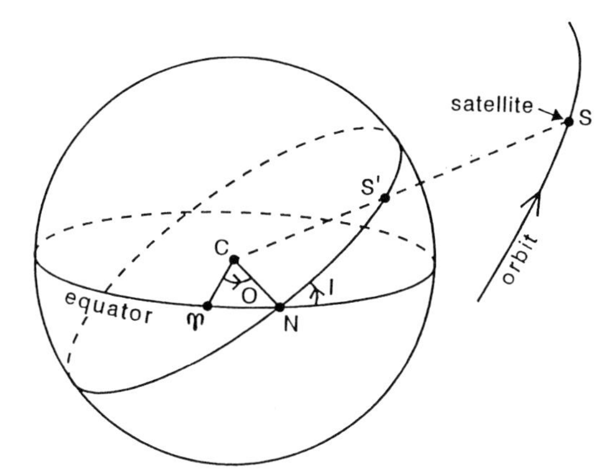

Figure 14.1. The basic geometry of a satellite's orbit with inclination I and (slowly regressing) nodal angle O relative to the equinoctial direction ♈. N is the *ascending node*. S′ is the nadir point beneath the satellite S; its highest latitude is I. The North Pole is at the top of the diagram. With the prograde orbital direction shown, dO/dt is negative; for $I>$ 90° dO/dt would be positive.

odd-order coefficients J_3, J_5, J_7 ... could be similarly derived from an analogous expression for oscillatory changes in the perigee distance $a(1-e)$.

The rocket of *Sputnik-1* lasted only 57 days before disintegrating in the atmosphere, too short a time for an estimate of J_2 of worthwhile accuracy. But *Sputnik-2* was brighter and survived for 162 days, enabling the RAE group to perform a pioneering analysis of the earth's gravity field from both orbits. The initial result,[4] $J_2 = 1084 \times 10^{-6}$, corresponding to $f = 1/298.1$, contrasted with the internationally accepted values dating from 1948,[3] namely 1091×10^{-6} and 1/297.1, causing consternation among traditional geodesists. However, the satellite results, slightly modified as more data were added, proved to be incontrovertible; the later accepted international standards were 1082.64×10^{-6} and 1/298.257.

Proliferation of satellites launched from 1958 onwards, chiefly by USA, USSR and France, was accompanied by growing refinement and numbers of instruments for measuring their position, from binoculars, kinetheodolites and cameras to lasers and electromagnetic-Doppler range recorders. The stream of data was absorbed by an increasing number of research groups in universities and national institutions, mostly in the USA, spurred on by a sudden interest in the gravity field expressed by US naval and military authorities. The objective soon became, not merely the zonal harmonics J_n, but the full expansion in tesseral harmonics of colatitude θ and longitude ϕ:

$$U(r, \theta, \lambda) = \frac{GM}{r} \sum_{n=0}^{\infty} \sum_{m=0}^{n} \frac{a_e^n}{r^n} P_n^m(\cos\theta)\,[A_n^m \cos m\phi + B_n^m \sin m\phi],$$

where the P_n^m are *associated Legendre functions*, suitably normalised, and the A's and B's are arrays of arbitrary coefficients to be determined from measurements.

With sufficient positional data to resolve the first $(N+1)^2$ harmonic elements A_n^m, B_n^m up to degree N, $U(r, \theta, \phi)$ was found to have a rich structure in a wide range of wavenumbers *n*, *m*. The potential *U* was usually translated into the height of the *geoid*, that is, the equipotential surface at about mean sea level,

$$U(a_e, \theta, \lambda) \;/\; (\delta U/\delta r),$$

measured relative to the ellipsoid which best represents the earth's Figure. The geoid was found to have spatial undulations with amplitudes up to about 100 m and fine structure related to geological features. The findings were of great interest to geophysicists. The most authoritative and widely used maps of the geoid were the series of *Goddard Earth Models* (GEM) computed by a large team of satellite geodesists (a new profession) at the Goddard Space Flight Center of the US National Aeronautic and Space Administration (NASA). An early GEM is mapped in the upper panel of Figure 14.5.

The pioneers of orbit analysis at the UK–RAE concentrated mostly on organising a large body of voluntary semi-professional observers, and on the study of a phenomenon first suggested by Sir Harold Jeffreys. An orbit slowly decaying under air drag will *resonate* with integer fractions of a nodal day (time between successive transits of the node *N* past an earth meridian). Analysis of orbit behavior around such resonances gave particularly precise estimates of the sets of coefficients A_n^m, B_n^m for certain harmonic orders *m*, mostly in the ranges $m = 11-16$ and $27-32$.[(1)] Such values calculated by RAE provided useful checks on the corresponding GEM harmonics, which had been derived from the inversion of massive matrices.

Tidal variations in satellite orbits

Towards 1965 some investigators noticed that orbital parameters, notably *I* and O, varied coherently with long periods related to the tidal periods. The major cause of a tidal variation in the gravity field as sensed at satellite height is the *earth tide*. Having the same spatial characteristics as the tidal potential, its wavelength is 180° in longitude and its effect reaches further out into space than the ocean tide with its relatively short length scale (large values of *n*), in virtue of the factor $(a_e/r)^n$ in the last formula. It will be recalled that the tide potential, and hence the earth tide, are resolvable into time-varying tesseral harmonics P_2^0, P_2^1, P_2^2. The relation between the intrinsic frequency σ of a tide constituent with spatial shape P_n^m and the frequency σ_s of its effect on a satellite of given orbital parameters is too complicated to explain here in general terms. It is greatly simplified in practice by the fact that the amplitude of the observed perturbation is inversely proportional to σ_s, so that only the longest of many periodic effects need be taken into account.

For the perturbation period σ_s^{-1} to be much longer than a day the relation may be reduced to:

$$\sigma_s = |\sigma + m\,(d/dt)(O - \Gamma)|$$

where Γ is the Right Ascension of the Greenwich meridian (equivalent to *sidereal time*) – Appendix A. For example, $\sigma_s = dO/dt$ for the tide constituent K_1, $2dO/dt$ for K_2, each reduced by 2 cycles per year for P_1, S_2, or by 2 cycles per month for O_1, M_2. dO/dt is typically of order -2 cycles per year for prograde orbits ($I < 90°$) or of order $+2$cpy for retrograde orbits ($I > 90°$) – see Figure 14.1. Typical periods for the prograde orbit of the satellite *GEOS-1*, whose node regressed by 2.28 cpy, (i.e. $-dO/dt$), are shown in the inset panel of Figure 14.2. Having identified the observed satellite frequencies with their respective tide constituents, spectral analysis of an orbital element at σ_s gives amplitude and phase lag which bear a known relationship to the appropriate constituents of the tide potential.

The earliest investigators of this effect[5,6] assumed that the earth tide was the *only* contributor, and hence they interpreted results as pertaining to the quantity $(1 + k_2)\Delta U$, where k_2 is the usual Love number for the added potential due to an elastically yielding earth (Chapter 10) and ΔU is the primary tide potential. (I have added the symbol Δ to distinguish the tide potential from the mean gravity potential U of the earth itself.) The resulting estimates of k_2 differed from the previously accepted value and suggested phase-lags and frequency-dependence. These apparent properties were out of keeping with ideas based on seismology.

A new wave of research into tidal perturbations was started towards 1974 by a group of earth scientists at the Institut de Physique du Globe (IPG) at Paris and at the Groupe de Recherches de Géodesie Spatiale (GRGS) at Toulouse.[7,8] They realised that the orbital perturbations are significantly affected by the oceanic tide as well as by the earth tide. They accordingly expressed the disturbing potential in its full spherical harmonic development (Appendix F):

$$(1 + k_2)\Delta U + g \sum_{n=0}^{\infty} \sum_{m=0}^{n} (1 + k_n{}') \frac{3\rho}{2n+1} \left(\frac{a_e}{r}\right)^{n+1} P_n^m(\sin\theta)$$

$$[{}^{+}C_n^m(t) + {}^{-}C_n^m(t)]e^{im\phi},$$

where $k_n{}'$ is the *loading* Love number of degree n, ρ is the density ratio (ocean:earth), and ${}^{+}C_n^m$, ${}^{-}C_n^m$ are complex coefficients of the respective eastward and westward progressing parts of the spherical harmonic expansion of the ocean tide. (Complex values are required to express the tide in-phase and in-quadrature with the tide potential; four arbitrary elements are thus needed for each pair (n, m) – see Appendix F.)

A spherical harmonic expansion of the oceanic tide is a new concept in this story. It was unthinkable at the time of the rudimentary and incomplete tide maps of the pre-computer age, but it arose naturally in the context of satellite orbitography as an extension to the expansion of the earth's gravity field. A similar expansion was assumed at about the same time in the extension to Laplace's equations to allow for loading effects (Chapter 12). Here was a method by which ${}^{+}C_n^m$, ${}^{-}C_n^m$ could in principle be directly measured by spectral analysis. It was unlikely to give reliable coefficients of degree high enough to reconstruct the details of the tide maps, but the low degree coefficients could be obtained with fair accuracy to provide useful constraints on dynamic tide solutions.

Figure 14.2, taken from a later study by a NASA group, well illustrates the character of the tide-induced oscillations in the inclination of the orbit of *GEOS-1*.[9] The upper panel shows the total oscillation over a 640-day span of data, with typical amplitude 1 arc-second. The lower panel shows the residual oscillation after subtracting a synthesis for the earth tide involving the constituents S_2, K_1, P_1, which had the longest periods for this satellite. (A standard value for the Love number k_2 was assumed; the period σ_s^{-1} corresponding to M_2 is only about 14 days. Since the amplitude of the perturbation is inversely proportional to σ_s, M_2 gives a much smaller perturbation.) The residual amplitude, which is due to the oceanic tide, is much less than in the total oscillation by a factor of about 5, but it is evidently still resolvable from the data.

However, the most radical innovation made by the IPG/GRGS group was to point out that the eastward progressing coefficients of degree 2, namely ${}^{+}C_2^m$, are sufficient to determine the global tidal dissipation, the deceleration ($-\mathrm{d}n/\mathrm{d}t$) of the moon, and the tide-induced retardation of the earth. The point is that the net work performed by the moon and sun to keep the tide going comes only from the part of the tide which is spatially correlated with the potential, which has degree 2 and travels eastwards. For example, the work done by the constituent M_2 is proportional to the part of ${}^{+}C_2^2$ which causes oscillations in a satellite orbit with the frequency $\sigma_s(M_2)$.

The concept, illustrated in Figure 14.3, was also alluded to in Chapter 10, being first introduced by Thomson and Darwin in the late 19th century. Figure 14.3 shows the profile of the ocean tide in the equatorial plane with greatly exaggerated vertical scale. For simplicity of illustration, the moon is depicted as moving in the same plane. Early forms of this diagram, which have frequently been copied, misleadingly depict the ocean tide as a smooth oval, stemming from Darwin's idea that all the dissipation occurs in the earth tide; in fact, the earth tide has negligible phase lag. As shown, the tide lags the moon *on average* by the geometrical angle denoted in the figure by $\varepsilon_{22}/2$, as a result of which the tidal forces acting on the tide itself – a second-order effect – constitute a torque T opposing the earth's rotation, proportional to

$$D_{22} \sin \varepsilon_{22},$$

where D_{22} is the amplitude of the tide's harmonic of degree and order 2. (Some geodesists use the symbol ε to denote $(450° - \varepsilon)$ in the present notation, in which case the cosine is involved.) The rate of working is $-T\Omega$, and the earth's deceleration, $-\mathrm{d}\Omega/\mathrm{d}t$, is T/C, where C is the usual symbol for the axial moment of inertia. Conservation of total angular momentum gives an expression for the moon's deceleration, $-\mathrm{d}n/\mathrm{d}t$, also proportional to T/C. In fact, D_{22} is directly proportional to $|{}^{+}C_2^2|$ and ε_{22} is the harmonic phase-lag of ${}^{+}C_2^2$ with respect to the tide potential.[7,8]

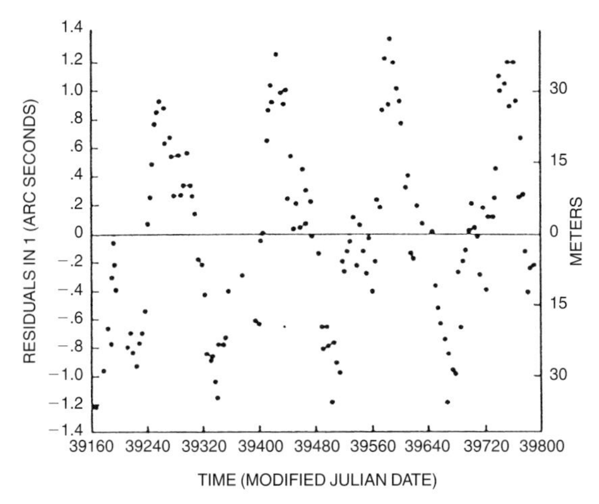

Perturbation of the Geos 1 orbit caused by solid earth and ocean tides.

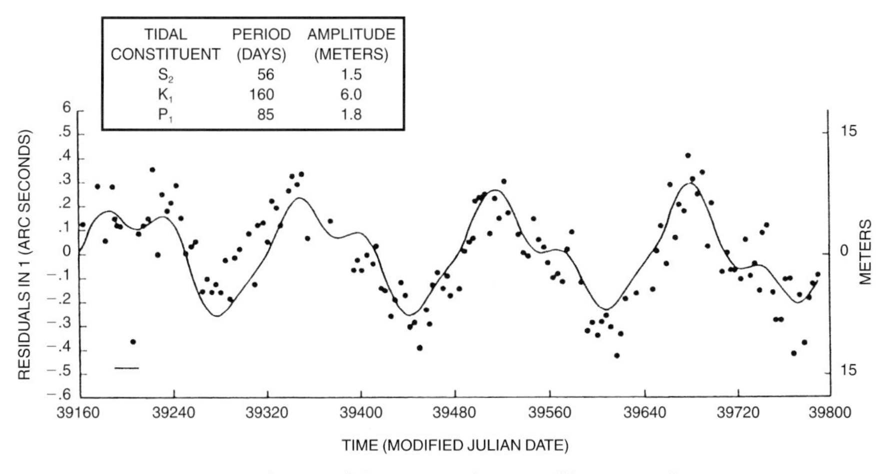

Perturbation of the Geos 1 orbit caused by ocean tides.

Figure 14.2. Two diagrams from ref. 9, showing tidal variations in the inclination of *GEOS-1* during 640 days in 1966–67. Upper panel shows the total observed variation; lower panel shows the residual (ocean-induced) variation after the effect of the earth body-tide was removed. The solid curve is a combination of three sinusoids with alias-periods listed in inset panel. Left-hand scales denotes the angular variation of I; right-hand scales denote the linear variation of the highest latitude on the earth's surface.

Thus, the astonishing fact emerged, that important geophysical parameters which had been sought for more than a century in terms of frictional drag on the sea bed, and currents and elevations of the sea surface, could be directly deduced from the variations in a satellite's orbit. The approach was quickly taken up by research groups in France and USA, and applied to a variety of satellites. In 1975, the French Comité National d'Etudes Spatiales (CNES) launched

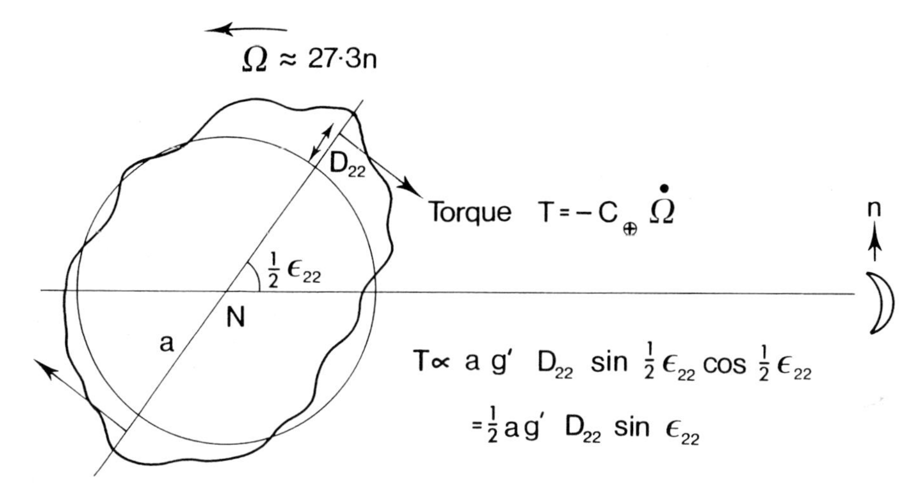

Figure 14.3. Diagram to represent the tidally induced torque on the mean P_2^2 bulge of the ocean tide, whose amplitude is D_{22} and phase lag (in time) is ε_{22}. n is the mean orbital velocity of the moon, Ω is the earth's angular velocity, $a = a_e$ is the equatorial radius. The quantity $g' = g \times$ mass ratio $\times$ cube of parallax $= O(10^{-7}g)$ represents the magnitude of the horizontal tidal acceleration due to the moon.

a spherical satellite named *Starlette* for the express purpose of precise orbit determination for the study of gravity and tides. *Starlette* was soon followed by a sequence of similar satellites launched by USA named *Lageos*-1 and *Lageos*-2.

I will not describe the many successive numerical results obtained from these satellites or attempt to account for small differences between them, because other methods of space geodesy were soon developed which could also be used to extract the same tidal parameters. An overall summary and comparison will be found at the end of this chapter.

Laser ranging to the moon

The American astronauts who landed on the moon in July 1969 did more than create a page of human history; they left on the moon's surface a retro-reflector facing the earth, (Figure 14.4). This passive device was designed to reflect a beam directed by a laser transmitter on earth at energy levels near 1 joule, in exactly the opposite direction from which it came. The reflected beam was detected at a level of one photon every few seconds, and was timed to an accuracy equivalent to a few centimeters of light travel. Other reflectors were later laid on the moon in space missions organised by USA and USSR.

At the level of a few centimeters precision, the distance between given points on the earth and moon is affected by a great variety of small motions in both bodies as well as the gross variations of the lunar orbit. Of greatest interest to us is the mean secular rate of increase of mean lunar distance due to the ocean tides, about 3.8 centimeters per year. By 1989, with data from more than a nuta-

Figure 14.4. Photograph of one of the lunar laser retro-reflectors placed on the moon in July 1969 by the *Apollo 11* astronaut Buzz Aldrin, whose footprints are visible in the lunar dust. (Courtesy of NASA-JPL.)

tional cycle of 18.6 years, it was possible to measure this rate directly with some precision.[11] In terms of the semimajor axis a_m of the lunar orbit, Kepler's 3rd Law, which states that $n^2 a_m^3$ is constant, where n as usual denotes the mean angular motion of the moon, we have

$$\frac{da_m}{dt} = -\frac{2a_m}{3n}\frac{dn}{dt}.$$

So measuring the rate of recession by lunar laser ranging is tantamount to measuring the tide-induced deceleration of the moon's orbital velocity, the subject of so many historical debates.

It should be noted that the above measure does not discriminate between the various tidal harmonic constituents M_2, O_1 etc but gives their total effect. Further, it is not affected by the solar tide. By contrast, the analysis of the perturbations of satellite orbits gives, in principle, the individual constituents due to both sun and moon. Both solar and lunar tides contribute to the tide-induced change in Ω, that is to the change in length-of-day.

Radar altimetry of the sea surface – *Skylab, GEOS-3, Seasat*

All the investigations described so far were based on viewing satellites (including the moon) from the earth. The reciprocal operation, viewing the earth from satellites, was recognised from the start to be potentially useful, but it took several years to be implemented. High resolution photography of land surfaces had priority for obvious reasons, both civil and military. Even this had to await either manned flights or digital transmission of images to a ground station, both realised during the 1960s. Visual images of reflected sunlight from the sea are not very informative, although the use of color filters enabled massive algal growths to be detected, with certain physical implications.

Infrared radiometry, as performed by the *Nimbus* series of satellites, was a distinct step forward since it gave meaningful measurements at night as well as day. Infrared emission is a useful measure of sea surface temperature, but it will not penetrate clouds. As their name implies, the *Nimbus* satellites were launched primarily for meteorology by monitoring the temperature of clouds.

The idea of using *active radar* transmitted from a satellite began to be discussed in the mid-1960s, chiefly in the context of backscatter from sea waves. Interest soon focussed on two possibilities: an instrument for measuring the backscatter from directionally tilted radar beams, known as a *scatterometer*, and an instrument for collecting the return signal from a vertically transmitted narrow-beam pulse, known as an *altimeter*. The scatterometer could be used to measure the directional characteristics of waves and wind. The altimeter could also give a measure of the roughness of the sea surface directly beneath the satellite, but more importantly to our subject, it would also measure the height of the satellite above the mean sea surface and hence the surface profile of the ocean. Both instruments could of course operate by day and night and through cloud cover. A third type of radar known as *Synthetic Aperture* (SAR), requiring much greater power and data storage, used multiple phase information to define a detailed radar image of the sea surface.

The potential benefits of a radar altimeter to geodesy and physical oceanography were fully discussed in an important pair of seminal papers in early 1969 by a group of oceanographers at New York University.[(12)] With a feasible precision at that time of 1m in about 700km height, the altimeter would easily resolve the undulations of the geoid over the oceans at shorter wavelength than the gravimetric analysis, including the surface signatures due to seamounts and trenches. Next in decreasing order of amplitude would be the large-scale undulations of the tides, mostly less than 1m on the ocean but detectable by averaging over several rapidly consecutive radar pulses. Finally, it seemed possible to detect the geostrophic rise of the surface across intense current jets such as the Gulf Stream. These and subtler features were all eventually resolved, with growing precision as altimeter technique developed.

The first satellite radar altimeter to be realised was an instrument known as 'S193' on board the manned NASA spacecraft *Skylab* in 1973–74.[(13)] The upper

Strategy B, at the opposite extreme, used no assumed physical formulations, but this fact could be seen by some as a virtue, in putting more trust in observations than in theory. Prior to satellite altimetry, the interpolation between tide-gage stations relied largely on inspired guesswork, or could be done reliably only in regions of closely spaced bottom-pressure stations (Chapter 13). With improvements in altimetry through *Seasat–Geosat–T/P*, a dense network of observations became available, limited only by the spacing of satellite tracks and by sampling errors due to noise. Both deficiencies could be greatly reduced by spatial smoothing. An effective variant of B was to use the altimetry to correct errors of long wavelength in another solution of type A or B.[25]

If method A may be criticised for placing too much reliance on approximate mathematical equations with arbitrary parameters, and method B for absorbing data-noise into its solutions, then strategy C is a logical compromise between the two. Inverse solutions for ocean tides based on tidal gravity and ground data were mentioned at the end of Chapter 13. The general technique, known as *data assimilation*, had been applied to ocean tides before altimetry was available.[26] One variant made use of a set of spatial *basis functions* originally postulated in theory as long ago as 1917 by J. Proudman (Chapter 12).[27a]

The most ambitious and effective of the assimilation methods, devised at Oregon State University (OSU), was to compute a series of 6355 spatial functions known as *representers*, each associated with a position where the ascending and descending earth tracks intersect ('crossovers'). The tide solution for a given frequency consisted of a sum of a dynamic solution of Laplace's tidal equations and a series of representers with coefficients adjusted to the 6355 data sets by least-squares. The representers required a great deal of computational effort, but once established could be used to adjust solutions to any quantity of data received at the same positions. Solutions from the OSU group to date are convincingly smooth and accurate with respect to *in situ* data.[28]

Figure 14.7 maps the principal solar constituent S_2, derived by a group at NASA–Goddard Space Flight Center by strategy C, using the Proudman expansion.[27b] It is just one example from the twelve 'best' tidal maps, all superficially very similar, produced around 1994–95.[29]

Assessments of relative merit between the various solutions have been made by comparing their constituent amplitudes and phases with those recorded at a set of 100–104 carefully chosen bottom pressure stations and small islands, reasonably well distributed over the oceans.[29] (Ref. 29 also gives detailed comparisons of the maps themselves.) The rms discrepancies of the various models differ by only a few millimeters. Typical average rms discrepancies are 20 mm for M_2 alone, 30 mm for the combination of four major constituents, M_2, S_2, O_1, K_1. Such levels of accuracy are comparable with the accuracy of some of the data sets themselves, and so have practically reached the limit of possibility to judge merit. Space technology has thus come near to bringing about the culmination of an era of research into mapping the tides which started in 1833 with the initial speculations of John Lubbock and William Whewell, (Chapter 9).

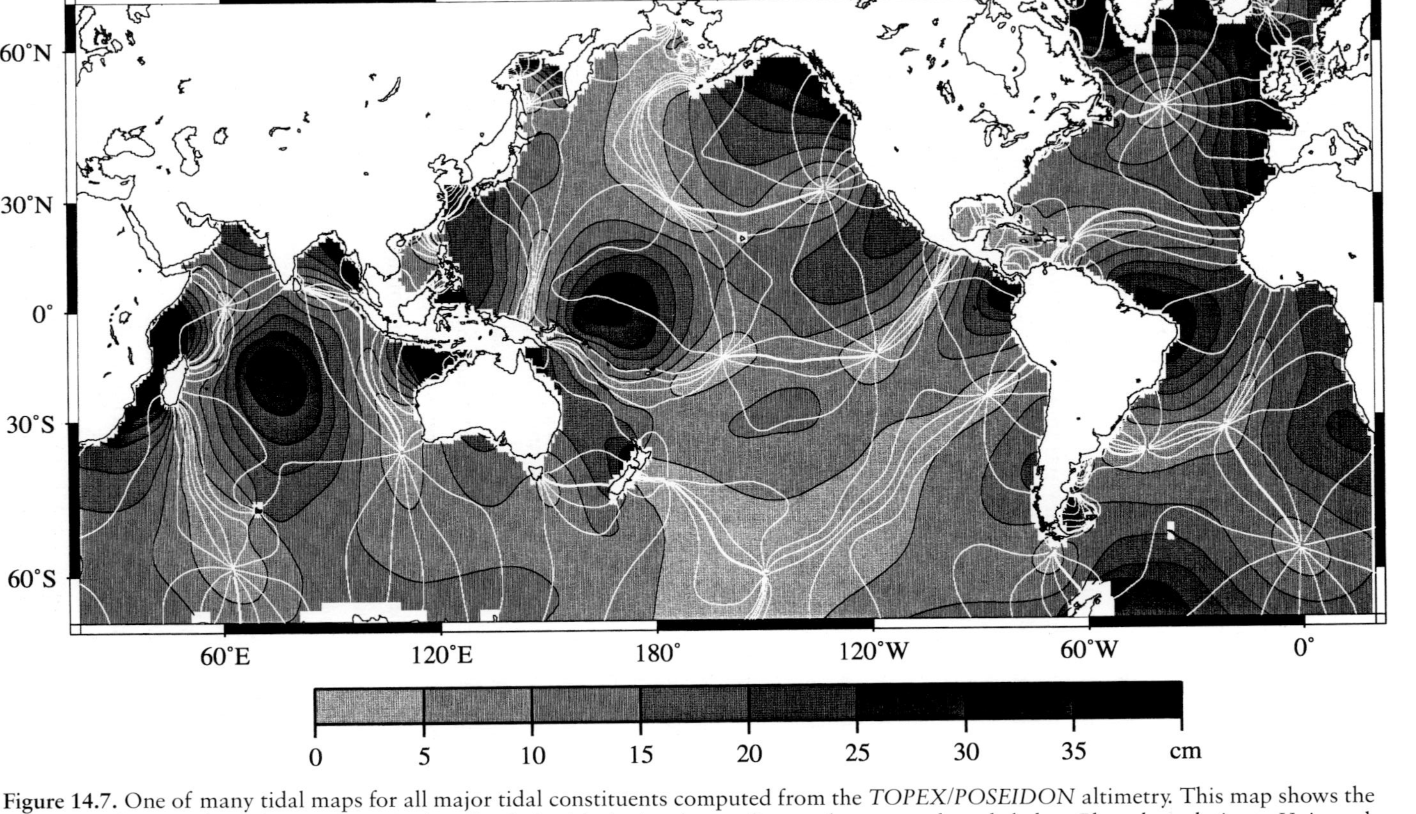

Figure 14.7. One of many tidal maps for all major tidal constituents computed from the *TOPEX/POSEIDON* altimetry. This map shows the principal solar semidiurnal constituent S_2. Amplitude (cm) is depicted according to the gray-scale code below. Phase lag relative to Universal (Greenwich) Time is constant along each white line, except for changes of 180° (6 hours) on passing nodal points. Thick lines denote zero phase; others progress around nodal points at 30° intervals. The sense of phase progression is not directly indicated, but it may be deduced from the facts that the three large amphidromic systems in the North Atlantic and those west of Africa and California all progress anticlockwise, while the systems off Western Australia, in the Arabian Sea and close to Antarctica progress clockwise. The depicted tidal field is a synthesis of more than 1000 *Proudman functions*, adjusted for an anomaly due to solar radiation, with coefficients determined for least-squares fit to the altimetry.[27b]

Global parameters for earth–moon dynamics

The advances in tide modelling described above naturally led to improved estimates of energy dissipation and of the decelerations of earth- and moon-rotations caused by tides. As standards of comparison one had the integral measures supplied by astronomical observation, by spectroscopy of satellite orbits, and by lunar laser-ranging.

As explained earlier in this chapter, the key parameters to the tidal influences on these integral properties were seen to be the amplitudes and phase lags (D_{2m}, ε_{2m}) of the second degree spherical harmonics $^{+}\mathbf{C}_2^m$ of the semidiurnal ($m=2$) and diurnal ($m=1$) constituents of the ocean tide. An important review of the subject in 1977, rather too early for good quality tide models to have been available,[10] quoted values of D_{22} for the M_2 constituent between 4.9cm and 3.6cm, and ε_{22} values between 126° and 105°. The 1991 model based entirely on *Geosat* altimetry[19b] gave (3.5cm, 133°), and finally the twelve models associated with T/P gave results closely clustered in the range (3.20 ± 0.05cm, 130 ± 1°), corresponding to a dissipation rate for M_2 of 2.45 ± 0.05TW. The last estimate contrasts strongly with the historical estimates: 1.4TW (Jeffreys – 1920 estimate), 2.7TW (Spencer Jones), roughly 3.7 ± 0.5TW from pre-1977 models[8], and 1.94 TW (Schwiderski).

Comparison with astronomical measurements entails summation over all tidal constituents, lunar and solar, for dE/dt, $d\Omega/dt$; only the lunar tides are needed for dn/dt. Representative figures are listed in the Table.

Tidally induced changes in the earth's (Ω) and moon's (n) rotation rates

Source	$-dE/dt$ TW	$d\Omega/dt$ $10^{-22}s^{-2}$	Δ(l.o.d.) ms/cy	dn/dt $''.cy^{-2}$
T/P tide model[25]	3.4	−5.9	2.3	−23.9
Orbit spectroscopy[10,30]	3.6	−6.4	2.4	−26.1
Gravity analysis[20]	3.5	−6.1	2.3	−25.3
Astronomical[31]	—	−6.4	2.4	−26
Lunar ranging[11]	—	—	—	−25.9 ± 0.5

The column headed $-dE/dt$ is by convention the rate of energy loss through dissipation, and therefore positive. The column headed Δ(l.o.d.), the change of length of day in milliseconds per century, is simply proportional to the figures for $d\Omega/dt$ by the factor 0.375.

The first row of figures was supplied by NASA (Richard Ray – personal communication) from a typical recent model based on T/P altimetry alone. Reference 25 is the only paper of the recent batch to quote figures for (D_{2m}, ε_{2m}); the others are preoccupied with rms differences from *in situ* data. The second row is mainly from a digest of the work of many authors on the spectral analysis of orbit inclination, eccentricity, and nodal progression. These authors had agreed closely on a figure of 2.19 TW for M_2 alone[30] which now appears too small, probably on account of small systematic errors in the spectroscopy method. The figures involving the solar tides were taken from earlier work by related methods.[10]

The third row of figures is from tidal parameters published with the GEM-T2 gravity model. Being based on the analysis of millions of ranging observations from 31 satellites, this model's method is closely related to the spectroscopy method, but it also involves the extraction of a huge array of Tesseral Harmonic coefficients for the earth's gravity field. Accuracies for the tide figures are claimed in the region of 1 percent.[20]

The 'astronomical' figures (fourth row) are based on authoritative studies of telescopic observations in the historical past of the moon's longitude and the earth's rotation.[31] The *tidal* part of $d\Omega/dt$ was deduced from long-term averages of the more recent data. Finally, the bottom row, containing only one figure with error estimate for dn/dt, is the latest result from many years of study of laser-ranging to the moon, using reflectors such as the one depicted in Figure 14.4. The figure is equivalent to a rate of recession of the moon of 3.82 ± 0.07 m/cy.

Some of these tabulated figures will probably be (and are already being) modified slightly as more data are acquired and analysed, but they summarise the remarkable degree of convergence of results from many independent methods. Like the tide maps described above, they mark a laudable climax to 20th century research on the ocean tides and their effects on earth–moon dynamics.

Notes and references

1. King-Hele, Desmond. *A Tapestry of Orbits*. Cambridge Univ. Press, 244pp., 1992.
2. (**a**) Staff of the Mullard Radio Astronomical Observatory, Cambridge. Radio observations of the Russian earth satellite. *Nature*, **180**, 879–883, 1957. (**b**) Staff of the RAE, Farnborough, Observations of the orbit of the first Russian earth satellite. *Nature*, **180**, 937–941, 1957.
3. Jeffreys, H. The Figures of the Earth and Moon (3rd paper). *Monthly Not. R. astr. Soc. – Geophys. Suppl.*, **5**, 219–247, 1948.
4. Merson, R.H. and D.G. King-Hele. Use of artificial satellites to explore the earth's gravitational field: *Sputnik-2 (1957β)*. *Nature*, **182**, 640–641, 1958.
5. Newton, R.R. An observation of the satellite perturbation caused by the solar tide. *J. Geophys. Res.*, **70**, 5983–5989, 1965.
6. Kosai, Y. Effects of the tidal deformation of the earth on close earth satellites. *Pub. Astron. Soc. Japan*, **17**, 395–402, 1965.
7. Lambeck, K., A. Cazenave, and G. Balmino. Solid earth and ocean tides estimated from satellite orbit analyses. *Rev. Geophys. & Space Phys.*, **12**, (3), 421–434, 1974.
8. Lambeck, K. Effects of tidal dissipation in the oceans on the moon's orbit and the earth's rotation. *J.Geophys.Res.*, **80**, 2917–25, 1975.
9. Felsentreger, T.L., J.G. Marsh, and R.W. Agreen. Analysis of the solid earth and ocean tidal perturbations on the orbits of *Geos-1* and *Geos-2* satellites. *J. Geophys. Res.* **81**, 14, 2557–63, 1976.
10. Lambeck, K. Tidal dissipation in the oceans: astronomical, geophysical and oceanographic consequences. *Phil. Trans. R.Soc. London*, **A**, **287**, 545–594, 1977.

11. Dickey, J.O. + 11 co-authors. Lunar Laser Ranging: A continuing legacy of the Apollo program. *Science*, **265**, 482–490, 1994.
12. Greenwood, J.A., A. Nathan, G. Neumann, W.J. Pierson, F.C. Jackson, and T.E. Pease. Radar altimetry from a spacecraft and its potential applications to (A) Geodesy, (B) Oceanography. *Remote Sensing of the Environment*, (Elsevier Pub.), **1**, (A) 59–70, (B) 71–80, 1969.
13. *Skylab EREP* Investigations Summary. NASA Special Pub.-399, Washington, D.C. 1978.
14. Gower, J.F. (Ed.) *Oceanography from Space*. Plenum, New York, 987pp., 1981.
15. Mazzega, P. The M_2 ocean tide recovered from *Seasat* altimetry in the Indian Ocean. *Nature,* **302**, 514–516, 1983.
16. Woodworth, P.L. and D.E. Cartwright. Extraction of the M_2 ocean tide from *Seasat* altimeter data. *Geophys. J. R. astr. Soc.*, **84**, 227–255, 1986.
17. Parke, M.E., R.H. Stewart, D.L. Farless, and D.E. Cartwright. On the choice of orbits for an altimetric satellite to study ocean circulation and tides. *J. Geophys. Res.*, **92**, (C11), 11, 693–707, 1987.
18. McConathy, D.R. and C.C. Kilgus. The Navy *Geosat* mission: An overview. *Johns Hopkins Tech. Digest*, **8**, 170–175, 1987.
19. (**a**) *Geosat*; sea level from Space. Various papers in *J. Geophys. Res.*, **95**, C3, 2833–3179, and **95**, (C10), 17865–18, 367, 1990.
    (**b**) Cartwright, D.E. and R.D. Ray, Energetics of ocean tides from *Geosat* altimetry. *J. Geophys. Res.*, **96**, (C9), 16, 897–912, 1991.
20. Marsh, J.G. + 16 co-authors. The *GEM-T2* gravitational model. *J. Geophys. Res.*, **95**, (B13), 22, 043–071, 1990. *Ibid*. Correction to Table 7 of above. *J. Geophys. Res.*, **96**, (B10), 16, 651, 1991.
21. Duchaussois, G. Status and future plans for *ERS-1*. Pp. 501–514 in T.D. Allan (Ed.), *Satellite Microwave Remote Sensing*. Ellis Horwood/John Wiley, Chichester (UK), 526pp., 1983. For later edition see: *ERS-1* System, European Space Agency, SP-1146, Noordwijk (Netherland), 87pp., 1992.
22. Fu, L-L, + 6 co-authors. *TOPEX/POSEIDON* mission overview. *J. Geophys. Res.*, **99**, (C12), 24, 369–381, 1994.
23. Papers by Tapley, B.D. + 14 co-authors, by Nouel, F. + 10 co-authors, by Nerem, R.S. + 19 co-authors, and by Bertiger, W.I. + 19 co-authors, in *TOPEX/POSEIDON*: Geophysical Evaluation. *J. Geophys. Res.*, **99**, (C12), 24, 383–485, 1994.
24. Le Provost, C., M.L. Genco, F. Lyard, and P. Canceil. Spectroscopy of the world ocean tides from a finite-element hydrodynamic model. *J. Geophys. Res.*, **99**, (C12), 24, 777–797, 1994.
25. Schrama, E.J.O. and R.D. Ray. A preliminary tidal analysis of T/P altimetry. *J. Geophys. Res.*, **99**, (C12), 24, 799–820, 1994.
26. Zahel, W. Modelling ocean tides with and without assimilating data. *J. Geophys. Res.*, **96**, (B12), 20, 379–391, 1991.
27. (**a**) Sanchez, B.V. and N.K. Pavlis. Estimation of main tidal constituents from T/P altimetry using a Proudman function expansion. *J. Geophys. Res.*, **100**, (C12), 25, 229–248, 1995.
    (**b**) Ray, R.D., B.V. Sanchez, and D.E. Cartwright. Some extensions to the response method of tidal analysis applied to T/P. (Abstract) *EOS*, Trans. A.G.U., **75**, (16), Spring Meeting Suppl. 108, 1994.
28. Egbert, G.D., A.F. Bennett, and M.G.G. Foreman. *TOPEX/POSEIDON* tides estimated using a global inverse model, *J. Geophys. Res.*, **99**, (C12), 24, 821–852, 1994.
29. Andersen, O.B., P.L. Woodworth, and R.A. Flather. Intercomparison of recent ocean tide models. *J. Geophys. Res.*, **100**, (C12), 25, 261–282, 1995.
30. Cazenave, A. Tidal friction parameters from satellite observations. Pp. 4–18 in: (P.Brosche & J.Sündermann, Ed.), *Tidal Friction and the Earth's Rotation -2,* Springer-Verlag, 345pp., 1982.
31. Stephenson, F.R. and L.V. Morrison. Longterm changes in the rotation of the earth: 700BC to AD 1980. *Phil. Trans. R. Soc. London*, **A**, **313**, 47–70, 1984.

15

Recent advances in miscellaneous topics, and final retrospect

The previous chapter brought the mainstream of tidal research up to the time of writing, but advances have also been made in subsidiary but relevant research topics which have not fitted conveniently under the chosen subject headings. In this last chapter I shall briefly review what seem to me to be the most important of these topics, without purporting to cover every published paper. We shall pay no further attention to questions of which features of modern technology are most relevant to each field of research. All modern aids to measurement and computation will be taken for granted.

The long-period tides

We considered the *pole tide* in Chapters 10 and 12, but I have said little about the tides of long period generated by the moon and sun since Darwin examined the harmonic constants of the monthly (Mm) and fortnightly (Mf) tides with a view to determining the earth's rigidity factor. As with the pole tide, reliable observational data have always been hard to come by, because of their weak amplitudes in spectral regions of high noise continuum. In the present context the long-period tides may be divided into three categories: the lunar harmonics Mm and Mf, the solar variations Sa and Ssa, and the very long-period term at 1 cycle per nodal period (18.6y).

The solar variations stand apart, because they are predominantly due to seasonal variations in the atmosphere and associated changes in ocean circulation and temperature. The observed spectral peak is at 1cpy (Sa), and is much greater than the annual component of the tide potential due to the ellipticity of the earth's orbit, which is quite negligible. The tide potential has a stronger line at 2cpy (Ssa), but this too is swamped by the second harmonic of the seasonal change. Analysis of the annual component of sea level variation at a worldwide distribution of 419 stations in the mid 1950's, based on the monthly averages compiled by the sea level service of IAPO (Chapter 12), showed it to be largely due to changes in the heat content of the top 100 meters, with some localised effects of atmospheric pressure and wind.[1] More recently, satellite altimetry has given new global views of seasonal changes in sea level, showing an annual shift of mass and heat between the two hemispheres with much regional and interannual variation.[2]

Dynamic study of the gravitational tides of long period must therefore rely on information at the relatively high frequencies of 1 and 2 cycles per month on the one hand, and on the extremely low frequency of the nodal cycle on the other. The main questions of interest are, as in Darwin's day: (1) how far are these tides affected by friction? and (2) how close are they to the 'equilibrium' form of the potential? These points were addressed in the last published paper by Joseph Proudman.[3]

Early in the century Darwin and Poincaré had assumed that friction was controlled by molecular viscosity, which they rightly concluded would have no sensible effect on even the fortnightly tide. Since G.I. Taylor's work on the Irish Sea (Chapter 11), it was generally recognised that friction is a turbulent process with much larger drag forces. Being quadratic in velocity, Proudman realised that turbulent friction would not be merely proportional to the square of the Mf tide velocity but to its product with the much larger velocities of the semidiurnal tides, a point which had been brought out by his departmental assistant at Liverpool, K.F. Bowden (1916–1989).[4] In brief, the frictional term in the momentum equations would be *linear* in the long-period velocity v, specifically of form kv where $k = 0.01V/\pi D$, V being a typical magnitude of the semidiurnal current in depth D.

A criterion established by Proudman for the effect of such friction on the tide was the magnitude of the ratio of the 'frictional period' $2\pi/k$ to the tidal period. Unfortunately, V/D varies so widely between ocean and shelf sea that he could only estimate that $2\pi/k$ should be between the limits of 5 days and 5 years, with 46 days as a rough estimate of its most likely global value. The frictional period being considerably larger than 14 days but small compared with 18.6 years, Proudman concluded that the tides Mf and Mm would be significantly modified by friction but the nodal tide should be close to the equilibrium form.

This rather bland conclusion hides the fact that Proudman employed some fairly sophisticated analysis involving current-intensive modes of second-class as well as gravity modes.[3] Evidence for second-class modes in the Mf tide came a few years later in the form of irregularities of short scale in tide constants derived from long series of sea level records from islands in the Central Pacific.[5] These were attributed to *Rossby* waves of short wavelength, as mentioned in Chapter 13.

The first computer model of the Mf tide[6] appeared to support the Rossby wave hypothesis, but later models on finer grid scales[7,8] gave preference to a slightly modulated phase-lagged equilibrium form for the surface displacement. The latter property was confirmed by analysis of the zonally-averaged altimetry from *Geosat–ERM* (Chapter 14).[9] The zonal averaging of course smooths out all short wave modes and so concentrates on the principal basin-scale modes. Average phase lags in the equatorial zones are typically around 15° for Mf, corresponding to a 'Q-factor' of about 4, roughly equivalent to a friction period of 55 days for Proudman's $2\pi/k$. With such a low value of Q, friction must play an important part in the dynamics of this tide. Results from Mm were similar but too affected by noise for a meaningful difference from Mf.[9]

At the lowest frequencies, 'stacking' monthly mean sea levels from some 720 worldwide stations between 1900 and 1979 led to the clearest demonstration yet achieved that, over the oceans as a whole, both the pole tide (1.2y) and the nodal tide (18.6y) with typical amplitudes about 15mm are indistinguishable from equilibrium.[10] It seems unlikely that this conclusion will be seriously challenged in the forseeable future.

Observational evidence for normal modes and Q of the ocean

There is a gulf of understanding between the computed normal modes of oscillation of the ocean and the real world of observation. The computed normal modes explain much of the spatial behavior of the ocean tides in a heuristic way, but individual modes do not appear as identifiable peaks in the non-tidal parts of the spectrum of sea level. It is therefore hard to assess the Quality factor Q (related to dissipation) of any particular mode, or indeed whether the computed resonant frequencies and their spatial forms are correct.

In 1983 this lacuna of understanding prompted Walter Munk to offer a small prize for the best essay from the international oceanographic community to shed more light on the problem. There were few competitors and the prize was awarded to the one serious entry.[11] The principal conclusion of the winning essay was that normal modes do not resonate to random excitation in the weather on account of a mis-match between the spatial scales of the weather and the oscillatory modes at the resonant frequencies. The problem of analysis was rendered more difficult by the dense spacing between the resonant frequencies and by a justified belief that the ocean's response is highly damped (low Q); any one mode could therefore be excited over a wide band of frequencies. Detection of a normal mode would require spatial as well as temporal filtering, for example by some form of 'stacking'. Data was not then readily available for such analysis.

Once again the tides had to be invoked for more definite results, in virtue of their unique and well defined structure in space and time. It is legitimate to include the tides of the Bay of Fundy (Nova Scotia) in this discussion, as a limiting case of a resonating arm of the North Atlantic Ocean. Because of its resonance, the semidiurnal tides in parts of the Bay are the highest in the world; they have been much studied on account of their potential for power extraction. The nature of the predominant mode in the Bay of Fundy provided an interesting exercise in analysis.[12]

As indicated in Chapter 12, the basic approach was to try to express the response at frequency σ to a single resonator of natural frequency σ_r in the form of an admittance:

$$Z(\sigma) = \left[\frac{\sigma_r - \sigma}{\sigma_r} - \frac{i}{2Q}\right]^{-1},$$

where Q is the Quality factor, defined as

$$Q = \sigma \times \text{Energy/Dissipation rate.}$$

In principle, measurements of amplitude $|Z|$ and phase Arg(Z) at two tidal frequencies σ should suffice to determine both σ_r and Q from the above equation. However, in the case of the Bay of Fundy the direct solution gave too large an amplitude ratio S_2:M_2 inside the Bay. In order to improve realism it was found necessary to modify the physical representation in two ways. In the first place, the dissipation within the Bay had to be distinguished from radiation into the Atlantic Ocean. Secondly, friction being proportional to the square of current amplitude had been shown by Jeffreys (*The Earth*, Art. 8.03, 5th and 6th Editions) to dampen minor constituents such as S_2, N_2 more vigorously than M_2. In effect, the frictional dissipative Q for M_2 could be realistically taken as $1.5 \times$ the Q for S_2.

With these refinements the observed tidal response at N, M and S gave a total Q of 5 for M_2, compounded of $Q = 7$ for external radiation and $Q = 18$ for friction within the Bay. The amplitude of S_2 was then satisfactory. The resonant period $2\pi/\sigma_r$ turned out to be 13.3 hours, a little greater than any of the semidiurnal periods.

Such an elaborate formulation could hardly be applied to the normal modes of the global ocean, whose resonant modes are clustered in close frequencies. However, the single-mode analysis has been successfully applied to the diurnal tides of the Southern Ocean.[13] Here, the Kelvin wave mode progressing westwards round the Antarctic continent is prominent and relatively isolated in frequency. A carefully computed set of diurnal tidal constants from a long-standing station on the Antarctic Peninsula was shown to accord extremely well with the response to a single mode of period 30.6h with $Q = 3.3$, (Figure 15.1). Dynamic analysis showed that, treated in isolation, the period of the Antarctic Kelvin wave is theoretically 30.7h, in close agreement with the modal analysis, but the corresponding mode coupled with the global ocean was 28.7h, a little less. Another single-mode analysis, based on satellite altimetry of the whole Southern Ocean, gave $2\pi/\sigma_r = 28.5 \pm 0.2$h with $Q = 3.8 \pm 0.4$, closer to the oceanic response.[14]

The very low Q's thus obtained for the diurnal tides may result from interaction of diurnal currents with the generally stronger semidiurnal currents to produce a quasi-linear damping, as had been assumed by Proudman for the long-period tides.[3] Experiments to determine Q from semidiurnal constants at Bermuda indicated a dominant natural resonance at 12.8h with a higher Q of

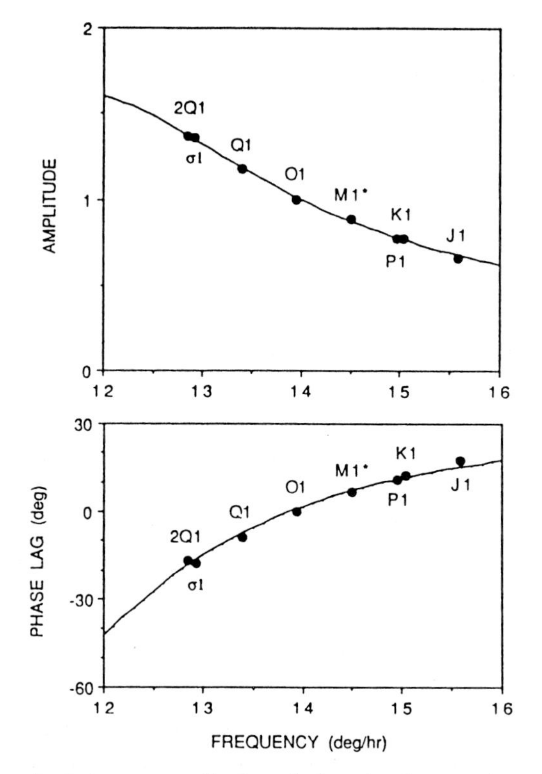

Figure 15.1. Diurnal admittance amplitude and phase lag for 'Faraday Station' (65° S, 64° W) on the Antarctic Peninsula, plotted relative to their values for O_1. The curves describe the response of a single resonator of period 30.6h (frequency 11.8deg/h) with $Q = 3.6$ (from ref. 13, courtesy of John Wiley & Sons.)

16–18. However, in that case the fit was not so good as in Figure 15.1 and suggested that at least three resonant modes with theoretical periods in the range 13.0–12.1h should be taken into account, with Q in the region of 5–8.[13]

In summary, the extraction of normal oceanic modes from direct observations of the tides is hindered by the multiplicity of modes and the relatively high rates of dissipation, but the existence of some prominent modes has been clearly indicated, with suggestive estimates of their Q-factors. Translation of Q into physical dissipation rate is limited by the uncertainty of estimating the *kinetic* contribution to the total mean energy. (The potential energy is easily obtained from models of surface elevation.) The directly estimated rates of global dissipation derived from the methods described at the end of Chapter 14 are more reliable.

Interactions between air tides and ocean tides

The development of the theory of air tides was treated in Chapter 10 as a process quite separate from the ocean tides. Separate treatment is justified by the differences between thermal and gravitational excitation and by the thermodynamics of the atmosphere, irrelevant to the ocean tides. However, in a broad sense the ocean and the air may be seen as a multi-layered fluid responding as a whole to tidal stresses. This view has been explored by a few scientists who have examined the ways in which tidal interactions are transmitted across the common interface at the ocean surface. It is natural to consider these interactions in two ways: air to ocean, and ocean to air.

Air-to-sea interaction in general occupies a great deal of modern oceanographic research, but in the context of tides, because of the dominance of the S_2 tide in the air, we need only consider the effect on the ocean of the thermally generated 12-hourly variation in surface air pressure. Empirical formulae for the global distribution of the mean S_2 tide in surface pressure (averaged over the seasons) have been known for many decades;(15) the most recent formula is from a digest of several hundred records from weather stations under the direction of Bernhard Haurwitz.(16a) Converted to the equivalent *static* elevation of the sea surface, (i.e. dividing the pressure amplitude by ρg and adding 180° to the phase), the principal term may be expressed:

$$12 \sin^3\theta \cos\,(30°t + 2\phi - 111°) \text{ millimeters.}$$

This is equivalent to a sort of solar 'equilibrium tide' travelling westward with the sun at a time-lag of 111/30 = 3.7 hours; the maximum pressure is lagged by 9.7 hours. The corresponding gravitational equilibrium tide, from the harmonic amplitude of S_2 in the gravitational potential divided by g, as obtained from standard tables, is

$$114 \sin^2\theta \cos\,(30°t + 2\phi) \text{ millimeters.}$$

Since $\sin^3\theta$ and $\sin^2\theta$ are not very different near the equator, ($\theta = 90°$), one might heuristically expect the air pressure to generate a tide in the ocean similar to the gravitational S_2 tide but with about one-tenth of its amplitude and a phase lag of about 111°.

The 'response method' of tidal analysis (Chapter 12) allows for a loosely-termed 'radiational' input to sea level, distinct from the gravitational input. An early application of response analysis to 19 years of hourly sea levels at Honolulu(17a) found the S_2 tide to include a radiational component of nearly a quarter the gravitational amplitude, with about 100° phase difference. It was concluded that there may be other radiational inputs to sea level such as heating and variable wind in addition to air pressure, but this was not substantiated in later analyses. There have since been several assessments of the radiational component of S_2. A recent comparison of harmonic analyses of 80 reliable tide stations in all oceans gave an average amplitude ratio (rad:grav) of 0.105 with phase lag 108°;(17b) the phase lag is remarkably close to the value derived from the Honolulu analysis. The net effect of the vectorially combined components is a reduction of S_2 amplitude by about 3 percent and an increase of phase lag of 5°.9 relative to the purely gravitational admittance. Results from individual stations show a variation about these values of typically ±10 percent in amplitude, ±5° in phase. The most noticeable result of the radiational anomaly is that the phase lag at S_2 is usually greater than that at K_2, whose frequency is slightly higher, whereas the general trend of oceanic admittance is for phase lag to increase with frequency.

Despite local anomalies from the mean, the above observational results suggest that the surface air pressure is the main agent for driving the interaction. However, a full analysis would have to take account of the empirical factor $\cos^3 \theta$ and a small additional standing-wave element of the air tide. (There is an alternative expansion in spherical harmonics.[16a]) The problem has apparently not yet attracted ocean tide modellers.

By contrast, studies of ocean-to-air tidal interaction have been mainly theoretical. The phenomenon in question is the effect on the atmosphere of the surface movement of the lunar oceanic tide, in addition to the very small air tide generated directly by the tide potential. The agent for the interaction is the vertical velocity field imposed at the interface by the complex structure of the oceanic M_2 tide, together with a small contribution from the earth tide of the land surface.

Laplace (*Mécanique Celeste*, Livre IV) included the motion of the interface in his formulation of air-tide dynamics, but he omitted it to simplify his computations. The driving potential of the ocean tide in relation to the air was revived in the mid 20th century, as knowledge of the ocean and earth tides and of the vertical structure of the atmosphere improved.[18,19] Some tentative calculations using first-generation oceanic tide models showed without doubt that the tidal movement of the interfaces exerts a strong influence on the air tide, including tidal winds.[20]

Figure 15.2 shows the striking result of the first fully numerical solution of the dynamic equations of the M_2 air tide driven by the tide potential alone (above) and by the combination of the tide potential with motion of air/sea and air/earth interfaces (below). The computations were performed in the 1970's by a tidal research group at the Shirshov Institute of Oceanology at St. Petersburg (then Leningrad).[21] Loading and linear friction have been allowed for and the ocean tide model was known to have modest realism within the limitations imposed by a 5° × 5° grid size.

With a rigid interface (Figure 15.2, above), the air tide has roughly zonal amplitude contours to a maximum of about 0.07 millibars (1mb = 100Pa) and meridional contours of Greenwich phase lagging the moon by about an hour (29°). Ocean and earth tides, by contrast, induce amplitudes up to 0.21 mb and a complex pattern of phase contours with nodes and antinodes. An intermediate result (not shown) obtained by omitting the earth tide is not markedy different, showing that the oceanic movement is largely responsible for the interaction.

On the observational side, the M_2 components of barometric records at 104 meteorological stations, each with average duration 21 years, have been arranged to define empirical global fields of amplitude and phase by Haurwitz

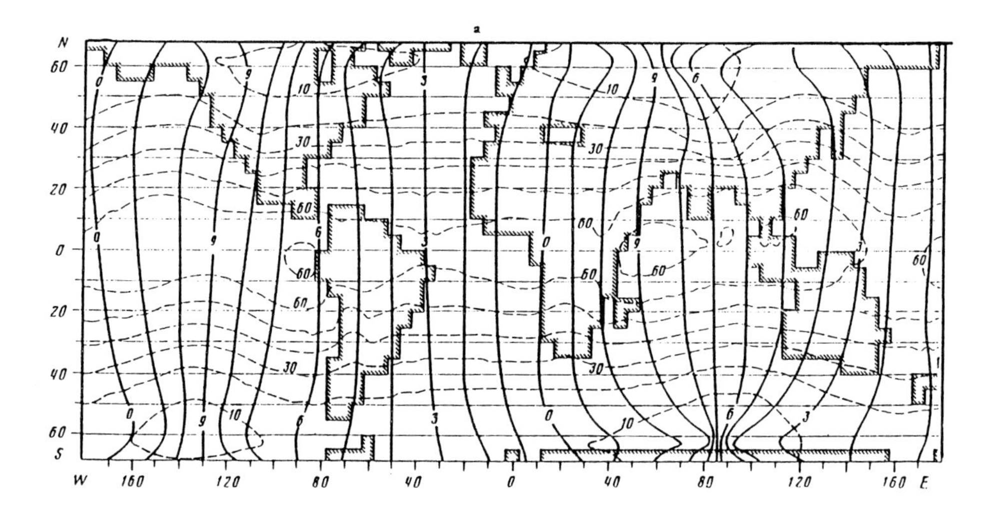

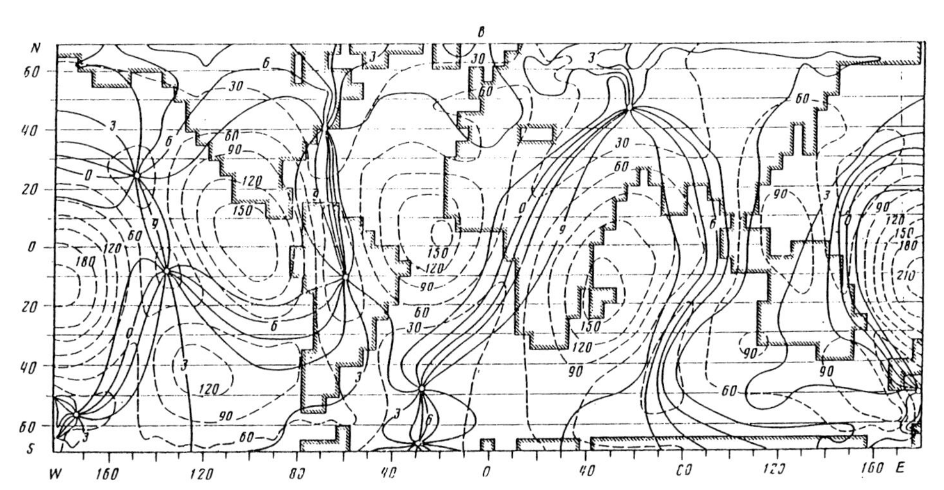

Figure 15.2. Numerically computed M_2 tide in the atmosphere driven by (a) the gravitational tide potential alone – upper panel; and (b) the combined tide potential, earth tide and ocean tide – lower panel. Phase contours (solid lines) are marked in degrees; amplitude contours (broken lines) are in microbars (10^{-1}Pa). (From ref. 21, by permission.)

and his colleagues.[16b] The distribution varies with the seasons; we shall assume the annual mean in the rest of this discussion. The mean picture shows amplitudes similar in magnitude to those of the lower panel of Figure 15.2, but only qualitative comparison is possible because nearly all of the stations used were necessarily situated on the continent. The data may have been over-smoothed spatially, and the computational model[21] could no doubt be improved.

A complementary question is: 'at what rate is gravitational tidal energy dissipated in the atmosphere in relation to the energy imparted by the sea and by the moon?' This question has been intensively studied by Professor G.W.

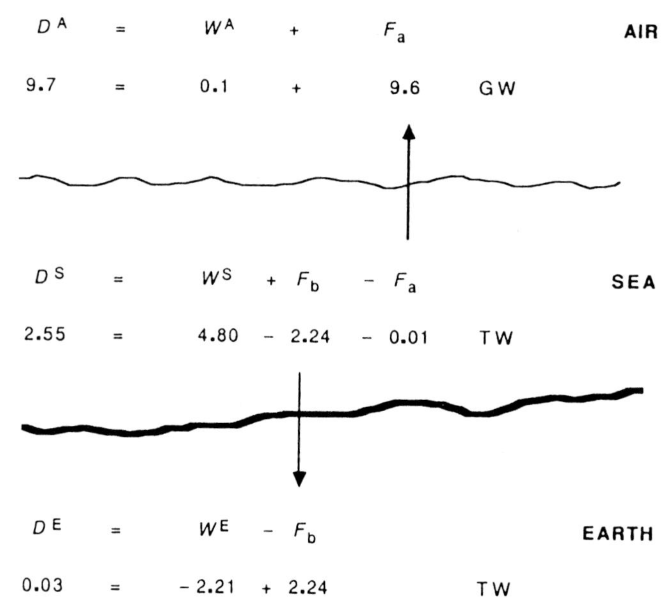

Figure 15.3. Schematic diagram of dissipation rates D in earth, sea and air, partitioned into work W performed by the gravitational tide potential at the M_2 frequency, and fluxes F transmitted across the common boundaries, as computed by Platzman.[22] The units are terawatts (10^{12}W) for earth and sea; gigawatts (10^9W) for the air. (Courtesy of G.W. Platzman.)

Platzman of Chicago.[22] The rate of tidal dissipation within each of the three zones of the geosphere (earth, sea, air) is the sum of the work done on that zone by the moon and the rate of energy transmitted into it from other zones. Both these rates of working may be computed explicitly in terms of integrals of products of vertical velocity and pressure at the interface and the total tide potential, without requiring specific knowledge of the frictional mechanism responsible for dissipation within the zone. (This simplifying property applies only to gravitational, not to thermal tides.)

Figure 15.3 succinctly summarises the results of Platzman's computations for the earth, sea and air, based on the best sources of data available in 1990. Chapter 14 showed a recent convergence of results for dissipation in the ocean-and-earth of about 2.5 terawatts, the earth being implicitly included through the Love number h_2. The work done on the ocean *alone* is considerably greater than the oceanic dissipation rate, 4.8TW according to Platzman's computations, but 2.2TW of this is transmitted across the yielding sea floor. Very little of this transmitted energy is dissipated in the earth itself; the 2.2TW is nearly balanced by work done by the earth tide against the gravitational forces, as illustrated in Figure 15.3.

The rate of energy transfer from ocean to air is seen to be almost negligible compared with the oceanic dissipation, but as Figure 15.2 suggests, it forms a very considerable part of the work budget of the atmosphere, 9.6 gigawatts

($1\,GW = 10^{-3}\,TW$) in Figure 15.3. This includes the small contribution from the earth tide, and allowance has been made for the *Lagrangian* nature of the recorded pressure at stations which move with the earth tide. The model of Figure 15.2 gave a dissipation rate of 17GW. In contrast with the tidal budget of the solid earth, however, the transferred energy is not used to work against the gravitational forces; it almost entirely accounts for the dissipation within the atmosphere. Thus, the energy budget of the lunar air tide is practically accounted for by input from the ocean and dissipation by turbulence in the higher levels of the atmosphere.

Oceanic tidal dissipation in the geological past

It will be recalled from Chapter 10 that Sir George Darwin's principal interest in tidal friction (i.e. dissipation) was to provide a time scale for the history of the moon since the catastrophic event which, in his view, caused the moon to separate from the earth. Related researches, some of them having little to do with tides, continued intermittently after Darwin's time. A major revival of interest in the subject started in the 1960s, partly from Munk and MacDonald's book *The Rotation of the Earth*, and most notably from the work of a German astronomer named Horst Gerstenkorn (1923–1981).

Darwin had imagined the moon to have emerged by fission from what is now the Pacific Ocean, roughly a billion years ago. This idea was made obsolete by the theory of plate tectonics. Gerstenkorn elaborated a different scenario: the moon may have been captured from remote parts of the solar system by a close encounter with the earth on a retrograde orbit (opposite to the earth's rotation).[23] This puts the moon on the left ($x<0$) of Darwin's diagram, Figure 10.5, where tidal friction causes it to *approach* the earth. It was then shown by calculations that the inclination of the orbit ($I>90°$) would steadily decrease, eventually to pass 90° to become a prograde orbit, ($x>0$ in Figure 10.5), when the distance would start to increase towards present conditions. The point of interest is: 'how close did the moon get before its orbit became prograde?' Gerstenkorn reckoned that it came close to the 'Roche limit' ($2.89 \times$ earth radius), where gravitational forces are so great as to cause general rupture, but the theory can also be adjusted to give minimum distances greater than the Roche limit.[24] At best, enormous heat would be generated in the earth and moon over a period of several hundred years.

In his 'Harold Jeffreys Lecture' to the Royal Astronomical Society in 1968, Munk[24] called this syndrome the 'Gerstenkorn Event' and invited his audience to consider: 'A heavy hot atmosphere over a darkened earth; giant tides on a 5-hour day, with steaming tidal bores following the moon on a 7-hour polar orbit'. This event would certainly be hostile to any existing forms of life, and would have left a recognisable mark in the geological record, then roughly estimated as 1–2 aeons ago (1 aeon $= 10^9$ years).

These ideas stimulated research in many diverse scientific disciplines. Some scientists looked at the geological and palaeontological (fossil) records; some sought to quantify the number of days per year in the geological past; some tried to estimate past rates of tidal dissipation or the rate of increase of the length of day. International Symposia and Workshops were organised to discuss and compare results. I shall not attempt to summarise all the arguments and counter-arguments, some of them still unresolved, and in any case too far from my sphere of competence to judge merit. Instead, I will comment on a few selected papers from the first two of a series of Workshops held at the 'Centre for Interdisciplinary Research' of the University of Bielefeld,[26a,b] and a few other papers. The published Proceedings of the second Bielefeld Workshop[26b] were formally dedicated to the memory of Horst Gerstenkorn, who had recently died.

In the first place, geologists could find no evidence for a global catastrophe in crustal or fossil records in the last 3.9 aeons.[27] The solar system is itself supposed to be 4.8 aeons old, so geologists could hardly be expected to find any evidence before 4 aeons. Mat-like structures caused by algal growth in marine inter-tidal zones, known as *stromatolites*, are found in pre-Cambrian rocks 3 aeons old,[28] showing that the oceans existed then and showed signs of lunar tidal activity. Now, calculations of the moon's history by Gerstenkorn[23] and others working on similar lines[24,25] arrived at their close-to-earth events around 1–2 aeons ago by assuming tidal friction to have been roughly at the present level throughout the period, (apart from calculable changes in the tidal forces due to changing distance). The geological evidence therefore demanded a rationale for much lower rates of dissipation in the remote past.

Whatever model of tidal friction is assumed, it is a straightforward matter to calculate the number of days per year (dpy) and of days per synodic month (dpm) at any past epoch.[29] Direct estimates of these quantities or their rates of change may therefore provide valuable constraints to a model, without going right back to the Gerstenkorn event. Estimates of the 'current rate' (geologically speaking) of decrease of dpy from telescopic and pre-telescopic accounts of solar eclipses and timed occultations of stars by the moon had improved during the 20th century with the accumulation of historic records translated from eastern languages and the modern use of atomic clocks. One result was that there have been fluctuations of rotation speed on a time scale of decades in addition to the steady decrease due to the tides. Many centuries of observations are required to smooth over the decadal fluctuations, and one also has to allow for a very long-term *nontidal* change in l.o.d. (an acceleration) due to glacial history and change of global sea level.[30] Despite these complications, it has been possible to estimate a basic *tidal* rate of change of l.o.d. (decrease of dpy) of about 2.0–2.5 milliseconds per century, or − 85 to − 105 dpy per aeon – cf. Table at the end of Chapter 14.

It had long been known that fossil and present day corals and other marine organisms show a clearly defined annual growth pattern due to seasonal changes in water temperature. The discovery in 1963 that corals also possess identifiable *daily* growth rates[31] suggested a means of measuring dpy in past epochs, and started a busy inter-disciplinary research activity. Another observed periodicity in the region of 29–33 days was interpreted as a measure of dpm, although the biological mechanism is unclear. The dpy counts gave better results, and together with similar markings in bi-valve shells could be relied upon back to the Lower Ordovician epoch, that is, about 0.5 aeons ago.[28] Pre-Cambrian stromatolites (1–3 aeons) were also used but their results seem to be controversial. On the whole, the resulting figures for dpy were indeed lower than one would expect from constant dissipation – typically 415 dpy at 0.5 aeons compared with 430 dpy from holding constant the lowest estimate of present-day dissipation.[28] Thus the fossil record was seen to support the hypothesis of lower rates of dissipation in the past, though at epochs long after the supposed close approach of the moon.

Dynamic computer models of the ocean tides were also used to estimate tidal braking in past epochs. Geologists had deduced rough maps of the shapes of continents and ocean basins in past epochs from paleomagnetism and plate tectonics. The Institut für Meereskunde at Hamburg applied their well-researched system for solving the nonlinear tidal equations in the present ocean to past oceans at defined epochs, using reasonable assumptions for the depth and extent of shelf seas. Their results, in brief, supported the view that the braking torque was considerably lower in the Cretaceous and Permian epochs (0.1–0.2 aeons), but the torque appeared to have something like the present level in the Silurian and Ordovician epochs (0.4–0.5 aeons), probably on account of some resonant property in the shape of the ocean at the earlier epochs.[32]

Sensitivity to oceanic tidal resonance was investigated by modelling the present oceans with a small perturbation in the position of the entire American continent.[33] The present configuration did indeed seem to be close to resonance, as others had suggested, again confirming that present friction is atypical of past epochs. Other investigators preferred to be less specific about details of past ocean shapes, and instead solved the tidal equations and the moon's orbital history for generalised oceans. A student of G.W. Platzman performed complete finite-element calculations with a single ocean and continent of present areas, in one case as a spherical cap centered on the North Pole, in another centered on the equator. Both cases were solved with two types of linear bottom friction, 'weak' and 'strong', and included a solution for the changing lunar orbital inclination.[34] None of the four combinations produced a separation less than 35 earth radii in the 4.5 aeons allowed. A closer approach in the time scale of the solar system would have required the M_2 tide to be permanently near resonance, a highly unlikely circumstance. Again the present system was shown to be atypical of the geological past.

Another model, explored at the UK–IOS, consisted of a hemispherical ocean whose center could be placed arbitrarily between pole and equator, but with the moon's orbit confined to the equatorial plane.[35] The tidal equations with linear friction were here solved by analytical expansion, and the results were averaged over all positions with respect to pole and equator. Because the higher tidal frequencies in the geological past excited shorter wavelengths which were less conducive to global resonance, allowance was also made for dissipation in the solid earth at physically appropriate rates. Total dissipation and braking torque were once more found to be much less than at present at all past epochs. Variations in parameters suggested close approach, effectively a 'Roche' limit, between 3.9 and 5.5 aeons ago, consistent with the absence of a recognisable geological record.

In summary, most of these varied investigations agreed that the moon is older than can be detected in the geological record, but whether it was captured as Gerstenkorn suggested or was accreted from a ring of planetary debris resulting from a planetary collision appeared to be insoluble. Researches on the origin and the age of the moon of course continue,[(45)] but further discussion would be outside the scope of this book.

Variable earth rotation at tidal frequencies

We have so far discussed only the very long-term or secular effects of the tides on the earth's rotation. As we have seen, these are entirely due to friction. The tides also have periodic effects which are independent of friction, and which are important in the accurate determination of Universal Time (commonly called UT1). The tidal effect is partly a result of the displacement of solid and fluid mass causing periodic changes in the axial moment of inertia, and in the higher frequencies also due to the exchange of angular momentum between the ocean and the lithosphere. Such effects had been predicted by Jeffreys as early as 1928,[(36)] but with amplitude less than a millisecond they were undetectable by the pendulum clocks of that time.

Seasonal changes in the l.o.d. were detected in 1936 by N. Stoyko at the 'Bureau International de l'Heure' (BIH), Paris.[(37a)] Quantification of the seasonal change as an annual and a semiannual term changed as methods of timekeeping improved, and was found to be slightly variable from year to year.[(37b)] Such matters have considerable geophysical interest, but I shall not discuss them here since the cause of the seasonal change in l.o.d. is largely meteorological, hardly tidal.

Tidal variations at monthly (Mm) and fortnightly (Mf) frequencies were first identified in the late 1960s, when the BIH had started to compile uniform series of UT1 to 0.01 ms at 5-day intervals by averaging daily visual and photographic star positions from a number of observatories.[(38a)] The basic cause of these variations is relatively simple. The long-period component of the tide-raising force being zonally uniform alternately draws the earth's material (including ocean and air) towards and away from the equator. Displacement of material towards the equator increases the axial moment of inertia and hence decreases the rate of rotation and *vice versa*. A harmonic increment of moment of inertia C with amplitude δC_0 and frequency σ induces a harmonic increment in UT1 with amplitude

$$\delta(\mathrm{UT1}) = \sigma^{-1}\,\delta C_0/C \text{ days}$$

and with a phase change of 90°.

Analysts treated δC as being primarily that of an elastic earth with small modifications due to the ocean, air and the earth's liquid core;[39,40]

$$\delta C = k\,Ma_e^2\,\frac{2m}{3M}\,\Pi^3 P_2^0(\cos\Theta),$$

where M,m are the masses of earth and moon, a_e the earth's equatorial radius, Π, Θ are the equatorial parallax and co-declination of the moon, respectively, and k is a constant to be determined. This form was chosen because it was shown that, for an oceanless solid elastic earth, k turns out to be identical to the Love number $k_2 = 0.30$ with possible frequency-dependence. The main part of the expression for δC is simply proportional to the zonal tide potential, as one would expect, so the constituent harmonic amplitudes of δ(UT1) are exactly similar to the low frequency harmonic expansion of the potential, weighted by period through the factor σ^{-1}.[39]

Early analyses of the UT1 data from BIH and from an independent source in the Soviet Union suggested that k was indeed frequency-dependent (e.g. 0.30 for Mm, 0.33 for Mf).[38b] After 1980, improvements in the BIH data and corrections applied from a new series for the zonal momentum of the atmosphere brought the estimated values of k for Mm and Mf closer together, within 0.32 ± 0.01.[40] It was then concluded that there were too many uncertainties in the data and in geophysical parameters, for the rotational response of the earth to tidal forces to be distinguished from equilibrium by this type of analysis.

From about 1981, measurements of earth rotation (i.e. UT1) were dramatically improved by the development of laser-ranging to satellites (SLR) and by the interferometry of distant galactic radio sources ('Very long baseline interferometry', or 'VLBI').[41] Series of UT1 at 1–2 hours interval and individual accuracy better than 0.1 ms became available, and from 1983 various techniques were monitored and compared through an international campaign known as MERIT,[42] leading to eventual collection on a regular basis by the BIH. Soon, short-period tidal variations in UT1 with amplitudes of order 10 microseconds were identified.[43a,b]

In contrast with the long-period components, the body-tide of the solid earth is practically irrelevant to diurnal and semidiurnal earth rotation because its zonal mean value at those frequencies is zero. The observed variation in UT1 is almost entirely due to the changing global inertia of the tidal surface displacement of the ocean and to the net change of zonal momentum in the pattern of oceanic tidal currents. These dynamic effects are conveyed to the solid earth partly by bottom friction, but to a greater extent by the zonally horizontal component of pressure force on the sloping sea floor, especially near the edges of

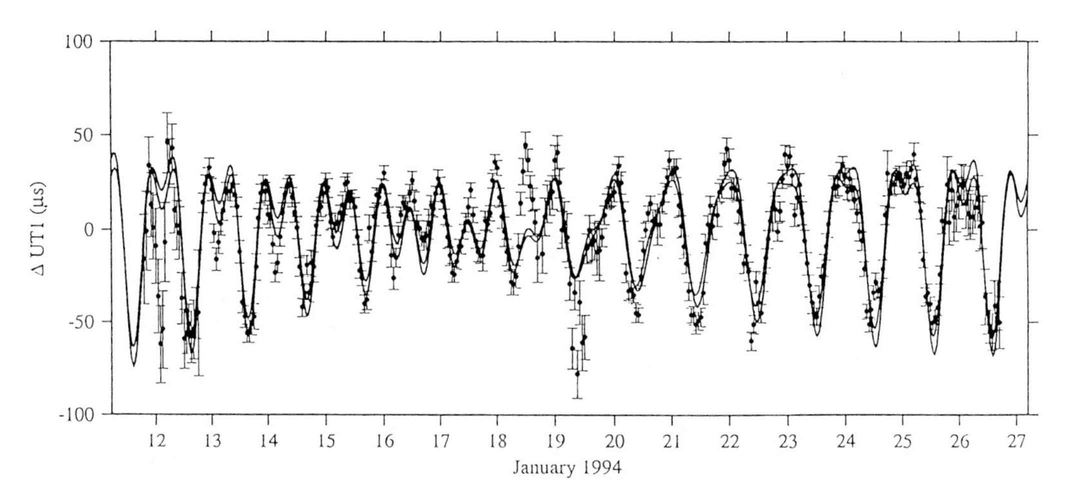

Figure 15.4. Fluctuations in Universal Time (UT1) recorded hourly by a VLBI network from 11 to 26 January, 1994 – spots with standard error limits – compared with two similar syntheses computed from independent models of four diurnal and four semidiurnal tide constituents – continuous lines. Vertical scale in microseconds. A low-order polynomial trend has been subtracted from all data. (Courtesy of R.D. Ray.)

continental shelves. The zonal momentum of the sea had been ignored in the long-period studies, (though mentioned by some as a possible correction), but it became clear that it is in fact the dominant agent in the short-period tidal variations.[43a]

Attempts to compute inertia and momentum exchange from successive models of ocean tides were very demanding of model accuracy, since they involve global integration of many positive and negative increments. Only surface elevations had been intensively modelled (Chapters 12, 14); models of barotropic tidal currents were rare and largely untested against measurements. Some comparisons of the tidal component of UT1 with model calculations were unconvincing, but eventually, comparison with the new generation of tide models derived from the TOPEX/POSEIDON altimetry (Chapter 14) showed agreement to about 10 microseconds.[44]

Figure 15.4 shows a detailed comparison of a 15-day period of UT1 measurements from VLBI with calculations from two independent ocean tide models, developed at NASA and at the University of Oregon, respectively. The predictions are based on models for the leading diurnal constituents (Q, O, P, K_1) and the semidiurnals (N, M, S, K_2). The contributions from all eight constituents have comparable magnitudes; the horizontal momentum terms exceed the inertia terms by factors of 2–5. As well as the apparent excellent agreement within the limits of data error, the authors[44] were able to detect a small discrepancy in phase lag at M_2, pointing to a subtle component in the rotation ascribed to 'spin libration', previously neglected. Thus, continued improvements in measurement and models reveal small hidden effects, and in turn call for still more accurate data to be recorded for longer periods.

Final retrospect

In these chapters, surveying the history of scientific research in a concentrated field, we have come a long way from the time four centuries ago (not to mention much earlier periods of history), when there were no measuring instruments beyond rudimentary aids to navigation, no aids to computation beyond the abacus, and significant writings on natural philosophy were published only three or four times in a century. The transformation from that intermittent trickle of scientific thought to the torrent of today's activity in research, open discussion and publication can seem bewildering, unless one examines the steps which brought it about.

Previously, the Church had stifled all enquiry into the nature of the universe which did not conform to religious dogma whose physical theory was largely based on the writings of Aristotle. The Church's violent antagonism to supporters of Copernican theory such as Bruno and Galileo heralded the end of their era of repression, and was followed by the emergence of freer thinkers such as Descartes and Kepler. Thus began the European scientific revolution of the 17th century.

The formation of the Royal Society in Britain and the Académie Royale in France in the 1660s (preceded by the 'Accademia del Cimento' in Florence) enabling like-minded intellectuals to meet and discuss their experiments and theories about natural processes and providing a medium for publication, was an important step towards modern scientific activity, soon imitated by the royal and imperial dynasties of other states. From these Societies emerged the great names of Newton and Laplace which have featured prominently in this history, together with many other influential scientists.

The Industrial Revolution relied on practical science for its commercial success, and encouraged careers in science and the introduction of science courses into university curricula previously confined to mathematics and astronomy. Soon, scientific societies proliferated with greater specialisation into specific branches of knowledge, each with its own journal of learned papers. The British Association for the Advancement of Science and its American counterpart encouraged a more popular dissemination of the latest advances in knowledge. The AAAS helped the US Coast Survey in providing a medium for publicity and discussion of their work. The BAAS fostered formal liaison between the universities and the Naval Hydrographic Office in Britain, and (later) helped to create the Liverpool Tidal Institute.

Geophysics emerged from geology in the late 19th century and physical oceanography emerged from marine biology in the early 20th century, both becoming established disciplines in their own right. University departments and institutions were founded to concentrate on research in these subjects. Both acquired international following, were supported by the IUGG and were the subjects of symposia in all parts of the world. The related subject of meteorology had its own parallel history of development. All had important applications to military activities on land and sea and in the air, and benefitted in various ways from their technology and funding.

Enough has been said to remind the reader of the progress of the earth sciences towards their establishment as part of modern international culture. Details describing how the subset of tidal studies has grown through the last few centuries have been described in the other chapters of this book. In Chapter 11, I quoted statistics for the numbers of published papers on 'tides' in 50-year blocks from roughly 1620 to 1970, and questioned whether the quasi-exponential rise would continue into the 21st century. I shall conclude with some remarks on this question from the viewpoint of 1996.

'Tidal research' has had different meanings in different eras. Before Newton, it meant explaining why the sea rises and falls twice per lunar day with the known monthly and annual inequalities. Methods to improve prediction of the times of High Water also proceeded, but on a more empirical basis. From Newton to Laplace the outstanding question was how to formulate the dynamics of the ocean tides. After Laplace, it was how to solve Laplace's dynamic equations, and how to describe the progression and standing oscillations of tidal waves across the world oceans. The practical problem of computing tidal ephemerides for chosen ports was essentially solved by Thomson (Kelvin) and Darwin towards the end of the 19th century with financial assistance from the BAAS, although improvements in accuracy continued into the 20th. Thomson and Darwin also introduced new problems for tidal research in the growing discipline of geophysics, and so created more areas of research than they completed.

In the accelerating pace of late 20th century science, the long saga of research to quantify the tides of the ocean was finally resolved by a combination of high-speed computers and satellite radar technology. The same activity provided definitive answers to the long-debated question of the rate of change of the earth's rotation due to tides. First-order solutions were also found to explain and quantify the tides of the solid earth and the atmosphere. The miscellaneous problems discussed in the previous sections of the present chapter have been well explored; one would not expect to see radical advances in them in the near future.

It seems clear, therefore, that most of the original mainstream problems of the tides are now part of the history of science. This was in fact the main motivation for writing this history, as mentioned in Chapter 1. Research will of course continue in response to the inexorable demand for higher accuracy in predictions; this in turn is liable to uncover niceties of theory previously neglected. The accuracy of tide-tables for most ports is limited by the less predictable effects of weather; hence more research of the last few decades has been directed toward modelling the response of sea level to wind than toward the astronomical effects. However, although tide-table production may now seem to be a routine computer operation, such work will always require expert supervision by a tidalist (as distinct from a computer technician), and a background of research into how best to extract prediction parameters from noisy data.

The most likely directions for new tidal research are in specialised applications of tidal theory to internal motions, mixing of density layers, air-sea interaction, earth-tide dissipation, and related subjects. All geophysical measurements include tidal variations, even if the tides are seen as a nuisance factor; these often generate research into ways of eliminating them to an acceptable level of accuracy. Such matters are far removed from the problems which were considered to be the most important fifty or more years ago.

Interest in merely counting research papers in a given span of time is now dubious. It is also made difficult today by the diversity of the subject matter, despite the availability of computerised library indexes. When Proudman started to compile the history of tidal publication on behalf of IAPO as described in Chapter 11, the subject was restricted to marine contexts; geophysical and astronomic implications were specifically excluded. Such exclusion would give a very distorted view of the recent research activity related to tides, as I hope I have made clear in the pages of this book.

As these pages go to press, publication of a symposium on ocean tides which was held in London in October 1996 is being prepared. The symposium focussed on the outlook for tidal science just before the start of the new Millennium. Its Proceedings should be consulted for the latest views of the international community of experts on the subject.[46]

Notes and references

1. Patullo, J.G., W. Munk, R. Revelle, and E. Strong. The seasonal oscillation in sea level. *J. Marine Res.*, **14**, 88–156, 1955.
2. Nerem, R.S., E.J. Schrama, C.J. Koblinsky, and B.D. Beckley. A preliminary evaluation of ocean topography from the TOPEX/POSEIDON mission. *J. Geophys. Res.*, **99**, (C12), 24, 565–583. 1994.
3. Proudman, J. The condition that a long-period tide shall follow the equilibrium law. *Geophys. J. R. astr. Soc.*, **3**, (2), 244–249, 1960.
4. Bowden, K.F. Note on wind drift in a channel in the presence of tidal currents. *Proc. R. Soc. London*, **A**, **219**, 426–446, 1953.
5. Wunsch, C. The long-period tides. *Rev. Geophys.*, **5**, (4), 447–475, 1967.
6. Kagan, B.A., V.Y. Rivkind, and P.K. Chernyayev. The fortnightly lunar tides in the global ocean. *Izv. Akad. Sci. USSR, Atmos. & Oceanic Phys.*, **12**, 274–276, 1976.
7. Schwiderski, E.W. Global ocean tides – X: The fortnightly lunar tide (Mf); atlas of tidal charts and maps. Naval Surface Weapons Center, Dahlgren, Virginia, NSWC TR 82–151, 1982.
8. Miller, A.J., D.S. Luther, and M.C. Hendershott. The fortnightly and monthly tides: resonant Rossby waves or nearly equilibrium gravity waves? *J. Phys. Oceanogr.*, **23**, (5), 879–897, 1993.
9. Ray, R.D. and D.E. Cartwright. Satellite altimeter observations of the Mf and Mm ocean tides with simultaneous orbit corrections. Pp. 69–78 in: *Gravimetry and Space techniques applied to Geodynamics and Ocean Dynamics.* AGU Geophys. Monograph no. 82, IUGG vol. 17, 1994.
10. Trupin, A. and J. Wahr. Spectroscopic analysis of global tide-gauge sea level data. *Geophys. J. Internat.*, **100**, 441–453, 1990.
11. Luther, D.S. Why haven't you seen an ocean mode lately? *Ocean Modelling*, no. 50, 1–6, + 2pp. Figures. 1983. (Unrefereed Manuscript)
12. Garrett, C. Tidal resonance in the Bay of Fundy. *Nature*, **238**, 441–443, 1972.

13. Platzman, G.W. Tidal evidence for ocean normal modes. Pp.13–26 in: B.B. Parker (ed.) *Tidal Hydrodynamics*, J.Wiley, New York, 883pp., 1991.
14. Cartwright, D.E. and R.D. Ray. Energetics of ocean tides from *Geosat* altimetry. *J. Geophys. Res.*, **96**, (C9), 16, 897–912, 1991.
15. Wilkes, M.V. *Oscillations of the Earth's Atmosphere*, Cambridge Univ. Press, 76pp., 1949.
16. Haurwitz, B. and A.D. Cowley. (**a**) The diurnal and semidiurnal barometric pressure oscillations, global distribution and annual variation. *Pure and Appl. Geophys.*, **102**, 193–223, 1973. (**b**) The lunar barometric tide, its global distribution and annual variation. *Ibid.* 77, 122–150, 1969.
17. (**a**) Munk, W.H. and D.E. Cartwright. See Ch. 12, ref. 33. (**b**) Cartwright, D.E. and R.D.Ray. On the radiational anomaly in the global ocean tide, with reference to satellite altimetry. *Oceanologica Acta*, **17**, (5), 433–459, 1994.
18. Chapman, S., S.K. Pramanik, and J. Topping. The worldwide oscillations of the atmosphere. *Beitrage Geophysik*, **33**, 246–260, 1931.
19. Sawada, R. The possible effect of oceans on the atmospheric lunar tide. *J. Atmos. Sci.*, **22**, (3), 636–643, 1965.
20. Hollingsworth, A. The effect of ocean and earth tides on the semidiurnal lunar air tide. *J. Atmos. Sci.*, **28**, (10), 1021–1044, 1971.
21. Kagan, B.A. and N.V. Shkutova. On the effect of ocean tides on gravitational tides in the atmosphere. *Oceanology*, **25**, 146–150, 1975.
22. Platzman, G.W. An observational study of energy balance in the atmospheric lunar tide. *Pure & Appl. Geophys.*, **137**, (1/2), 1–33, 1991.
23. Gerstenkorn, H. Über die Gezeitenreibung beim Zweikörperproblem. *Zeitschr. Astrophys.*, **36**, 245–274, 1955. Also: On the controversy over the effect of tidal friction upon the history of the earth-moon system. *Icarus*, 7, 160–167, 1967.
24. Munk, W. Once again – tidal friction. See Ch. 12, ref. 42.
25. Alfvén, H. and G. Arrhenius. Two alternatives for the history of the moon. *Science*, **165**, 11–17, 1969.
26. (**a**) Brosche, P. and J. Sündermann (Ed.) *Tidal Friction and the Earth's Rotation*, Springer-Verlag, Berlin, 242 pp., 1978. (**b**) *Ibid.* TFER-2, 345pp., 1982.
27. Piper, J.D.A. Geological and geophysical evidence relating to continental growth and dynamics and the hydrosphere in Precambrian times. Pp. 197–241 in ref. 26a.
28. Scrutton, C.T. Periodic growth features in fossil organisms and the length of the day and month. Pp. 154–196 in ref. 26a.
29. Lambeck, K. The earth's palaeorotation. Pp. 145–153 in ref.26a.
30. Stephenson, F.R. and L.V. Morrison. History of the earth's rotation since 700 BC. Pp. 29–50 in ref. 26b.
31. Wells, J.W. Coral growth and geochronometry. *Nature*, **197**, 948–950, 1963.
32. Krohn, J. and J. Sündermann. Paleotides before the Permian. Pp. 190–209 in ref. 26b. Also: Sündermann, J. and P. Brosche. Numerical computation of tidal friction for present and ancient oceans. Pp. 125–144 in ref. 26a.
33. Brosche, P. and W. Hövel. Tidal friction for times around the present. Pp. 175–189 in ref. 26b.
34. Hansen, K. Secular effects of tidal dissipation in the moon's orbit and the earth's rotation. *Rev. Geophys. & Space Physics*, **20**, (3), 457–480, 1982.
35. Webb, D.J. On the reduction in tidal dissipation produced by increases in the earth's rotation rate and its effect on the longterm history of the moon's orbit. Pp. 210–221 in ref. 26b. (Includes references to related papers by the author.)
36. Jeffreys, H. Possible tidal effects on accurate timekeeping. *Monthly Not. R. astr. Soc., Geophys. Suppl.*, **2**, 56–58, 1928.
37. (**a**) Stoyko, N. Sur la périodicité dans l'irregularité de la rotation de la terre. *Comptes Rendus Acad. Sci.* **205**, 79–81, 1937. (**b**) Stoyko, A. and N. Stoyko. Les fluctuations périodiques de la rotation de la terre pendant 1933–40. *Bull. Acad. R. Belgique*, ser. 5, **42**, 693–702, 1956.
38. Guinot, B. (**a**) Short-period terms in Universal Time. *Astron. & Astrophys.*, **8**, 26–28, 1970. (**b**) A determination of the of the Love number k from the periodic waves of UT1. *Ibid.* **36**, 1–4, 1974.

39. Yoder, C.F., J.G. Williams, and M.E. Parke. Tidal variations of earth rotation. *J. Geophys. Res.*, **86**, (B2), 881–891, 1981.
40. Merriam, J.B. Tidal terms in Universal Time: effects of zonal winds and mantle *Q. J. Geophys. Res.*, **89**, (B12), 10, 109–114, 1984.
41. Carter, W.E. + 6 authors. Variations in the rotation of the earth. *Science*, **224**, 957–961, 1984.
42. Wilkins, G.A. International cooperation in the monitoring of the rotation of the earth. *Vistas in Astronomy*, **28**, 329–335, 1985.
43. (**a**) Baader, H.-R., P. Brosche, and W. Hövel. Ocean tides and periodic variations of the earth rotation. *J. Geophys.*, **52**, 140–142, 1983.
(**b**) Brosche, P., J. Wünsch, J. Campbell, and H. Schuh. Ocean tide effects in Universal Time detected by VLBI. *Astron. & Astrophys.*, **245**, 676–682, 1991.
44. Chao, B.F., R.D. Ray, and G.D. Egbert. Diurnal/semidiurnal oceanic tidal momentum: T/P models in comparison with earth rotation rate. *Geophys. Res Letters*, **22**, (15), 1993–1996, 1995.
45. Hartmann, W.K., R.J. Phillips, and G.J. Taylor, (Eds.) *The Origin of the Moon*. Lunar & Planetary Institute, 759pp., Houston, Texas, 1986.
46. Tidal Science – 1996. Proceedings of a symposium held in London, 21–22 October 1996. *Progress in Oceanography*, Elsevier Press. ('Special Issue', vol. 40, Issue 1–4, 1998.)

Appendix A

Common astronomical terms

Age of the moon: The number of days (0–29.5) since the last 'Change' to New Moon.

Apparent Time: Local time as would be measured by a sundial, with noon corresponding to Upper Transits of the sun. *Mean* Apparent Time differs from **Universal Time** (q.v.) by 4 minutes per degree of longitude, (later, east; earlier, west of the zero meridian). Before about 1850, all times were recorded in Apparent Time.

Declination: The angle of the direction of a celestial object above the equatorial plane; north positive, south negative.

Ecliptic: The plane of the earth's orbit round the sun, to which the equator is inclined by 23°.4, (obliquity slowly decreasing).

Equinox, Equinoctial point, (usual symbol *♈*): The **ascending node** ('vernal equininox') of the earth's orbit round the sun, i.e. the point on the equatorial plane where it meets the ecliptic plane around 21 March. The **descending node,** ('autumnal equinox') is passed similarly around 23 September. For tidal calculations, the equinox is taken as a fixed reference point in space, but it actually precesses westward at about 50″ per year relative to fixed stars.

Evection and **Variation**: The largest two periodic terms in the perturbation of the moon's orbit by the sun. Evection has amplitude 72′, period 31.8 days, argument = moon's anomaly (angular distance from perigee) minus 2× (distance from sun). Variation has amplitude 39′, period 14.8 days, argument = 2× (moon's angular distance from sun). The corresponding principal perturbations to the semidiurnal tide are: Evectional, ν_2, λ_2; Variational, μ_2; see Figure 8.3.

Luminary (mainly archaic): The sun or the moon.

Month: synodic – the mean time interval between Full Moons (29.53 d); sidereal – the mean time interval between the moon's passing a fixed star (27.32d);

anomalistic – the mean time interval between the moon's passing perigee (27.55 d).

Nodal cycle: The cycle of westward precession of the moon's ascending node relative to the equinox (−18.61 years).

Parallax (equatorial): The angle subtended by the earth's equatorial radius from a celestial body within the solar system; an inverse measure of the body's distance. The moon's mean parallax is 3422″.5; the sun's mean parallax is 8″.794.

Perigee/Apogee: The point in a satellite's (e.g the moon's) orbit where it passes closest to/furthest from the earth.

Perigee cycle: The cycle of eastward progression of the moon's perigee relative to the equinox (8.85 years).

Perihelion/Aphelion: The point in the earth's orbit where it passes closest to/furthest from the sun. The date of perihelion is at present about 2 January, but it was earlier in past centuries by about 1.74 days per century.

Quadrature (in the context of Chapters 4–7): The configuration of the First and Last Quarters of the moon, when the Right Ascensions of the sun and moon differ by 90°.

Right Ascension: The angle of the direction of a celestial body measured along the equatorial plane eastward from the equinox. Right Ascension (RA) and declination (Dec) are analogous to east longitude and north latitude in defining the position of a celestial body relative to the earth's equator. The RA of the Greenwich meridian, expressed as a time at 1 hour per 15°, is known as '**Sidereal Time**'.

Solstice (summer- or winter-): The time of (highest – around 21 June, or lowest – around 22 December) declination of the sun.

Southing (archaic): Upper Transit of the moon as seen from a place with latitude greater than the moon's declination, (e.g. Europe).

Syzygy (in the context of Chapters 4–7): The configuration of Full or Change of the moon, when the right ascensions of the sun and moon differ by 0° or 180°.

Transit: The passage of a celestial object (usually sun or moon) past the meridian of the place of observation – 'Upper Transit' – or the local meridian + 180° – 'Lower Transit'. (See also **Southing.**)

Tropical year: The mean interval between the sun's passages of the equinox, (365.2422 days).

Universal Time (previously 'Greenwich Mean Time'): The time system such that Upper Transits of the sun at the Greenwich meridian occur (on yearly average) at 12 hours (noon). UT is not quite uniform compared with atomic time on account of non-uniformity of the earth's rotation; it is occasionally adjusted by a 'leap second' in order to maintain the above definition. (See also **Apparent Time.**)

Variation (of the moon's orbit): See **Evection.**

Appendix B

Terms commonly applied to tides

Age of the tide: The time delay of highest spring tides at a place after the corresponding syzygy (Full or New Moon); typically 1–2 days, but zero or negative in certain locations.

Annual inequality (archaic): The twice-yearly variation of spring tide amplitude between the equinoxes and the solstices.

Cotidal map (mainly 19th to mid-20th century): A map of an ocean or sea showing contours – 'cotidal lines' – joining places of equal mean lunitidal interval or equal phase of a harmonic constituent, usually in steps of one-twelfth of the period or 30°, (e.g. Figure 9.2). The more general map, containing both iso-phase and iso-amplitude contours, is simply called a 'tidal map' in this book, (e.g. Figure 12.6).

Diurnal inequality: The difference in HW (or LW) heights or times between two consecutive semidiurnal tides.

Diurnal (semidiurnal) tides: Tides with 1 (2) maxima per lunar day. Since Laplace, this has usually referred to the respective harmonic **species**, in which case 'terdiurnal' (3), 'quarter-diurnal' (4), or 'sixth-diurnal' (6) may also be used. Laplace's original specification was 1st, 2nd, 3rd and 4th species (*espèces)* for long-period, diurnal, semidiurnal, terdiurnal tides, respectively.

Earth tide: The vertical tidal variation of the surface of the solid earth, frequently split into the *body tide* – direct yielding to the tidal potential, and the *load tide* – additional deformation due to the ocean tide. The earth tide also has horizontal tilting and horizontal strain.

Equilibrium tide: A hypothetical global form of the sea surface elevation which would be in equilibrium with the tide-generating stresses in the absence of inertia and currents. It is commonly approximated by the primary tide potential multiplied by (0.69/g), but modifications are required if the shapes of the oceans and continents are taken into account.

Establishment (somewhat archaic): The mean time of flood tide (HWT) at Full Moon or Change, at a particular place. Whewell (1830's) called this the 'Vulgar Establishment', and defined a 'Corrected Establishment' as the mean of all lunitidal intervals in a spring-neap cycle at the place. The CE is usually about 30 minutes less than the VE (From French 'Etablissement'.)

Geocentric tide: Tidal variation of the sea surface relative to the earth's center – equal to the sum of the **ocean tide** and the **earth tide** (q.v.)

Harmonic constituent: One of many spectral components of the tide separated in frequency by at least 1 cycle per year, first identified by Thomson and Darwin in the late 19th century. See Figure 8.3 for details.

High Water (Low Water) height: (HWH) The height of the maximum (LWH – minimum) vertical elevation of the tide in a 12h or 24h cycle, relative to an arbitrary fixed datum. HWT (LWT) the clock time of High (Low) Water.

Long period tides: The harmonic components of the tide which have periods much longer than a day (Laplace's '1st species'), typically 9.1 days to 18.6 years. They are essentially due to the zonal harmonics of the tide-generating potential.

Lunitidal interval: The difference between a HWT and the time of the previous lunar transit.

Monthly inequality (archaic): The twice-monthly variation of tidal amplitude between spring and neap tides.

Neap tides: The tides of least amplitude in a 15-day cycle.

Ocean(ic) tide: Conventionally defined and measured as the tide in sea surface elevation relative to the local sea floor. (See also **Geocentric tide**.)

Range (tidal): Height difference between consecutive High and Low Waters. (Usually applied to the greatest spring tides, or as the double-amplitude of a harmonic constituent.)

Seiche: A non-tidal oscillation of sea level, reflecting local resonances of enclosed seas, bays and islands, excited by weather or seismic disturbance. Typical periods of seiches are 10–60 minutes. The name 'seiche' was coined by F.A. Forel (1841–1912) to describe oscillations of Lake Geneva.

Spring tides: The tides of greatest amplitude in a 15-day cycle. *Perigean* spring tide: a spring tide coinciding with Perigee, and therefore with greater amplitude than average.

Tide (-generating) potential: (Primary): The scalar function whose gradients define the tidal forces at any place. (Secondary): Modifications to the primary potential necessary to allow for the earth's elasticity and the self-attraction of the tidal deformation itself.

Appendix C

Development of the tide-generating potential

Under Newtonian attraction, the *gravitational potential* at a distance R from an attracting body of mass S is $V=\gamma S/R$, where γ is the Universal Constant of Gravitation. ($\partial V/\partial R$ gives the inverse-square law.)

Referring to Figure 5.1(c), where SP $=R$, ST $=d$, PT $=a$, KP $=x$, the *tide-generating* potential at the earth's surface is

$$U=\Delta V=\gamma S\,(R^{-1}-d^{-1}-xd^{-2}),$$

where the term proportional to x is the potential of a constant force parallel to TS, balanced by the orbital acceleration of T about S.

In terms of the zenith angle PTS $=Z$ of the sun above P, we thus have $x=a\cos Z$, and

$$U=S[\,(r^2+d^2-2rd\cos Z)^{-\frac{1}{2}}-d^{-1}-xd^{-2}]_{r=a},$$

which on expansion in powers of the small quantity $q=r/d$ becomes:

$$\frac{\gamma S}{d}\left[\frac{q^2}{2}(3\cos^2 Z-1)+O(q^3)\right]_{r=a}.$$

The leading term of the tide-generating potential is therefore

$$U(a,Z)=\frac{\gamma S}{2d^3}a^2(3\cos^2 Z-1),$$

or, in terms of the *equatorial parallax* ($\varpi=a/d$) of the luminary S, (Appendix A):

$$U(a,Z)=\frac{3\gamma S}{2a}\varpi^3(\cos^2 Z-\tfrac{1}{3})+O(\varpi^4). \tag{C1}$$

It is easily verified that the gradients of $U(a,Z)$ correspond to the force vectors indicated in Figure 5.1(c); in the vertical direction,

$$\frac{\partial U}{\partial a} = \frac{\gamma S}{d^3}(3x\cos Z - a),$$

and in the eastward horizontal direction,

$$a^{-1}\frac{\partial U}{\partial Z} = \frac{\gamma S}{d^3} = 3x\sin Z.$$

If we now express the zenith angle Z in terms of the geographical coordinates of P, namely (θ, ϕ), and the corresponding coordinates of S, (Θ, Φ), the spherical triangle PNS with apex at the North Pole N gives

$$\cos Z = \cos\theta\cos\Theta + \sin\theta\sin\Theta\cos(\Phi - \phi - \Omega t).$$

So finally, to order q^2:

$$U(\theta, \phi, \Theta, \Phi) = \frac{3\gamma S}{2a}\varpi^3[\{\cos\theta\cos\Theta + \sin\theta\sin\Theta\cos(\Phi - \phi - \Omega t)\}^2 - \tfrac{1}{3}]. \quad \text{(C2)}$$

With translation to Laplace's notation, in which $a=1$ and the gravitational constant γ is omitted, C2 differs from Laplace's 'R' (Figure 7.1 – below equation 9) by a numerical constant, which is immaterial since only the gradients of U are involved in LTE.

Algebraic expansion of the square-bracketed expression in C2 gives the sum of three terms:

$$f_1 = \cos^2\theta\cos^2\Theta + \tfrac{1}{2}\sin^2\theta\sin^2\Theta - \tfrac{1}{3}$$

$$= \tfrac{1}{24}(1 + 3\cos 2\theta)(1 + 3\cos 2\Theta),$$

$$f_2 = \tfrac{1}{2}\sin 2\theta\sin 2\Theta\cos(\Phi - \phi - \Omega t),$$

$$f_3 = \tfrac{1}{2}\sin^2\theta\sin^2\Theta\cos 2(\Phi - \phi - \Omega t),$$

corresponding to Laplace's three *species* (espèces). f_1, being independent of Ωt, varies only with the orbital periods of 1 month, 1 year, etc. f_2, being proportional to $\cos(\Phi - \phi - \Omega t)$, has periods of 1 day with orbital modulations. f_3 has, similarly, periods of a half-day with orbital modulations. The functions of θ and Θ may be recognised as the *associated Legendre functions* of degree 2.

In practical applications all these functions are additionally modulated by variations in the parallax factor, ϖ^3. Further expansion into harmonic terms requires expressions for the orbital motions in their respective planes inclined to the equator.

The small term of order ϖ^4 in C1 is quite negligible for the sun, but its lunar part generates detectable secondary tides. It may be expanded in similar terms, and produces in addition to tides of species 1,2,3, a term f_4 proportional to $\cos 3(\Phi - \phi - \Omega t)$ of terdiurnal periodicity.

Appendix D

Internal tidal waves in a flat rotating sea of uniform depth

As an approximation to global coordinates in a localised region, let (x,y,z) be a right-handed set of axes with x,y horizontal and z vertically upwards, in a region where the *Coriolis* or *inertial* frequency, $2\omega = 2\Omega\cos\theta$ (θ=colatitude), and the depth D may be considered uniform. (This is known technically as an 'f-plane', since 'f' is a symbol often used for the inertial frequency.) In some contexts the geographical orientation of the x,y axes is arbitrary, but we shall assume that x points southwards and y eastwards unless otherwise stated. The origin of z is taken as the mean air/sea interface, whether the motion is considered as taking place in the sea ($-D<z<0$) or in the air ($z>0$). In the latter case, $z=0$ at the level of dry land also applies.

Let $u(x,y,z,t), v(x,y,z,t), w(x,y,z,t), p(x,y,z,t)$ be the oscillatory components of fluid velocity in the directions x,y,z and of pressure, respectively. Nonuniform density ρ is assumed:

$$\rho(x,y,z,t) = \rho_0(z) + \rho'(x,y,z,t), \quad (|\rho'| \ll \rho_0),$$

where ρ_0 is the mean density at height z and ρ' its oscillatory part at a fixed position. The fluid is assumed incompressible and isothermal in the present treatment. Therefore ρ' and w are related through:

$$\rho'_t = -w(\rho_0)_z = w\,\rho_0 N^2/g, \tag{D1}$$

where $N(z) = \sqrt{[-g(\rho_0)_z/\rho_0]}$ is known as the *buoyancy frequency*, and suffixes z, t represent derivatives. The wave equations therefore require only the four independent variables u,v,w,p.

The 'extended Laplace' tidal equations read:

$$u_x + v_y + w_z = 0 \tag{D2}$$

$$u_t - 2\omega v = -(1/\rho_0)\, p_x + U_x/g \tag{D3}$$

$$v_t + 2\omega u = -(1/\rho_0)\, p_y + U_y/g \tag{D4}$$

$$w_t = -(1/\rho_0)\,(p_z + g\rho') \tag{D5}$$

where $U(x,y,t)$ is the tide-generating potential at the earth's surface. Combining (1) with (5),

$$w_{tt} + N^2 w = -(1/\rho_0)\, p_{zt}. \tag{D6}$$

With time-dependence $\exp(-i\sigma t)$ at a tidal frequency σ, D3–D6 become, on ignoring U_x, U_y:

$$\mathrm{i}\sigma u + 2\omega v = (1/\rho_0) p_x, \tag{D7}$$

$$\mathrm{i}\sigma v - 2\omega u = (1/\rho_0)\, p_y, \tag{D8}$$

$$\mathrm{w}(N^2 - \sigma^2) = (1/\rho_0)\, \mathrm{i}\sigma p_z. \tag{D9}$$

Using D7, D8 to express u and v in terms of p_x, p_y, and substituting the result, with D9, into D2 gives:

$$w_{zz} = \frac{N^2 - \sigma^2}{\sigma^2 - 4\omega^2}(w_{zz} + w_{yy}). \tag{D10}$$

If the waves are supposed to progress in the y-direction (say) with wavenumber κ, D10 becomes:

$$w_{zz} + \frac{N^2 - \sigma^2}{\sigma^2 - 4\omega^2}\kappa^2 w = 0. \tag{D11}$$

In the simplified case of uniform $N(z)$, D11 has a general solution of the form:

$$w = A e^{\mathrm{i}\nu z} + B e^{-\mathrm{i}\nu z} \tag{D12}$$

with constants A and B to be determined by boundary conditions at two levels z. (For the atmosphere, and with slight approximation also the sea, $A + B = 0$.)

Now in the tidal context, $N >> \sigma$, so real ν requires $\sigma > 2|\omega|$. Semidiurnal tides satisfy this condition almost everywhere except very near the poles, but diurnal tides, for example the principal harmonic K_1, for which $\omega = \Omega$, satisfy it only at latitudes between 30° N and 30° S. Outside the *critical latitudes* where $\sigma = 2\omega$, ν must be complex and hence w is exponential in z. Waves propagate vertically only when ν is real; otherwise the wave is *trapped* at the height where it is energised. This proved to have radical consequences for the air tides, since the thermal excitation occurs high in the atmosphere. Semidiurnal tides can propagate the localised energy down to the surface $z = 0$ but diurnal tides cannot do this except between the critical latitudes. This heuristically explains why the tide in surface air pressure is predominantly semidiurnal, without requiring resonance.

In the ocean, the horizontal wavenumber κ is not given *a priori*. The usual approach to solution of D3–D6 is to separate the horizontal and vertical variations:

$$u = u'(x,y,t)\,F_z(z), \quad v = v'(x,y,t)\,F_z(z),$$

$$w = z_t'(x,y,t)\,F(z)/D_n, \quad p = \rho_0 g z'(x,y,t)\,F_z(z),$$

where D_n is a length parameter analogous to depth, which will later be shown to have a sequence of possible values ordered by n. With these substitutions the continuity equation (2) gives

$$D_n(u_x' + v_y') = \mathrm{i}\sigma z';$$

and the momentum equations (7, 8) give

$$\mathrm{i}\sigma u' + 2\omega v' = g z_x',$$

$$\mathrm{i}\sigma v' - 2\omega u' = g z_y'.$$

These last three equations are identical in form to the classic LTE in an ocean of depth D_n with surface displacement $z'(x,y,t)$. They include classic solutions like the Kelvin wave, (Chapter 7). D9 gives an additional equation for the vertical variation $F(z)$:

$$F_{zz} + \frac{N^2 - \sigma^2}{g D_n} F = 0. \tag{D13}$$

Neglecting σ/N for convenience, and again assuming N to be independent of z for simplicity, the solution of D13 which has $F(0) = 0$ is

$$F = A \sin N (g D_n)^{-\frac{1}{2}} z.$$

The boundary condition $F(-D_n) = 0$ then gives D_n as a sequence of *eigensolutions*

$$D_n = g^{-1}(ND/n\pi)^2, \quad n = 1,2,3, \ldots$$

The corresponding *eigenfunctions* $F_n(z)$ have $(n-1)$ zeroes between $z=0$ and $-D$. A solution corresponding to $n=0$ is the quasi-barotropic solution $F(z) = A(z+D)$, with horizontal velocity components uniform in the vertical. An internal tide of Kelvin-type (exponential in one horizontal direction) would travel with phase speed $c_n = \surd(g D_n)$ and horizontal wavenumber $\kappa_n = \sigma/c_n$, very much larger (i.e. shorter wavelength) than the barotropic Kelvin wave in depth D. Similar but more complicated results obtain when N is not assumed uniform, as in reality.

Extension to air tides

The dynamics of air tides have features similar to the case discussed above, but their analysis is very much complicated by the essentially spherical geometry, compressibility of the medium, and the thermodynamics of adiabatic processes.

The assumed horizontal waveform is $e^{i(m\phi-\sigma t)}$, $m=0$, 1, or 2, and density ρ' and temperature variation T' have to be included as independent variables in addition to u,v,w,p; the divergence is no longer zero but has to be taken as another independent variable; the thermal driving potential has a complicated vertical distribution.

Equations representing the physical relationships between these variables are in principle separable in the horizontal and vertical as in the simpler case discussed above. For example,

$$w(\theta,\phi,z,t) = Y^{(w)}(\theta,\phi,t)\,\Theta(z),$$

and again the function Y may be shown to satisfy a set of equations analogous to LTE with eigenvalues D_n. The eigenfunctions Y_n associated with D_n are the *Hough Functions* mentioned at the end of Chapter 7. The vertical functions Θ_n satisfy an equation:

$$\frac{\partial^2\Theta_n}{\partial x^2} + \left[-\frac{1}{4} + \left(\frac{1}{D_n}\,kH(x) + \frac{\partial H}{\partial x}\right)\right] + \Theta_n(x) = \frac{\kappa J_n(x)}{gD_n}e^{-x/2} \qquad \text{(D14)}$$

where $x=\ln[p_0(0)/p_0(z)]$, an artificial variable; $k=0.2857$ is a thermodynamic constant. $H(x)$ is a variable proportional to static temperature T_0, and the right hand side involves the thermal driving potential.

As in the simpler case, D_n appears in both the horizontal and vertical equations; its values are conventionally known as *equivalent depths* by analogy with the classic marine case. The most radical discovery in the late 1960's was that for 'completeness' (in the mathematical sense) the set of Hough Functions has to admit *negative* equivalent depths. With $D_n<0$, the coefficient of Θ_n in D14 is strictly negative, so that vertical wavelike solutions are impossible, as we found in the simple case above when $\sigma<2\omega$. In fact, calculations show D_n to be predominantly negative for $m=1$ (diurnal tides), so that energy from the heating source does not propagate to the earth's surface. For $m=2$, (semidiurnal tides), D_n is predominantly positive and a more uniform distribution of energy with height results. The horizontal distribution with respect to the critical latitudes is reflected in the shape of the Hough Functions $Y_n(\theta,\phi,t)$ themselves. See Chapter 10, ref.13 for a full discussion.

Appendix E

Some simplified cases of barotropic waves of second class – Rossby waves and Continental Shelf waves

Eschewing mathematical complications, the essential characteristics of these waves may be seen in a limited flat sea with rectangular coordinates x(east), y(north) and Coriolis term approximated by $f = f_0 + \beta y$, where β is everywhere positive. It may be shown that they are close to the condition of zero horizontal divergence:

$$(Du)_x + (Dv)_y = 0,$$

where $D(x, y)$ is the depth and suffixes denote derivatives. Solutions of this equation (including surface elevation) are conveniently expressed in terms of a *stream function* $\psi(x, y)$ defined to satisfy

$$Du = \psi_y, \quad Dv = -\psi_x.$$

Elimination of the pressure gradients between the two momentum equations (Chapter 7) yields:

$$(\nabla^2\psi)_t + \beta\psi_x - D^{-1}[(D_y\psi_y + D_x\psi_x)_t + f_0(D_y\psi_x - D_x\psi_y)] = 0, \qquad \text{(E1)}$$

where $\nabla^2\psi = \psi_{xx} + \psi_{yy}$ is the *vorticity* of the motion. The surface elevation may be obtained from the pressure gradients, which are simply expressible in terms of ψ.

In the case of *uniform depth* D, equation (E1) reduces to

$$(\nabla^2\psi)_t + \beta\psi_x = 0. \qquad \text{(E2)}$$

This is satisfied for waves of form $\exp[i(kx + ly - \sigma t)]$, provided

$$(k^2 + l^2)\sigma + \beta k = 0.$$

Since $\sigma > 0$, $\beta > 0$, k must be negative, so waves of this type always travel with a westward component of phase velocity. These are the *Rossby* or *Planetary* waves, first identified by the meteorologist Carl-Gustav Rossby (Chapter 13, ref. 34).

Turning now to the case where the gradients of D are more important than those of f, consider for simplicity a long uniform shelf parallel to the x-axis (which may now take any orientation), defined by $D(x,y) = D(y)$, a monotonic function from shelf depth $D(0)$ to an abyssal depth $D(\infty)$ at large y. The stream function y must now satisfy

$$(\nabla^2 \psi)_t - (D_y/D)\,(\psi_{yt} + f_0\,\psi_x) = 0, \tag{E3}$$

with a null-flow condition $\psi_x = 0$ at $x = 0$. For an assumed depth variation $D(y)$, substituting the wave function

$$\psi\,(x, y) = F(y)\,\exp[i(kx - \sigma t)]$$

leads to a second-order differential equation for $F(y)$ with eigen-solutions $F_n(y)$ corresponding to a discrete set of wavenumbers k_n, determined by the boundary conditions $F(0) = 0 = F(\infty)$. It may be shown that all k_n have the same sign as σ/f_0, that is, all waves travel with shallow water to the right in the northern hemisphere, (left in the southern), like the classic Kelvin wave. They differ from the Kelvin wave, however, in having much larger wavenumbers k_{n} (i.e. much shorter wavelength) and in being more closely trapped to the shelf edge. Also, whereas Kelvin waves can exist at any frequency σ, these waves, like Rossby waves, can exist only where $\sigma < |f_0|$. Such waves are known as *Continental Shelf waves* or *Topographic Rossby waves* .

Appendix F

Spherical harmonic expansion of a globally defined tidal constituent

A constituent of frequency σ at colatitude θ and east longitude ϕ may in general be expressed:

$$\begin{aligned}\zeta(\theta,\phi,t) &= H(\theta,\phi)\cos[\,\sigma t - G(\theta,\phi)] \\ &= H_1(\theta,\phi)\cos\sigma t + H_2(\theta,\phi)\sin\sigma t \qquad \text{(F1)}\end{aligned}$$

where H is the amplitude and G is the Greenwich phase lag as usually mapped over the ocean, and

$$H_1 = H\cos G, \quad H_2 = H\sin G$$

are the *in-phase* and *quadrature* parts of the tide, which may also be mapped. For (θ, ϕ) over land, $H = H_1 = H_2 = 0$ in an oceanic context.

If H and G are assumed to be known globally, then both $H_1(\theta,\phi)$ and $H_2(\theta,\phi)$ may be expanded in spherical harmonics with real coefficients a,b,c,d in the following standard form:

$$H_1(\theta,\phi) = \sum_{n=0}^{\infty}\sum_{m=0}^{n}{}'\,(a_{nm}\cos m\phi - b_{nm}\sin m\phi)\,P_n^m(\theta)$$

$$H_2(\theta,\phi) = \sum_{n=0}^{\infty}\sum_{m=0}^{n}{}'\,(c_{nm}\cos m\phi - d_{nm}\sin m\phi)\,P_n^m(\theta)$$

where the P_n^m are *associated Legendre functions* and the primes $'$ denote that the terms with $m = 0$ are halved.

Substituting these expressions into equation F1, we obtain:

$$\zeta(\theta,\phi,t) = \frac{1}{2}\sum_{n=0}^{\infty}\sum_{m=0}^{n}{}'\,\big[(a_{nm} + d_{nm})\cos(m\phi + \sigma t) - (b_{nm} - c_{nm})\sin(m\phi + \sigma t) +$$

$$(a_{nm} - d_{nm}) \cos(m\phi - \sigma t) - (b_{nm} + c_{nm}) \sin(m\phi - \sigma t)\big] P_n^m(\theta). \tag{F2}$$

The first two of the four terms inside the brackets represent a wave travelling in the direction $-\phi$, that is westward; the third and fourth terms represent a wave travelling eastward.

Dropping the summation signs and the suffixes n,m for convenience, equation F2 may be equivalently be expressed as the 'Real Part' of the complex expression

$$\frac{1}{2}\big[\{a+d+\mathrm{i}(b-c)\} + \{a-d+\mathrm{i}(b+c)\}\mathrm{e}^{-\mathrm{i}\sigma t}\big] P_n^m(\theta)\mathrm{e}^{\mathrm{i}m\phi}$$

$$= \big[{}^{+}\mathrm{C}n^{m}(t) + {}^{-}\mathrm{C}n^{m}(t)\big] P_n^m(\theta)\mathrm{e}^{\mathrm{i}m\phi},$$

as appears in the text.

Author index

Only authors mentioned by name in the text are included

Subject index

p48 - lunitidal interval re full/new moon

semidiurnal [illegible] - tides, and regional modulation and random element

p100 - periodogram

p167 - amphidromic systems - p.170

94 Kelvin / Darwin symbols

p.100 G.H. Darwin - citation

p.191 Darwinian 'Harmonic constants'

p.179 equilibrium tide and its component "species"

T/P [illegible] elements

dictionary: oblate, amphidromic tesseral geostrophic

p.238 - surface profile of ocean

247 interpolations between tide stations
OSU global tidal map

p272 declination - north → positive
south - negative

p275 - established - mean time of flood spring tide

p35 - monthly inequality - spring-neap tide cycle

110 - progression of high water, crest of tide wave
tide hour of a port
1831 Lubbock paper on high tide time

establishment = harmonic phase lag

Diurnal inequality especially large in Pacific and Indian Ocean